Wendelin Wichtmann, Christian Schröder & Hans Joosten (eds.)

Paludiculture – productive use of wet peatlands

Wendelin Wichtmann, Christian Schröder &
Hans Joosten (eds.)

Paludiculture – productive use of wet peatlands

Climate protection – biodiversity – regional economic benefits

with contributions from 73 authors (see list of contributors on page 263)

 Schweizerbart Science Publishers · Stuttgart, 2016

Wendelin Wichtmann, Christian Schröder & Hans Joosten (eds.)
Paludiculture – productive use of wet peatlands
Climate protection – biodiversity – regional economic benefits

Editors:	Wendelin Wichtmann, Christian Schröder & Hans Joosten
	Greifswald Mire Centre
	Ellernholzstraße 1/3
	17487 Greifswald
	Germany
	info@greifswaldmoor.de

This book is a product of the joint research project 'VIP – Vorpommern Initiative Paludiculture' funded by the Federal Ministry of Education and Research of the Federal Republic of Germany within the 'Sustainable Land Management' programme (FKZ 033L030) and the 'Wetlands Energy' Project, funded by EU Aid (DCI-ENV/2010/220-473). The book project was coordinated by the Greifswald Mire Centre (GMC). The GMC is a cooperation of the Ernst-Moritz-Arndt University of Greifswald, the Michael Succow Foundation for the Protection of Nature and the Institute DUENE eV.

Improvement of illustrations:	Stephan Busse
Reference management:	Annie Wojatschke
Translation:	Jenny Schulz, Fernando Andreu, Carl Skarbek, John Couwenberg, Hans Joosten
Proof-reading/review:	Almut Mrotzek, Alma Rodriguez-Tsuda, John Couwenberg
Front cover:	Summer harvest of reed in Peene Valley, Germany. Photo by Christian Schröder

Disclaimer: The authors take full responsibility for the technical and scientific content of this publication. Opinions expressed are attributed to the authors alone, and not to the institutions or personages supporting the publication.

ISBN 978-3-510-65283-9
Information on the English title: www.schweizerbart.com/9783510652839

This title is also available in German:
W. Wichtmann, C. Schröder & H. Joosten (Hrsg.): Paludikultur – Bewirtschaftung nasser Moore
Informationen zur deutschen Ausgabe: www.schweizerbart.de/9783510652822

© 2016 E. Schweizerbart'sche Verlagsbuchhandlung (Nägele u. Obermiller), Stuttgart, Germany

All rights reserved. No part of this publication may be reproduced, stored in a retrieval system, or transmitted, in any form or by any means, electronic, mechanical photocopying, recording, or otherwise, without the prior written permission of E. Schweizerbart'sche Verlagsbuchhandlung, Stuttgart.

Publisher:	E. Schweizerbart'sche Verlagsbuchhandlung (Nägele u. Obermiller)
	Johannesstr. 3A, 70176 Stuttgart, Germany
	mail@schweizerbart.de
	www.schweizerbart.de

♾ Printed on permanent paper conforming to ISO 9706-1994

Layout: DTP + TEXT Eva Burri, Stuttgart
Printed in Germany by Gulde Druck, Tübingen, Germany

Preface by the laureate of the Right Livelihood Award, Michael Succow

Paludiculture – Sustainable use of Wet Peatlands

It is the wet nature of mires, which explains why they were the last natural ecosystems to be taken into intensive agricultural use. This happened to the fens in our Central Europe no longer than 50 years ago. Till this day the vital water is – with large efforts – drained out of the peatlands. All over the world deep drainage continues to cause degradation of organic soils. The rapid depletion of the peat resource inevitably leads to a total loss of the functions of peatlands in the balance of nature, comparable to the loss of soils by desertification in hot climates.

Mainstream drainage-based peatland use has to be challenged fundamentally, because of the ensuing environmental damage and to ensure long-term preservation of agricultural land. In this context one must realise that current unsustainable peatland use in Germany and Europe can only be upheld by large agricultural subsidies. Already long ago I recognized the possibility to utilise biomass of mires without damaging their self-regulative properties and functionality by building upon traditional types of land use such as harvesting reed and using litter from wet meadows.

These ideas were taken up more than 20 years ago, scientifically studied and conceptually tested in various research projects. This led to the emergence of the concept of 'paludiculture', mainly coined by Hans Joosten and Wendelin Wichtmann, environmental scientists from Ernst-Moritz-Arndt-University of Greifswald.

Paludiculture is not merely a word: it is a principle, a rethinking of how to deal with peatlands in agricultural use. This book outlines this new approach to sustainable utilisation of fens for biomass production for the first time in all its complexity.

Paludiculture enables to maintain the peatland carbon stock, whilst at the same time using the land. Paludiculture is about establishing productive, possibly peat accumulating mire-typical plant communities on hitherto deeply drained agriculturally used peatland sites. This environmentally compatible, sustainable land use is urgently demanded as the only future-oriented way for our civilisation.

The prerequisites for these changes are particularly favourable, especially in countries with a high-intensity, peat degrading agricultural peatland use, provided that sufficient water for rewetting is available. An urgent need for action exists for Germany, where drained peatlands annually emit double the amount of CO_2 compared to the entire incoming and outgoing air traffic.

Since my biology studies in Greifswald more than 50 years ago, I am deeply attached to peatlands. I had still the fortune and opportunity to experience mire landscapes in Central and Eastern Europe under peat-preserving low-intensity land use. For me, it is a delayed gratification, a relief to see a new start being made, after 30 years of industrial agricultural peatland use. The end of abuse and the start of new sustainable ways of peatland use – which can only be paludiculture – are long overdue on the way towards a responsible way of managing our ecosystems.

In this pioneering spirit I experience highly motivated, well-educated and well-equipped research teams, that in international networks fathom the new, important knowledge on the functioning and functionality of peatlands in our landscape and that prepare the practical implementation of the results. For "without using nature we cannot exist, but by misusing nature we will perish!"

Gratitude is owed to all those who devote themselves with heart and mind to make amends to the damage to peatlands and who brought together their insight and vision in this book.

Let us give a new future to our peatlands, by restoring the role Nature had meant them to perform in the balance of nature: for the sake of their and our future.

Greifswald, Germany, 2015

Professor Emeritus Michael Succow
Chairman of the Board of the Michael Succow Foundation for the Protection of Nature

Foreword by the Food and Agricultural Organisation (FAO), Martin Frick

The 21st Conference of the Parties of the UN Framework Convention on Climate Change, COP21, concluded in December 2015 with the historic Paris Agreement. As outlined in the Agreement, the international response to climate change should combine efforts to reduce greenhouse gas emissions with those aimed at building resilience, fighting poverty and eradicating food insecurity. The agricultural sectors (crops, livestock, forestry, fisheries and aquaculture) are vital to accomplishing this.

Agriculture, forestry and other land use are responsible for more than 20 percent of all human-induced greenhouse gas emissions. Without concerted and immediate action, this figure will continue to rise in tandem with income and population growth.

The agricultural sectors also provide livelihoods and food security for the vast majority of the world's poor. These sectors are among the most vulnerable to a changing climate. Low-emissions and climate-resilient agricultural growth strategies are therefore vital to effectively tackle climate change, reduce poverty and feed a growing global population.

Peatlands have a particularly important role to play in the fight against climate change. They cover only three percent of global land area, but store 30 percent of the world's soil carbon. When managed unsustainably, peatlands can be a major source of greenhouse gas emissions.

For instance, unsustainable agriculture and forestry activities can drive peatland drainage, which can in turn lower the water table and result in significant greenhouse gas emissions. Drained peatlands are also more prone to fires that emit greenhouse gases and undermine human health and biodiversity.

Conservation and responsible management of peatlands should be top international priorities. Paludiculture is particularly important in this respect. Paludiculture offers a means of cultivating on rewetted peatlands, which reduces greenhouse gas emissions while providing environmental and socio-economic benefits such as reduced frequency of fires, as well as improved livelihoods and food security. Paludiculture can also halt land subsidence, thereby reducing land loss, flood and fire frequency as well as salt-water intrusion.

FAO has long supported better management of peatlands, including through paludiculture. As part of the Mitigation of Climate Change in Agriculture series, FAO and its partners have published guidebooks such as "Towards climate-responsible peatlands management" and "Peatlands – guidance for climate change mitigation through conservation, rehabilitation and sustainable use". Meanwhile, the FAO online collection of peatland management practices provides many examples of projects that are effectively using paludiculture for biomass and energy production.

This new book provides users with detailed and comprehensive information on paludiculture practices. It includes hands-on guidance related to the harvesting, logistics and economics of paludiculture at farm and regional levels. It highlights the need to promote and further develop paludiculture for more widespread adoption.

This publication provides an important basis for more responsible global peatlands management which can be an important element to fighting both climate change and hunger.

Martin Frick
Director
Climate and Environment Division (NRC)
Food and Agriculture Organization of the United Nations (FAO), Rome

Table of Content

Preface by the laureate of the Right Livelihood Award, Michael Succow.. V
Foreword by the Food and Agricultural Organisation (FAO), Martin Frick... VI

1 Paludiculture as an inclusive solution 1

2 The limits of drainage based peatland utilisation 3
2.1 Fen peatland use in Northeast Germany 3
2.2 Drainage induced peat degradation processes 7
2.3 Impact of drainage on productivity 9
2.4 Ecosystem services of peatlands 13

3 Production and utilisation of paludiculture biomass 21
3.1 Promising plants for paludiculture 22
3.2 Edible and medical plants from paludiculture 38
3.3 The production of fodder in paludiculture 39
3.4 Material use of biomass from paludiculture 43
3.5 Solid energy from biomass 44
3.6 Liquid and gaseous biofuels 54

4 Harvest and logistics 59
4.1 Trafficability of wet and rewetted fens 59
4.2 Agricultural machinery for wet areas 64
4.3 Logistics of biomass production on wet peatlands 70
4.4 The feasibility of biomass harvest from paludiculture 76

5 Ecosystem services provided by paludiculture 79
5.1 Greenhouse gas emissions 79
5.2 Biodiversity 94
5.3 Local climate and hydrology 102
5.4 Nutrient balance and water pollution control 106

6 Economics of paludiculture 109
6.1 Economic aspects of paludiculture on the farm level 109
6.2 Certification of biomass from paludiculture 120
6.3 The creation of regional value 132
6.4 Welfare aspects of land use on peatland 134

7 Legal and political aspects of paludiculture 143
7.1 The legal framework 143
7.2 Agricultural policy 149
7.3 Control mechanisms and incentives for paludiculture 152

8 Social aspects of paludiculture implementation 157
8.1 The relationship between humans and mires over time 157
8.2 The integration of stakeholders and the public 162
8.3 Acceptance and implementation at the producer level 168
8.4 Transfer of knowledge 171

9 Sustainability and implementation of paludiculture 175
9.1 Sustainable land use 175
9.2 Availability of suitable areas 178
9.3 The decision-support tool TORBOS 185
9.4 Technical measures for implementing paludiculture 188
9.5 Implementation and administrative approval in Germany 194

10 Paludiculture in a global context 197
10.1 Global demands and international commitments 199
10.2 The global potential and perspectives for paludiculture 200
10.3 Germany – Rewetting and paludiculture in Mecklenburg-West Pomerania 204
10.4 Belarus – Biomass from rewetted peatlands as a substitute for peat and for promoting biodiversity 205
10.5 Poland – Paludiculture for biodiversity and peatland protection 207
10.6 Indonesia – Paludiculture as sustainable land use 217
10.7 China – Paper from the water 223
10.8 Canada – Harvesting Typha spp. for nutrient capture and bioeconomy at Lake Winnipeg 226

11 The way out of the desert – What needs to be done 229
11.1 Problems of peatland management and the necessity of paludiculture 229
11.2 Challenges for practice 230
11.3 Awareness raising and communication 231
11.4 Politics and society 231
11.5 Research questions 232
11.6 Outlook 233

References 235

List of contributors 263

Index 265

1 Paludiculture as an inclusive solution

Wendelin Wichtmann, Christian Schröder & Hans Joosten

Drainage-based land use of peatlands for agriculture and forestry causes many problems all over the world. Peat soil degradation and subsidence, which are the direct consequences of peatland drainage, progressively decrease the yields and increase the costs to the extent that lands have to be abandoned. Greenhouse gas emissions, loss of biodiversity, and catastrophic peat fires are other consequences of drained peatland use. The massive emissions from drained peatlands worldwide have put them in the international agenda to mitigate climate change (Chapter 10.1; Joosten et al. 2012). Almost all peatlands in Western Europe have been transformed to degrading systems which are now sources of greenhouse gas emissions. Similar processes of peatland degradation are happening in Eastern Europe, America and Asia.

There is an increasing acknowledgment of the fact that problems caused by peatland drainage can only be addressed and solved through raising the water table. Therefore, rewetting and restoration measures have been implemented around the world with the aim to reduce greenhouse gas emissions and to avoid peat fires, as well as to conserve or re-establish peatland biodiversity (Kratz & Pfadenhauer 2001, Kowatsch 2007, Rieley & Page 2008, Tanneberger & Wichtmann 2011).

With rewetting, however, conventional drainage-based agriculture loses the peatland area as productive land. The arisen question is whether degraded peatland can be rewetted and simultaneously be used for production, possibly even with renewed peat accumulation. Management practices have to be found and implemented in a way that links comprehensive peat conservation with the production of renewable raw materials. This is exactly what paludiculture implies (Box 1.1).

Depending on the climate zone, a wide spectrum of plant species can be cultivated in wet peatlands (Box 3.1). Paludiculture includes the traditional uses of semi-natural peatland sites, such as commercial reed cutting or utilisation of litter as bedding material for cattle. Examples of new forms of paludiculture in Central Europe are the use of biomass from reed beds, sedge vegetation, or wet meadows for energy genera-

---

Box 1.1: What is paludiculture?

Wendelin Wichtmann, Christian Schröder & Hans Joosten

Paludiculture is the agricultural or silvicultural use of wet and rewetted peatlands. Paludiculture uses spontaneously grown or cultivated biomass from wet peatlands under conditions in which the peat is conserved or even newly formed (Wichtmann & Joosten 2007).

Paludiculture differs fundamentally from drainage-based conventional peatland use, which leads to huge emissions of greenhouse gases and nutrients and eventually destroys its own production base through peat degradation (Joosten et al. 2012).

Paludiculture allows the re-establishment and maintenance of ecosystem services of wet peatlands such as carbon sequestration and storage, water and nutrient retention, as well as local climate cooling and habitat provision for rare species (Chapter 5; Joosten et al. 2012, Wichtmann et al. 2010).

Paludiculture implies an agricultural paradigm shift. Instead of draining them, peatlands are used under peat-conserving permanent wet conditions. Deeply drained and highly degraded peatlands have the greatest need for action from an environmental point of view, and provide the largest land potential. The implementation of paludiculture is the best choice for degraded peatlands.

Paludiculture is a worldwide applicable land management system to continue land use on rewetted degraded peatlands. Various plants can be cultivated profitable under wet conditions.

Paludiculture is also a land use alternative for natural peatlands particular for regions where the increasing demand for productive land drives the drainage. Because of their vulnerable ecosystem services, pristine peatlands should best be protected entirely. If land use on pristine mires is unavoidable, paludiculture should always be given preference over drainage-based land use (Joosten et al. 2012).

tion. Additional examples include the plantation of alder on rewetted fens for the production of timber or veneer (Schäfer & Joosten 2005), as well as the cultivation of peatmosses on rewetted bog grassland as a raw material for high quality horticultural growing media (Gaudig & Wichmann 2011, Gaudig et al. 2014) (Chapter 3.1).

For paludiculture to be an alternative to agriculture or forestry on drained peatlands, economic efficiency is of prime importance. Income from paludiculture is generated by the utilisation and sale of biomass, but may also be derived from the provision of other services (e.g. agri-climate and agri-environmental funding programmes, framework contracts for nature conservation, water abstraction charges and carbon credits) and premiums (e.g. direct payments, support for organic farming). Furthermore, biomass from semi-natural vegetation in protected areas can be used if biomass removal is necessary or desirable for habitat and species conservation. In this case, the main emphasis is on conservation, while the harvest of biomass is ancillary.

This book starts with elucidating the negative effects of drainage on the production function of peat soils and their ecosystem services (Chapter 2). With regard to the globally increasing demand for agricultural land to secure food and biomass supply, the world can no longer afford the degradation of peatlands. Faced with an increasing shortage of productive land, we must maintain the productive capacity of the soil and its multifunctionality, and keep options open for the future. The use of above-ground biomass from wet peatlands offers a wide array of perspectives. Valuable raw biomass materials or energy crops can be produced on sites that are not or less suitable for food production. Thus, paludiculture can take away pressure from those areas needed for food production – a potential that is not yet fully quantified. New cultural plants in paludiculture may offer new utilisation opportunities (Chapter 3). New, innovative technical solutions are needed to master the logistic challenges of wet land use and to mobilize the potentials of paludiculture (Chapter 4). Cultivation of wet peatlands stops further peatland degradation, reduces greenhouse gas emissions, supports wetland biodiversity and secures or restores various important functions within the regional water and nutrient cycle (Chapter 5).

Profitability is of crucial importance for the implementation of paludiculture as a new land management concept. For business and regional development, the utilisation of biomass from wet peatlands offers various interesting economic prospects (Chapter 6). Since a change to paludiculture implies a paradigm shift in land management, the framework of legislation and agricultural policy must be adapted to this new land use concept. The recognition of paludiculture as a new form of land management is paramount for its large-scale implementation in the European Union. Ideas how to overcome existing obstacles and how to provide incentives for the implementation of paludiculture are proposed in Chapter 7.

When implementing paludiculture, the interests of the local population must be taken into consideration, as the landscape may change substantially in appearance when peatlands are rewetted. Information provision and goal-directed communication between all parties involved are urgently needed prior to the implementation of paludiculture. Conflicting interests can be identified through new participatory approaches, and solutions to problems and conflicts can be developed together with the local stakeholders (Chapter 8). Furthermore, the book contains advice on how to implement paludiculture in practice and what steps must be considered during planning (Chapter 9). Case studies are presented to show how the decision to implement paludiculture can arise from various motives (Chapter 10). Last but not least, it is important to learn from existing paludiculture projects, because we are just entering a new era of peatland management (Chapter 11).

This book shows the need for a broad, interdisciplinary approach to facilitate a land use change on peatlands. Expert knowledge from various fields must be combined with practical experience. Paludiculture has many facets, and the most important ones are described and discussed in this book. Other questions will only be touched upon superficially, as much experience is still lacking. Doubtlessly, paludiculture remains a big challenge for scientists, decision makers and practitioners alike. But the first important steps have been set.

2 The limits of drainage based peatland utilisation

Knowledge, as well as the quality and availability of means of production, determine the nature and intensity of all land use (Jutta Zeitz, 2015).

In the past, human impact on peatlands has varied depending on existing technology, human needs and financial means. Peatlands were exploited in order to extract peat for fuel but the main interest was to turn them into agricultural use. Often, the reclamation of peatlands was linked to political changes – e.g. to win new land for settling refugees. Large-scale cultivation efforts were driven and funded by state policy – for instance, the Cultivation Act (Urbarmachungsedikt) of Frederick the Great and the complex amelioration (Komplexmelioration) in the German Democratic Republic (Chapter 2.1). However, regardless of motivation and techniques, the precondition for peatland cultivation was always to drain the water away from the naturally wet peatlands.

Drainage leads to alteration and intensification of the transformation processes in the upper peat layer (Chapter 2.2), which, dependent on their nature and intensity, leads to changes in soil productivity (Chapter 2.3). This 'secondary pedogenesis' in drained peatlands is similar in all climate zones. In the beginning and with shallow drainage, soil productivity may initially even improve. On the longer term, however, deep drainage, frequent ploughing and the sowing of cash crops result in such profound changes of the physical, chemical and biological properties of the soil, that the peat soil degrades. Not only the conditions for biomass production deteriorate, but also other ecosystem services of mires, such as water retention and the sequestration and storage of nutrients and carbon are negatively affected (Chapter 2.4). As these changes are largely irreversible, drainage based land use of peatlands is neither ecologically nor economically sustainable. Long term peatland utilisation is only possible when agriculture is adapted to the natural site conditions and not the other way around, as it was the case during past centuries.

2.1 Fen peatland use in Northeast Germany

Jutta Zeitz

Peatlands form by the accumulation of peat, when the production of biomass is larger than its decomposition. This is the case when decomposition is retarded by oxygen depletion caused by a permanently water-saturated environment.

Whereas in many regions mires were the last areas to be taken into cultivation, archaeological evidence shows that already Palaeolithic hunter-gatherers made use of peatlands. However, this usage did not significantly impact peat formation and the ecosystem services mires provided. Before humans started to change the water balance of mires intentionally, they used them, and particularly their margins, in several ways: as pasture, for hay making or for gathering litter, especially in dry years (Fischer 1999). The biomass was mown by hand and transported onto mineral ground, whereas animals were driven onto the mires for grazing if the load-bearing capacity of the ground allowed it. Later, peatlands also became providers of fuel peat and a variety of raw materials, including salt (extracted from saline peat in coastal mires), bog iron ore, and bog lime (Succow & Jeschke 1986, Succow 1988, Lehrkamp & Zeitz 2014). The deforestation of large areas, as well as pondage for watermills, changed the regional hydrology and even stimulated the growth and expansion of mires.

In Northeast Germany and other parts of West and Central Europe mires have increasingly been modified by agricultural use in the course of the last 300 years. Mires were drained to be used as pasture, hay meadow or arable land. Depending on the nature and intensity of the practices the effects on the mires range from small to irreversible (Table 2.1).

Drainage improves the trafficability of the ground, whereas the altered microclimate reduces the frequency of early and spring frosts. Peat decomposition and mineralisation result in the release of plant nutrients, especially nitrate. Fertilisation and management enable the cultivation of crops that are specifically bred, highly productive, but thus also more demanding to site conditions.

2 The limits of drainage based peatland utilisation

Table 2.1: History of agricultural use of fen peatlands in Northeast Germany (changed after Succow 1988 and Zeitz 2003, Zeitz 2014). Floodwater and water rise mires were in many cases the first mires to be cultivated, followed by percolation mires.

Time	Typical utilisation/Example
6th century	• Very low intensity pastoral use of sedge vegetation
13th century	• Mowing of graminoid vegetation of eutrophic floodwater mires • Use of litter from sedge vegetation • Beginning drainage and establishment of hay meadows (1 cut per year) by Cistercian monks
18th century	• Extension of fen grassland by drainage and clearance of swamp forests • Water regulation of rivers and drainage of large valley mires • First establishment of polders • 1718 drainage of the Havelländisches Luch, initiated by Frederick William I • 1765 Cultivation Act (*Urbarmachungsedikt*) for peatland cultivation in Prussia issued by Frederick II
19th century	• Drainage using fascines (from the mid-19th century) • Land gain by lowering the water table of lakes • Valley mires are largely converted to humid meadows • Development of humid *Molinia* (Purple Moor Grass) meadows caused by removal of plant nutrients • Start of fen cultivation by covering the peat with sand (sand cover cultivation, *Sanddeckkultur*, 1817, C. Pogge) • Beginning of Rimpau sand cover cultivation (*Moordammkultur* 1887)
20th century, until the mid 1960s	• Application of artificial fertilisers • Development of humid meadows that are cut twice per year (hygrophilous tall herbaceous communities with *Cirsium oleraceum*) • Development of high quality cultivated grassland • Yields of up to 8.5 t dry mass ha^{-1} for fen grassland
mid 1960s	• Complex amelioration: renewed and very deep drainage by lowering the groundwater table to 50–80 cm and more below the ground surface • Intensive grassland use with high rates of fertilisation (100–200 kg N ha^{-1}) • Tilling and renewed grass cultivation every 4–5 years. Yields reach 10 t ha^{-1} dry mass ha^{-1} a^{-1} with 3–4 cuts per year • Partially arable use on deeply ploughed areas
Since the mid 1970s	• Increasing peat degradation on intensively used grassland • Technical upgrade of drainage infrastructure for water regulation in both directions (drainage and irrigation)
1990 until the end of the 20th century	• No further large scale ameliorations • Peatland use becomes increasingly problematic due to waterlogging, wind erosion and the development of microrelief • Increasing conflicts with respect to water regulation (usage rights for water abstraction, high costs for pumping, insufficient maintenance, soil wetness) • Abandonment for economic reasons • Rewetting for nature conservation
21st century	• Partly renewed intensification to secure peatland utilisation • Minimum maintenance, peatlands become areas of marginal agricultural revenues • Intensification of problems related to peatland utilisation • Rewetting in order to reduce greenhouse gas emissions

In the following sections we present some common practices of peatland utilisation in Northeast Germany.

2.1.1 Niedermoorschwarzkultur (black fen cultivation)

Black fen cultivation is the oldest form of intensive land use on peatlands in Germany, being widely practiced since the 13th century until now. The German name 'Niedermoorschwarzkultur' (Niedermoor = fen, schwarz = black, kultur = cultivation) derives from the black top soil which results from peat oxidation after drainage. This form of land use involves drainage of the peatland, removal of the natural vegetation and ploughing of the top peat layer. Subsequently, the soil is fertilised: in earlier times, solid and liquid manure were used whereas in modern times, mineral fertiliser is utilised. After fertilisation, specially bred crops are sown. In this form of land use the soil profile stays largely intact (Figure 2.1).

Fens were initially drained by shallow ditches that were dug manually and quickly collapsed. Later, ditches with a wide range of depths and widths plus subsurface drainage tubes were used. Up until the end of the 19th century all drainage ditches and channels were dug by hand (Figure 2.2). Clay drainage pipes from mecha-

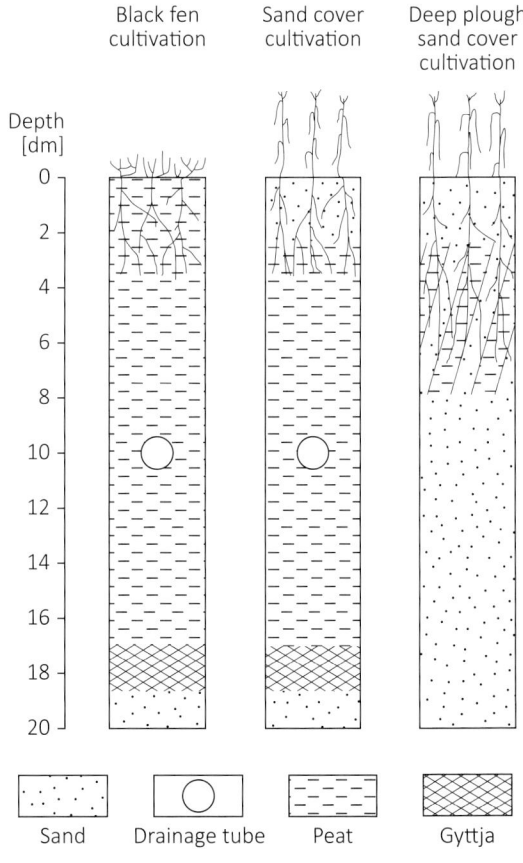

Figure 2.1: Soil profiles and root penetration in various fen cultivation alternatives (after Göttlich & Kuntze 1990).

2.1 Fen peatland use in Northeast Germany

Figure 2.2: Adolescents digging a drainage ditch in the Friedländer Große Wiese, Germany (Zentralbild Schwankler, 11.8.1958).

nised mass production were available and used in fen drainage since the mid 19th century. From the 1960s onwards the clay pipes were replaced by plastic pipes. In Northeast Germany, subsurface drainage of fens was also accomplished by cutting rectangular (18 cm x 14 cm) tunnels below ground (Figure 2.3), a practice called 'Maulwurfsfräsdränung', (Maulwurf = mole, Fräsen = cutting, dränung = drainage). These pipe-less drains were functional for about 10 years (Scholz 1986).

Currently, many of the formerly laid drainage pipes lay close to the ground surface, which illustrates the large peat losses associated with land use based on drainage. Colour picture 1 shows an example where 50 years of drainage has caused the loss of 100 cm of peat (as a consequence, drainage tubes originally installed at 120 cm depth lay now close to the surface) and a (black) degraded top soil is overlying the originally less decomposed bog peat.

2.1.2 Sanddeckkultur (sand cover cultivation)

In the 19th century, farmers observed that the soil properties of cultivated fen peatlands improve when the peat is covered with sand. For such 'Sanddeckkultur' (sand cover cultivation), sand is spread on top of the fen peat without mixing sand and peat. The stratigraphy of the underlying peat profile itself is not altered (Figure 2.1). A special technique applied in Mecklenburg, was 'Poggeln', named after its inventor Carl Pogge (1763–1831). In 1817, Pogge applied this technique for the first time on a 10 ha peatland meadow of his manor, near the town of Güstrow. The method involved covering the peatland meadow with sand brought from the margins of the peatland by cart. A sand layer of about 12–15 cm was spread onto the turf of the shallowly drained peatland. The weight load of the sand layer compressed the underlying peat, kept it

Figure 2.3: 'Maulwurfsfräsdränung' (mole pipe drainage) in the Friedländer Große Wiese 1971 (Zentralbild Bartocha, 26.4.71).

moist and reduced peat mineralisation (Ratzke & Mohr 2005). Later research has shown that in comparison to the aforementioned 'Niedermoorschwarzkultur' peat mineralisation rates at a mean water level 90 cm below ground are reduced approximately by half when the peat is covered with a sand layer (Mundel 1976).

The initial sandy cover layers were often heterogeneous in thickness. In the course of time, they got mixed with the underlying peat through very deep ploughing, and changed into a mixture of sand and peat. This process offset the positive effects of the 'Sanddeckkultur' in regard to soil productivity, workability, and peat mineralisation.

An improved form of Pogge's method was developed by Theodor Herrman Rimpau on his manor in Cunrau (Saxony-Anhalt) in 1862. The practice was called 'Moordammkultur' (Moor = peatland, damm = dam, kultur = cultivation, Figure 2.4). In this method, the fen peatland was deeply drained by a dense network of ditches. The area was divided in 25 m wide fields on which the peat from the ditches was evenly spread. Lastly, the fields were covered with a 10–12 cm layer of mineral soil (sand) material, which in shallow peatlands was taken from the ditches whereas in deep peatlands, it was taken from the surrounding mineral area. Around 1900 this technique was considered an excellent method to transform grassland areas with poor soil productivity into higher yielding arable land. Only shallow tilling was allowed to prevent the mixing of the mineral soil layer with the underlying peat. However, even with this method, the drainage was too deep to conserve the peat layer (Massenbach 1887: 'It is sufficient to lower the water table to 1 m below ground'...), resulting in shrinkage and peat degradation (Göbel 2000).

2.1.3 Tiefpflugsanddeckkultur (deep plough sand cover cultivation)

The 'Tiefpflugsanddeckkultur' (Tiefpflug = deep plough, sanddeck = sand cover, kultur= cultivation) form of peatland cultivation exists since the 1950s. A giant plough was used to plough shallow and already drained fen peatlands, so that the peat layer was mixed with the underlying sand (Figure 2.1, Colour picture 2). In the same action the fields were covered with an approximately 25 cm thick sand layer and subsequently used for mainly arable farming. The peat stratigraphy was

Figure 2.4: Rimpau sand cover cultivation (German: 'Rimpausche Moordammkultur) around 1900. https://upload.wikimedia.org/wikipedia/de/1/1a/Rimpauische_Moordammkultur_um_1900.jpg

completely disturbed by the ploughing, which strongly improved vertical water exchange in the profile, particularly on sites with impervious, water stagnating lake sediments. Compared with the 'Niedermoorschwarzkultur', the soil achieved a higher load-bearing capacity and fields were less prone to the proliferation of weeds. 'Tiefpflugsanddeckkultur' was applied to approximately 2,500 ha of land in Northeast Germany and has altered these former peatlands structurally in such a way that they cannot be classified as peatland anymore, although they remain groundwater dependent (Zeitz 2003, Zeitz 2014).

2.1.4 Komplexmelioration (complex amelioration)

'Komplexmelioration' was applied in East Germany from the 1960s until the 1980s, when many formerly shallowly drained fen peatlands were further 'ameliorated' by a variety of interventions. Part of the existing drainage channels were deepened until 2.5 m, whereas the fields were enlarged by filling in the remaining ditches between the large and the deep ones. Since the 1950s, national engineering standards guaranteed the standardization of ditch distances, depths and slope gradients, taking into account the hydraulic conductivity of peats and lake sediments, the presence of impervious layers, as well as the requirements of future land use. Smaller fields were merged to larger ones because large coherent fields are favourable in technical and economic terms. This also reduced the number of vehicle crossing points between fields thus reducing their high maintenance costs.

Several extreme summer droughts in the 1970s and the first signs of peat degradation on the overly drained fen peatlands lead to the installation of adjustable weirs in the ditch system, which allowed drainage, water retention and irrigation. Water retention by elevated weir levels made it possible to keep the water table high until May, that is to say, four to six weeks longer than before. However, these high spring levels could not prevent desiccation in summer (Box 9.4). Thus, irrigation was widely practiced but required additional water supply by pumping water from higher laying water courses. For the large Oberes Rhinluch peatland area in the federal state of Brandenburg, water was transported from lake Müritz via a chain of smaller lakes. This resulted in significant ecological damage due to falling lake water levels. The effect of irrigation was very limited, especially on degraded peatlands. Even though shrinkage cracks had developed in the drained peat, irrigation had little effect, probably because sealing (colmation) the ditch walls and compaction prevented water from entering the peat body (Hennings 1995).

2.2 Drainage induced peat degradation processes

Jutta Zeitz

For over 200 years the processes that follow on peatland drainage have been observed and documented. Initially, this monitoring was driven by concerns as to whether and how the newly drained areas would become utilisable. During the 1960s and 1970s, research was encouraged because secondary soil genesis in drained peatland prevented efficient drainage and required recurring intervention. Since the early 1980s, conventional drainage-based land use of fen peatlands became subject of critical public debate (Schmidt et al. 1981).

Drainage interrupts the primary process of soil genesis in peatlands, namely peat formation. Instead, peat accumulation is replaced by secondary pedogenetic processes including consolidation, compaction, shrinkage, humification and mineralisation (oxidation), as well as dislocation, leaching and accumulation of soil substances. The continuously decreasing soil productivity resulting from these processes challenges the long-term utilisation of peat soils under conventional agriculture (Sauerbrey & Zeitz 2003, Stegmann & Zeitz 2001, Zeitz 2001, Ilnicki & Zeitz 2003, Oleszczuk et al. 2008, Kalisz et al. 2010, Zeitz 2014).

While initially (i. e. in the first years) consolidation is the main process causing the lowering of the peatland surface upon drainage, subsequently other processes, including shrinkage, compaction and mineralisation (peat oxidation) contribute more significantly. Subsidence is the sum of all processes that lower the peatland surface.

2.2.1 Peatland consolidation

Pristine mires consist of peats and organic sediments that are, in general, extremely porous. In pristine state, peatlands are – with the exception of seasonal water level fluctuations, saturated with water up to the surface. Peatland drainage empties the pores and causes a purely mechanical setting of the peat because the buoyancy is lost. This loss leads to the compression of the formerly spongy peat and to a reduced peat porosity (Table 2.2). The load of the drained peat at the surface also causes compression of the underlying peat strata that are still saturated with water. As a result of these processes that occur immediately after drainage, the surface of the peat strongly subsides. The more porous and spongy the peat in pristine state is, the smaller the volume of the drained peat will be. Similarly, the thicker the peat layer and the deeper the drainage, the more the surface will be lowered. For example, when a typical percolation mire with a dry matter content of 7.5% and a thickness of 5 m is drained to a water level of 1.1 m below the original surface, subsidence may reach 0.9 m (Eggelsmann 1981). Any further drainage leads to renewed consolidation, though less dramatically, as the peat layer has already lost large amounts of water and therefore has a higher content of dry matter. In order to calculate the magnitude of consolidation, empirical formulas were developed and the depths of drainage ditches and drain pipes were accordingly positioned (Segeberg 1960, Eggelsmann 1981).

Table 2.2: Change of soil properties as a result of drainage and intensive agricultural use of fen soils and their effects on provisioning services (after Zeitz 2014).

Trend	Soil property	Direct and indirect effects
Increasing	Bulk density	• Improved load-bearing capacity and trafficability • Restricted movement of water and nutrients • Formation of plate-like aggregates impermeable to water, which cause flooding after heavy rain and may induce vegetation dieback
	Air filled pore space, mainly as cracks and tears in the top soil	• Strong heating up caused by increased insulation and possibly resulting vegetation dieback
	Concentration of nutrients in the top soil	• Uneven nutrient supply and shallow root penetration with higher risk of drought
	Microrelief	• More difficult movement of vehicles • Strong alterations in soil moisture • Reduced possibilities for water regulation
	Vulnerability to wind erosion	• Loss of nutrient rich top soil • Increase in fine dust emissions (particulate suspended matter, PSM) • Soiling of cultivated plants, water bodies and roads (on-site and off-site damage by erosion)
Decreasing	Saturated and unsaturated water conductivity	• Restricted water regulation • Decreased capillary water rise • Increased risk of droughts
	Storage capacity of plant available water	• Reduced yield
	Rewettability	• Top soil becomes hydrophobic • Microerosion and particle translocation after precipitation • Increasing heterogeneity of the soil
	Ground surface height	• Gravity drainage becoming increasingly difficult • Polders (with dikes and pumps) become necessary • Increased flood risk; in coastal areas also intrusion of and flooding by salt or brackish water
	Peat thickness	• Impeded water regulation and storage • Decline of filter and storage capacity for pollutants • Disappearance of peat layer and exposure of underlying unproductive soil (gyttja, acid sulphate soils, hard pan, quartz sand)

2.2.2 Shrinkage

Peats and organic lake sediments are porous and elastic substrates that shrink when they lose water. In pristine mires these processes are – within the range of natural water level fluctuations, reversible, i. e. a rise in the water level causes the peat layers to swell back to their original volume. The consequent oscillation of the mire surface is called 'Mooratmung' (German for 'bog breathing'). If desiccation exceeds a certain threshold (depending on the type of peat), shrinkage becomes largely irreversible. The dry matter content of the peat increases, and cracks and clefts appear in the peat. Depending on the peat type and the original degree of peat humification, vertical and horizontal fissures of varying dimensions develop. The more porous the peat was in pristine state, the stronger its volume reduction by drainage. Repeated shrinking and swelling lead to the development of characteristic segregation structures (Zeitz 1992). Long-term and intensive desiccation – reinforced by mechanical disturbance by tillage, results in a very fine grainy, dusty structure in the top soil, with high water repellency (Zeitz & Velty 2002).

2.2.3 Humification and mineralisation

Simultaneously with consolidation, compaction and shrinkage, also microbiological decomposition starts to take place in the oxygenated peat layers. During humification, degradation-resistant organic substances are converted into humus, whereas during mineralisation, decomposition to simpler anorganic substances takes place. Fertilisation, in particular with nitrogen, accelerates humification and mineralisation ('priming effect', Paepke 1992). Organic matter is converted into end products with simpler molecular structure, which are released from the peatland as gases (CO_2, N_2O) or as solutes (nitrate, dissolved organic carbon – DOC), which leach out or are transported into deeper peat strata. Mineralisation results in loss of organic matter and contributes to the continuous subsidence of the peatland surface (Leifeld et al. 2011).

2.2.4 Translocation, leaching and concentration of organic material

The enhanced break down of organic matter by soil macrofauna, followed by humification and mineralisation, enable the translocation and – depending on the thickness of the peat layers, also the leaching of smaller particles and soluble substances. As a result of subsidence, shrinkage, compaction and mineralisation, easily adsorbable plant nutrients such as phosphorus accumulate in the top soil. The translocation of soil particles, the cycle of thawing and freezing, as well as soil compression by heavy agricultural machinery may result in soil compaction and the formation of stagnating layers in the top soil that cause waterlogging (Zeitz et al. 1987).

2.2.5 Soil development after drainage

Drainage results in surficial peats that are highly decomposed by secondary decomposition, in which the original plant structures are no longer recognisable. The processes described above lead to the development of typical soil horizons, such as the shrinkage-induced 'peat crumb horizon' in the deeper soil strata. In case of shallow drainage, the top soil is 'earthified' (secondarily humified), with the peat structure consisting of crumbles that are greasy when wet, and hardly dusty when dry. Deep drainage causes the peat to become 'strongly earthified', fine granular, greasy-granular when wet and dusty-granular when dry. The German Soil Classification distinguishes between these soil types in drained peatlands as 'Erdniedermoore' and 'Mulmniedermoore', respectively (Sponagel 2005, Figure 2.5).

2.3 Impact of drainage on productivity

Jutta Zeitz

Soil productivity of drained fen peatlands is determined by a number of site conditions that in mutual interdependency determine trafficability, water retention and effective yield, nutrient retention and supply, and soil aeration and thermal regulation. Secondary pedogenesis, which begins with drainage and intensive agricultural use, influences these site conditions and consequently soil productivity. The impact of the interrelated processes resulting from drainage leads to considerable land management problems, and questions the productive land use on these sites (Table 2.2).

2.3.1 Consequences of peatland drainage

Peatland subsidence resulting from consolidation, shrinkage, compaction and mineralisation significantly complicates water management. Subsidence in temperate regions is 2–25 mm per year, depending on site conditions (peatland type, peat type, hydrology and climate); drainage intensity (time, duration and depth of water level drawdown); and land use (grassland or arable use) (Mundel 1976, Schothorst 1977, Lehrkamp 1987, Eggelsmann 1990a, Fell et al. 2015). In Southern Europe or Southeast Asia, subsidence can reach up to 70 mm per year (Hooijer et al. 2012) (Box 2.1). In some locations, subsidence has been systematically monitored by means of stakes placed in the ground to mark the former land surface.

2 The limits of drainage based peatland utilisation

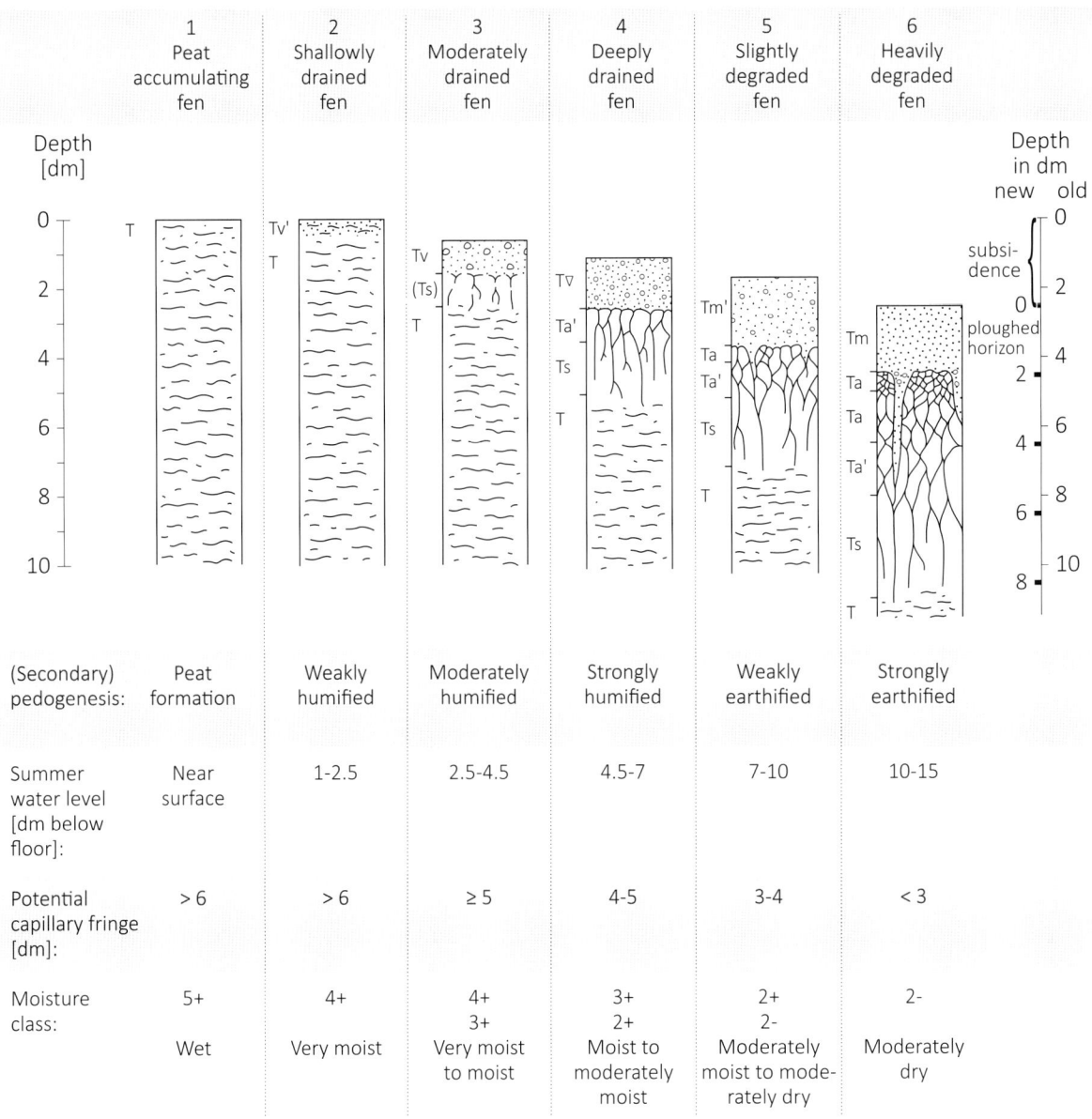

Figure 2.5: Pedogenesis on fen peat as a function of drainage depth and time. Consolidation at the beginning of drainage not included (after Stegmann & Zeitz 2001).

In other cases, subsidence is visible on the foundations of buildings, which are exposed by the disappearance of the peat (Colour picture 3). Progressing subsidence enforces a regular deepening of the drainage infrastructure, which leads to ever increasing drainage costs. Kuntze (1984) coined this process 'the vicious cycle of peatland utilisation' (Figure 2.6). When the costs of drainage exceed the revenues from agriculture, land use is usually stopped and the peatlands are completely abandoned. If subsidence has been severe, the stopping of the drainage pumps may even lead to the formation of shallow water bodies on those former agricultural fields.

The formation of stagnating soil layers creates a soil profile that contains two water levels. Near the surface, the soil is waterlogged, then a second groundwater level is situated significantly deeper and, in between both waterlogged horizons, a well-aerated soil horizon is located (Zeitz et al. 1987, Colour picture 4). Soil density measurements in fen peatlands have shown that surface waterlogging is caused by a plate-like soil structure that develops on a depth of approximately

2.3 Impact of drainage on productivity

Box 2.1: Consequences of drainage for the productivity of tropical peatlands in Southeast Asia

René Dommain

The conventional cultivation of plants in tropical peatlands requires a lowering of the groundwater table to 25–100 cm below the surface (Ambak & Melling 2000). Oil palm and acacia plantations need rather low groundwater levels of 50–80 cm (Hooijer et al. 2012). In smallholder plantations, water levels may even be 1m below ground level as they lack sophisticated adjustable drainage systems. Agricultural use of nutrient-poor acidic raised bogs with a pH of 3–4 requires regular liming and fertilisation (Andriesse 1988, Ambak & Melling 2000), which accelerates peat mineralisation. Furthermore, the application of liquid manure and mineral based nitrogen fertiliser results in extremely high nitrous oxide emissions of 3–40 g N_2O m^{-2} a^{-1} = 10–120 t CO_2e ha^{-1} a^{-1} (Takakai et al. 2006). Despite high fertilisation, yields of many crops such as rice remain lower than on mineral soils and even decrease the longer is the land cultivated (Limin et al. 2007).

The biggest problem of cultivated tropical peatlands is subsidence (Couwenberg et al. 2010, Couwenberg & Hooijer 2013). Subsidence can amount to more than 1 m during the first year after drainage as a result of consolidation (Den Haan et al. 2012, Chapter 2.2.1). After this initial phase, shrinkage and peat oxidation cause further subsidence. In peatlands that are deeply drained for the cultivation of oil palm, peat oxidation causes subsidence of about 5 cm per year (Couwenberg et al. 2010, Couwenberg & Hooijer 2013, Hooijer et al. 2012, Jauhiainen et al. 2012). A common problem in drained oil palm plantations is therefore the exposure of the palm roots, resulting in loss of stability up to the point that the trees fall over. The reason for this is that during the typical operating time of an oil palm plantation of 25 years, more than 1 m of peat can disappear (Hooijer et al. 2012).

The emissions that result from oxidative peat mineralisation are about 10 t CO_2 $ha^{-1}a^{-1}$ at a drainage depth of 10 cm and rise linearly with increasing drainage depth (Couwenberg et al. 2010, Hooijer et al. 2012). A typical drainage depth of -70 cm in oil palm and acacia plantations causes annual GHG emissions of 70 t CO_2 ha^{-1} a^{-1} (Hooijer et al. 2012, Couwenberg & Hooijer 2013). The continuous peat loss may lead to the complete disappearance of the peat. In coastal peatlands, this process can lead to the exposure of acid sulphate soils, which are useless for agriculture. The enormous peat subsidence in coastal tidal areas results in regular flooding with salt or brackish water, which can reach up to 70 km inland and drastically impacts on crop cultivation (Silvius et al. 1984). In order to avoid coastal flooding, tide-steered drainage systems have to be installed. Continuing peat losses require establishing polder systems, which will cause extremely high pumping costs because of the heavy tropical rainfall of more than 2000 mm per year. Under these conditions and for the background of globally rising sea water levels, it is to be expected that large areas of the Indonesian and Malaysian coastal peatlands will literally drown in the near future.

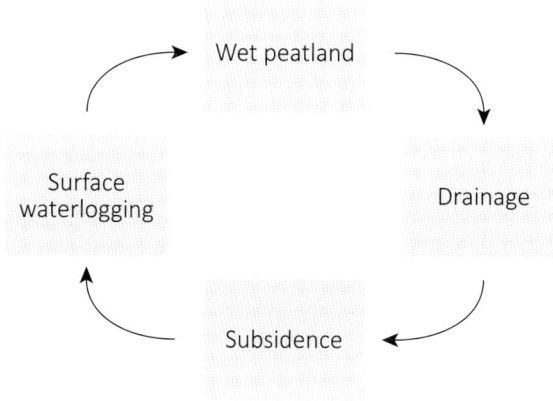

Figure 2.6: Vicious cycle of peatland utilisation (after Kuntze 1984).

30 cm. It is assumed that heavy agricultural machinery, translocation of soil particles and freezing play a role in the formation of this impervious soil layer (Zeitz et al. 1987). This layer is also the reason that capillary groundwater no longer reaches the top soil layer and plants wilt more quickly. Grassroots cannot or hardly penetrate this compacted layer. Consequently, they grow only parallel to the layer and cannot utilise space and nutrients of the soil layers below. It may be expected that, with climate change, the frequency and intensity of heavy precipitation events during summer will increase. Impounded water in summer will cause die off of the sward, resulting in subsequent problems such as wind erosion (Table 2.2). This risk increases further if the crumbled top soil degrades into a very fine dusty, grainy 'mulm' substance that is very difficult to wet.

Especially in Eastern Europe, Southeast Asia and Africa, soil productivity of deeply drained and strongly desiccated peatlands is strongly reduced by peat fires (Colour picture 5). After months of smouldering, the fires leave a bare peat that is largely water repellent and hardly wettable. The reclamation of these burned peatland areas for the reestablishment of soil productivity requires highly sophisticated technical input, as well as large financial expenditure. Additionally, peat fires cause enormous greenhouse gas emissions and produce haze most of the year, which poses a serious health risk to the people in the region because of its toxic fumes and fine dust particles.

2.3.2 Management problems resulting from peatland drainage

The load-bearing capacity of peat soils increases initially with drainage as a result of a higher density of the peat, which allows the use of heavy agricultural machinery. Through progressing secondary pedogenesis and the formation of granulose (mulmified) horizons in the topsoil, load-bearing capacity and accessibility decline again (Colour picture 6). The causes for this development are complex and are influenced by several factors (Table 2.2).

The altered physical soil properties reduce water supply to the crops, specially during the summer months. In contrast to pristine mires, seasonal differences in water availability are no longer balanced out via negative feedback mechanisms. Due to drainage, the pore volume is reduced and an oscillation of the peat body is no longer possible. The capillary rise of groundwater is restricted by the above mentioned impermeable soil layers and the hydrophobia of the peat. As the degree of subsidence varies, a microrelief develops on the land, which in combination with the hydrophobic character of the grainy peat leads to a strongly fluctuating moisture content of the top soil (Zeitz 2001). After heavy rainfall, a fine-scale pattern of alternating waterlogged, shallowly inundated, as well as drier spots, forms (Colour picture 4). Periodical water shortages and inundation after heavy rainfall cause that parts of the sod die off. As the bare peat heats up much more than the vegetation covered soil, extreme temperature fluctuations prevent the regrowth of vegetation. The patchy sod reduces the trafficability due to the decreased mechanical support of the grass sod and the grainy, strongly earthified significantly less firm structure of the underlying peat.

2.3.3 Productivity of drained peatlands

Very limited data exist in literature about grassland yields as a function of the progressing alteration of the physical soil properties of drained fen peatlands. Research on this topic has stopped and former experimental sites have been abandoned. However, field experiments under the climatic conditions of Eastern Germany have shown that agricultural yields depend on drainage depth and the progression of secondary pedogenesis (Schmidt et al. 1981), whereas also peat depth influences yields (Figure 2.7). Yields tend to decline with increasing groundwater depth. Along with secondary pedogenesis, yields initially rise (when the peat is only 'earthified', in German 'Erdniedermoor'), but later they decline when pedogenesis has reached the stage of (German) 'Mulmniedermoor'. Shallow peatlands – such as the large fen peatlands in the East German valley depressions, experience a much stronger decline in yield than deep peatlands. Additional nitrogen fertilisation has less effect on more deeply drained and more strongly degraded sites.

In the long term, these changing soil properties lead to a decline in fen productivity. Chronologically three phases of grassland use on peatlands can be distinguished:

In the first phase (which lasted until the middle of the 20th century) shallow drainage and nitrogen fertilisation led to a quantitative and qualitative increase in agricultural yields.

During the second phase, new technical innovations, progress in plant breeding and intensive grassland management, in combination with renewed and deeper drainage, resulted in maximum yields and high fodder quality. However, this stage did not last – in Germany it presumably lasted less than 20 years.

The current third phase is characterised by extremely declined yields combined with yield insecurity on the progressively very degraded peatlands. Peat subsidence has often made gravity drainage impossible. In order to continue land use, larger investments in water regulation infrastructure become necessary, including the reconditioning and maintenance of bridges and vehicle crossing points. In polder areas, the necessary deeper drainage would require the permanent operation of pumps, which cannot be financed with current water management fees. Moreover, deeper drainage would cause further peatland subsidence. Considering the predicted sea level rise as a result of global climate change, the continuation of peatland drainage is very counterproductive, in particular with respect to coastal protection in the coastal areas of the Baltic Sea and the North Sea (Trepel 2013). In recent years, the problems associated with drainage based peatland utilisation have come notably into the focus of public discussion, which in Germany led to the adoption of state peatland conservation programmes (Box 2.2).

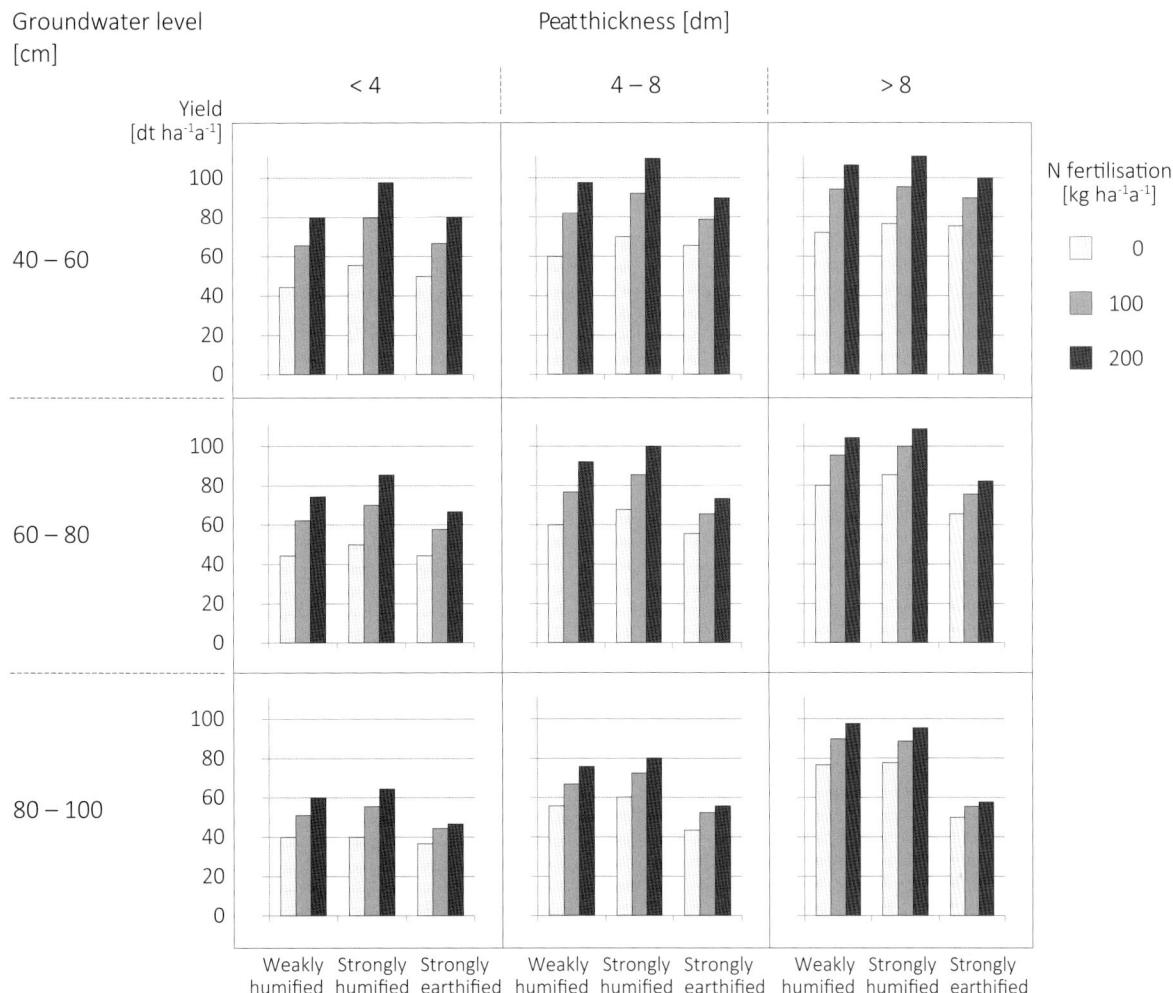

Figure 2.7: Yields of fen grasslands as a function of peat thickness, secondary pedogenesis, groundwater depth and nitrogen fertilisation (after Schmidt et al. 1981).

2.4 Ecosystem services of peatlands

Vera Luthardt & Sabine Wichmann

Awareness about ecosystem services that peatlands provide has grown significantly in recent decades. As semiaquatic ecosystems that mediate between terrestrial and aquatic ecosystems, pristine mires fulfil important functions in the nutrient, carbon and water balance of the landscape. With their specific features, mires furthermore offer refuge to highly specialised species and constitute the last remaining wilderness areas in many parts of the world. The utilisation of peatlands by humans has a large impact on these ecosystem services. Peatlands are drained to optimise provisioning services, which negatively affects many other ecosystem services. By peatland rewetting, part of these ecosystem services can be restored. If rewetting is combined with paludiculture, important regulating services can be restored while maintaining provisioning services of the peatland.

2.4.1 The changing appreciation of peatlands

For many years, mires were avoided as inaccessible wildernesses and wastelands (Chapter 8.1). They only gained importance when they were reclaimed by drainage and were used for agriculture, forestry, fuel peat extraction or as settlement area. The main focus of attention was their production function. As peatlands usability and productivity declined (Chapter 2.3) and knowledge of the ecological functioning of pristine and drained peatlands increased, questions arose about how to deal with peatlands in a sustainable way. Societal priorities widened in the face of biodiversity loss, eutrophication of water bodies and climate change. In

Box 2.2: Peatland conservation in Germany

Simone Witzel

The total area of peatlands including organic soils in Germany is approximately 1.3 million hectare. Some 940,000 ha (68%) are used for agriculture (Röder & Grützmacher 2012). Although peatlands only constitute 6% of the agricultural area, they are responsible for 99% of the CO_2-emissions from all agricultural land (UBA 2013). In total the agricultural peatlands emit 41 megaton of CO_2-e per year, which is 4.3% of the total emissions of Germany (Osterburg et al. 2013).

Due to Germany's federal structure, no coordinated national peatland conservation programme exists; some political declarations have, however, federal relevance. For instance, the national biodiversity strategy (2007) contains aims and objectives with regard to peatland conservation. The responsible authorities of the federal states with the highest peatland coverage (Schleswig-Holstein, Mecklenburg-West Pomerania, Brandenburg, Lower Saxony and Bavaria) released in 2011 a conjoint position on the potentials and the aims for the protection of peatlands and the climate (Jensen et al. 2012). Based on that position, the German Advisory Council on the Environment suggested a federal initiative for peatland conservation in its environmental assessment for 2012 (SRU 2012). All these initiatives aim for the reduction of greenhouse gas emissions, the conservation of biodiversity and the stabilisation of the local and regional hydrological balance. The federal states of Schleswig-Holstein, Mecklenburg-West Pomerania, Brandenburg, Lower Saxony and Bavaria each have their own peatland conservation programmes, which were introduced at different times and have different lifespans (Kowatsch 2007). Lower Saxony adopted its first peatland conservation programme in 1981, whereas Bavaria introduced its peatland development plan in 2005. In Brandenburg, a 10 point plan has been developed as a framework for a peatland conservation plan, similar to that of Mecklenburg-West Pomerania, which was already introduced in 2000 and was updated in 2009. Schleswig-Holstein introduced a programme for the conservation of fens in 2002, which was extended in 2008/09 by including raised bogs. In 2011, the state parliament merged these two programmes into a full peatland conservation programme (Ullrich & Riecken 2012).

The aims and objectives of these programmes are largely similar, but they differ in the degree of importance that is given to specific aims by each of the federal states. For instance, in Mecklenburg-West Pomerania priority is given to the reduction of greenhouse gas emissions. In Schleswig-Holstein, water management plays an important role and the fen programme is closely associated with the ecological improvement of water courses (aiming to support the implementation of the Water Framework Directive, Directive 2000/60/EC; (Chapter 5.4). In Lower Saxony, where the largest areas of raised bogs in Germany occur, peat extraction still takes place over more than 10,000 ha. These areas and their biodiversity are supposed to be largely restored after peat extraction. Recently new priority areas for peat conservation and peatland development are being designated, whereas the programme 'Lower Saxony Mire Landscapes' aims at a far-reaching peatland conservation strategy, which besides bogs, also includes fen peatlands. It also focuses on the reduction of greenhouse gas emissions, the phasing out of peat mining and an adjustment of agricultural use.

Within the peatland conservation programmes, land acquisition and rewetting are generally financially supported. Rewetting measures are implemented in agreement and with consent of the land owners and other stakeholders, who are being compensated financially. The projects are financed by a combination of funds allocated at EU, national, and federal state level or implemented within the framework of a compensation scheme (Intervention and Compensation Regulation, Chapter 7.1). The restored peatland areas are excluded from land use or the targeted water levels are defined in the land charge register. The complementation of peatland conservation with paludiculture is only included in the programme of Mecklenburg-West Pomerania but is discussed for Brandenburg and Lower Saxony as well. In future, paludiculture can contribute to peatland conservation at a larger scale while, at the same time, preserve the production function of peatlands.

2.4 Ecosystem services of peatlands

Germany and many other countries, the last remaining pristine mires became legally protected and peatland degradation with its complex effects increasingly gained attention.

Since the Millennium Ecosystem Assessment (MEA 2005), efforts are intensifying to consider not only the provisioning services of ecosystems but to take all other 'ancillary' ecosystem services also into account when deciding about land use (Box 2.3). Politics stresses the importance of a transparent presentation and monetary evaluation of ecosystem services (TEEB 2010, EC 2011a), but these requests can only partly be satisfied. Relevant ecosystem services include provisioning, regulating and cultural services (Haines-Young & Potschin 2011), which depend on specific ecosystem functions and processes (Boyd & Banzhaf 2007) and can be assessed by indicators (Schröder et al. 2013). The benefit that stakeholders derive from a peatland process makes that process become a 'service' (Chapter 6.4.1). An example of such functional chain is given in Figure 2.8 for the ecosystem service 'local and regional climate regulation'.

Peatland ecosystem services are more far-reaching than those of other ecosystems, in particular regarding climate regulation. The chain of effects range from the global via the regional level up to the individual farmer or hunter. Qualitative and quantitative assessments of ecosystem services – and possibly also the monetisation and commodification of changes in their provision, are only possible for specific areas and single services . Appropriate methods are currently being developed at both national and international level.

A comprehensive evaluation of peatland ecosystem services requires the assessment of not only all benefits but also of the damage that results from negatively impacted ecosystem functions. The different capacity of natural, drained and rewetted peatlands to provide various ecosystem services is generalised and compared in table Table 2.4 and discussed in detail in the following sections.

2.4.2 Ecosystem services of pristine mires

Pristine mires offer important provisioning services by supplying local people with natural produces, including Cranberries (*Vaccinium oxycoccos*), Cloudberries (*Rubus chamaemorus*), mushrooms, medical plants such as Marsh Labrador Tea (*Ledum palustre*) or Bogbean (*Menyanthes trifoliata*), as well as fodder, litter, and construction materials (Joosten & Clarke 2002). Furthermore, mires provide substantial regulating and cultural services. They have a stabilising effect on local hydrology, attenuate the effects of peak discharge during flooding events, and exert a cooling effect on local climate through evaporation and cloud formation during heat waves and drought periods. An anaerobic environment combined with the filtering characteristics of the peat body results in the conversion of soluble nutrients and pollutants (denitrification) and/or the sequestration of substances in the peat, thereby removing them from the nutrient or carbon cycle (Joosten & Clarke 2002). These functions promote high water quality of adjacent lakes and rivers and thus the supply of drinking water. Continuous peat formation leads not

---

Box 2.3: Ecosystem services of peatlands

Hans Joosten

In many cases, the utilisation of peatlands leads to irreversible destruction of the ecosystem. Therefore, we distinguish between peat-conserving services that are compatible with the formation and preservation of peat, and peat-consuming services, which eventually destroy the prime feature of peatlands: the presence of peat. Table 2.3 presents an overview of peatland ecosystem goods and services that builds on the Common International Standard for Ecosystem Services (CICES, Haines-Young & Potschin 2011). Unlike CICES, which does not include the supply of subterranean goods such as coal or crude oil in its classification, we consider the supply of peat as an (unsustainable) service of peatlands. In contrast to fossil coal and crude oil, which bear no functional relationship to the covering ecosystems, peat is a functional and defining part of the peatland ecosystem. Like wood in a forest, peat originates within the peatland ecosystem and contributes to its self-organisation and self-regulation. Whereas coal forms by a chemical process that merely requires specific physical conditions (high temperature and pressure) and proceeds in the absence of life, peat formation is a biological process, which is bound to the peatland ecosystem.

The supply of space is nowadays a further distinctive feature of peatlands. Most peatlands are difficult to access and not inhabitable, which makes them in many countries the last spaces unoccupied by humans. Therefore, peatlands often fall victim to the expansion of infrastructure. The common occurrence of peatlands in valleys and on wind-exposed places predestines them for the installation of water reservoirs or wind farms. The value of peatlands in providing these services does not derive from their characteristics as peatlands but from the parallel demands on the characteristics of the landscape.

Table 2.3: Peatland ecosystem goods and services (cf. Joosten & Clarke 2002) according to the Common International Standard for Ecosystem Services (CICES, Haines-Young & Potschin 2011).

Section	Division	Group	Subgroup	Examples of goods and services provided by peatlands	
				(Potentially) peat sequestering (undrained)	Peat degrading (drained or deeply flooded)
Provisioning services	Nutrition: Food and fodder	Natural		Wild game and fowl, fish, berries, mushrooms, sago, honey	
		Supported	Managed game	Meat of reindeer, deer or ptarmigan	Idem from high density populations that degrade peat by trampling, overgrazing or fire management
			In situ fodder	Fodder for livestock grazing wet peatlands (e.g. Water Buffalo)	Fodder for livestock grazing drained peatlands (e.g. high productivity dairy cattle)
			Ex situ fodder	Hay and silage from wet fen plant material	Hay and silage from drained and fertilised peatland
		Cultivated		Oil from Shorea-species, starch from sago	Carrots, potatoes, palm oil, maize and so on
	Water	Drinking, irrigation, industrial and cooling water		Outflowing (surplus) water	Withdrawn surface and groundwater
	Materials	Medicine and delicacy	Pharmaceuticals	Medicinal plants (and animals) e.g. Drosera, Menyanthes, Ledum	Humic preparations, peat baths and poultices, peat based fungi- and bactericides, active coal from peat
			Flavours	Plants for flavouring drinks (e.g. Menyanthes, Acorus, Hierochloe)	Peat for flavouring whiskey
		Fibres	Construction materials	Plants (z.B. Phragmites, Typha) for thatching, insulation, building, wattling and veneer	Peat as foundation, building and insulation material; wood from drained peatland
			Clothing and textiles	Fur, leather, wool	Cottongrass peat fibre, hemp, wool from high intensity sheep grazing
			Pulp for paper and cellulose	Biomass from Phragmites, Phalaris, Papyrus, Typha	Wood from Pinus, Picea, Acacia
			Absorption, filter and bedding materials	Litter from biomass	Peat for litter in stables, filters, active coal, oil spill absorbent, diapers
			Growing media, potting soils	Peatmoss biomass, biomass compost	Peat as constituent of horticultural growing media
		Fertilisers	Nutrient enrichment	Compost of fen biomass	Peat ash as potassium fertiliser, fen peat as nitrogen fertiliser
			Improvement of soil structure	Biomass compost	Peat for improving soil structure
		Chemicals	Raw materials for chemistry	Refined plant sap, latex (jelutung)	Peat waxes and dyes, active coal made from peat

2.4 Ecosystem services of peatlands

Section	Division	Group	Subgroup	Examples of goods and services provided by peatlands	
				(Potentially) peat sequestering (undrained)	Peat degrading (drained or deeply flooded)
Provisioning services (cont.)	Fuel	Fossil fuel		Marsh gas (methane)	Peat and peat-derived fuels
		Biomass based fuel		Reed, sedges, wood	Palm oil, maize for biogas production, wood, sugar cane for alcohol production
	Space	... for biomass provision		(See nutrition, materials and fuel)	(See nutrition, materials and fuel); fish ponds
		... for urban, industrial and infrastructural development		Space for some wind farms, some transport infrastructure	Space for settlements, harbours, airports, industry complexes, hydro-electricity reservoirs, landfills
		... for defence and isolation		Space for low intensity military training grounds	Space for high intensity military training grounds
				Little managed defence and border lines	Intensively managed defence and border lines
				Space for prisons and labour camps	Associated peatland drainage and reclamation
Regulating services	Regulation of waste	Bioremediation		Denitrification, nutrient retention and sequestration in plants and peat	Wastewater treatment, intensive denitrification
		Dilution and sedimentation		Clean water supply to dilute downstream pollution, filtering out of pollutants	-
	Regulation of flows	Regulation of water flow		Attenuation of run-off and discharge rates, mitigation of downstream floods	
				Maintenance of base flow, coastal protection	Rapid discharge and increased buffer capacity after drainage
		Regulation of mass flow		Erosion control	
	Regulation of the physical environment	Global climate		Carbon sequestration and storage in peat	Idem in biomass and litter in some boreal peatland forests (temporarily)
		Local and regional climate		Evapotranspiration cooling	
		Water quality		Nutrient retention, denitrification	Waste treatment, denitrification
		Soil conditions		Peat accumulation, initiation and conservation of permafrost	Improved soil structure through secondary pedogenesis, conservation of permafrost
	Regulation of the biotic environment	Life cycle maintenance and habitat protection		Pollination, seed dispersal	
				Wildfire control	
		Pest and disease control		Control of pathogens and invasive species	
		Gene pool protection		Rare and specialised mire and wetland species	Rare species of (slightly) drained fen meadows

2 The limits of drainage based peatland utilisation

Section	Division	Group	Subgroup	Examples of goods and services provided by peatlands	
				(Potentially) peat sequestering (undrained)	Peat degrading (drained or deeply flooded)
Cultural services	Symbolic	Aesthetic appreciation and inspiration		Areas of Outstanding Natural Beauty, mire patterning	Use of peat and fossil bog wood for artisan objects
		Heritage		Themes for arts and literature	
				Tradition, history and notions of cultural continuity, sense of place	Traditional peat extraction and land use, sense of place
		Symbols and mascots		Hunting trophies, Canadian beaver and Japanese crane as national symbols	
		Reflection and spiritual / religious enrichment		Wilderness, naturalness, quietness, solitude	Wide open spaces, wide horizon
				Notions of ecological and evolutionary connectedness, timelessness and naturalness	Sense of control over the landscape
				Sacred places and species	
	Intellectual and experiential	Recreation and community activities	Recreation and stress mitigation	Tranquility and scenery for tourism and outdoor activities, opportunity for hunting/angling and wildlife watching	
			Social amenity	Employment and volunteering in mire conservation and research	Employment in peat extraction and processing and in drainage-based agriculture and forestry
		Information and knowledge	Cognition and satisfaction of curiosity	Stratigraphical archives (palaeo record, preservation of archaeological artefacts)	
				Extreme habitat conditions and special adaptations of mire organisms, (reference for) self organisation – and regulation	Cultural land use history and sociology, behaviour of disturbed systems
			Indication	Palaeoecological record, indicator organisms	
			Education	Subject matter for educational literature, field excursions, presentations	Idem with respect to peat extraction, agriculture, forestry, water management and road building
	Transformation	Character development		Options for development of new tastes, moral and social skills, and growing awareness of evolutionary and ecological connectedness	
	Option and bequest	Continuous provision of ecosystem services		Benefits that still have to be discovered	Benefits that still have to be discovered

2.4 Ecosystem services of peatlands

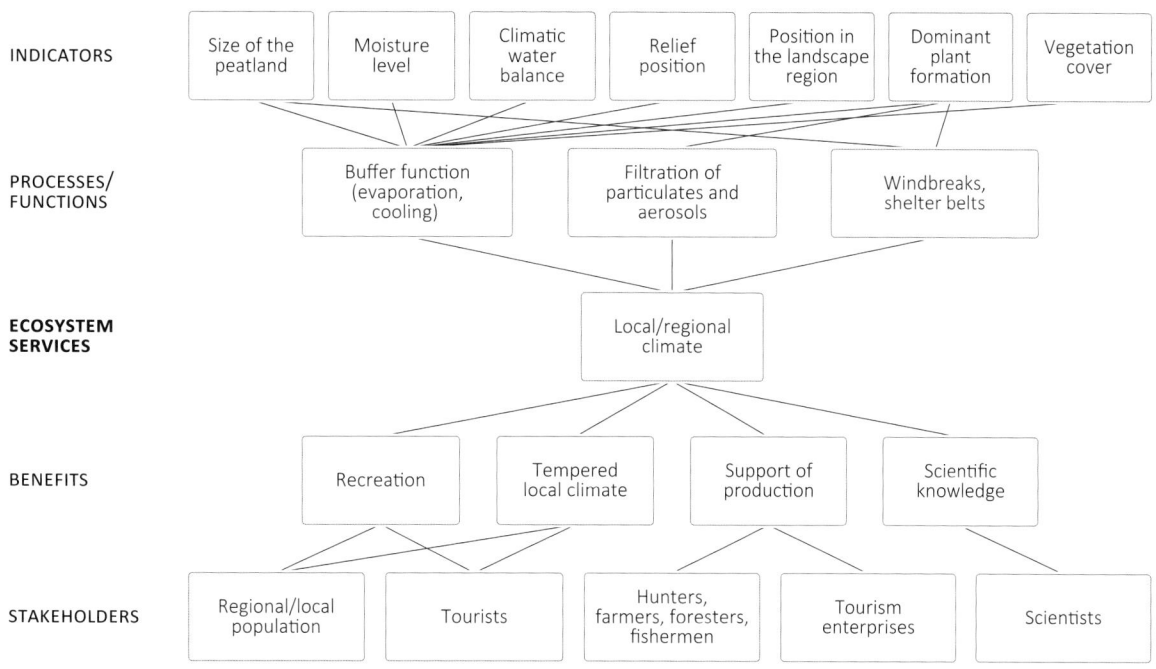

Figure 2.8: Factors determining the performance of the peatland ecosystem service 'regulation of the local/regional climate' (after Schröder et al. 2013).

only to the sequestration of carbon and mineral components of plants but also to the production of a raw material.

Mires are habitats with very specific characteristics such as excess water availability, a cool and moist microclimate and a continuously upward growing surface, whereas some mire types are extremely nutrient poor and acidic. They can only be populated by organisms that are adapted to these extreme conditions. In the course of time, very special biotic communities have developed in mires that with their food webs, flow paths of matter and energy and other interrelations represent particularly valuable ecosystems. Thus, a considerable part of biodiversity is found in these special habitats, even when they cover only a small area of the landscape (Luthardt & Zeitz 2014).

In these ecosystems, which often have a history of development that may exceed 10,000 years, natural process dynamics – both of the present and of the past, can be observed and studied. As places of wilderness, mires play a prominent role in science, as well as for recreation, experiencing nature, inspiration and environmental education.

2.4.3 The change of ecosystem services by drainage

Up until now, peatlands had to be drained in order to generate provisioning services. Drainage induces a wide range of processes (Chapter 2.2 and 2.3) and particularly affects the regulating services of peatlands: Peat accumulation stops and irreversible peat degradation sets in, nutrients sequestered in the peat are released, the hydrological balance is severely disturbed, and the habitats of specialised mire organisms disappear to make place for those of more ordinary species. Carbon sinks turn into sources that very rapidly release considerable amounts of carbon and other substances that had been sequestered over a very long time span. The filter and buffer function is dramatically reduced.

In Northern Germany, productive use of drained peatlands is largely directed at fodder production, peat extraction, and recently also at the production of biomass for energy generation. The landscape does not differ any more from that of other terrestrial areas, so that the special ambiance of mires can no longer be experienced. Instead, accessibility and cultural components start to characterize the landscape. Only the open space with its wide horizon remains. Science has to redirect its research focus on conversion processes, the behaviour of disturbed systems and succession.

Table 2.4: Selected ecosystem services of peatlands in dependence of water management and utilisation
(Positive impact: +++= strong, ++= medium, += existing; negative impact: ---= strong, --= medium, -= existing; 0= no influence: ~= changing over time).

Section	Group	Pristine/ near natural	Drained with land use	Drained & abandoned	Rewetted & unused	Paludiculture
Provisioning services	Food and fodder	+	+++	0	0	++
	Plant fibres (building material, litter, substrate)	+	++	0	0	++
	Biomass fuel	0	++	0	0	++
Regulating services	Climate regulation (global & local)	++	---	---	+++	+++
	Water purification/ nutrient retention	++	---	---	+~	++~
	Regulation of the water cycle	+++	---	---	++	++
	Habitat for specialised species/gene pool	+++	+	--	++~	++~
Cultural services	Experience of nature, recreation	+++	+	+	++	+
	Information and knowledge (processes, archive)	+++	+	+	++	++

Regional value creation used to be positive over long time as it was based on local resources. However, the growing inputs of material and energy that drainage demands make the balance increasingly negative. Eventually, conventional drainage based generation of provisioning services on peatlands is a cul-de-sac (Chapter 2.3).

2.4.4 Ecosystem services of rewetted peatlands

The rewetting of degraded peatlands focuses in particular on the re-establishment of mire typical biodiversity, on the improvement of water quality, on the reduction of greenhouse gas emissions and on the restoration of other regulating services. Some of these aims can be achieved almost instantly; other functions may take an extended time span to fully recover. Rewetting also secures cultural services by preserving the peat layer as an archive of climate, landscape and cultural history. Up until now, however, rewetting of peatlands largely implied the renouncement of the production capacity of peatlands.

Paludiculture is the only form of land use that allows using the provisioning capacity of peatlands without substantially compromising the supply of regulating or cultural services. In comparison to drained peatlands the latter services improve considerably (Table 2.4, Chapter 5).

3 Production and utilisation of paludiculture biomass

The aim of paludiculture is to produce plant biomass on wet or rewetted peatlands. This results in a range of services, whose renumeration allows an economically profitable use of the peatlands. This Chapter discusses the production and utilisation aspects and introduces the most important plant species for paludiculture (Chapter 3.1). A wide range of applications, such as food, medicine (Chapter 3.2), fodder (Chapter 3.3), industrial raw material (Chapter 3.4), and fuel for energy generation (Chapter 3.5 and Chapter 3.6) is discussed.

> **Box 3.1: Database of Potential Paludiculture Plants (DPPP)**
>
> Susanne Abel
>
> Apart from those described in Chapter 3.1, there are many other plant species that are currently not used on a regular basis but whose cultivation is possible at peat preserving water levels. On the one hand, many of these wetland plants have a long history of use, which is now partially forgotten or regionally limited. In any case, this results in an increasing loss of traditional knowledge. On the other hand, many new forms of plant utilisation – such as the use of biomass for energy production or the production of natural construction materials, provide new exploitation perspectives.
>
> The 'Database of Potential Paludiculture Plants' (DPPP) gathers information on plant species appropriate for paludiculture from all over the world (Abel et al. 2013). For each relevant species a 'plant portrait' is made (Figure 3.1, Table 10.6), which summarises, standardises and analyses the gathered information in a systematic way. To date (March 2016) the data base contains information on 1128 plant species, including 346 herb and forb, 410 tree, 145 grass, 184 shrub and 41 fern species, as well as one genus of moss. The species are divided into seven utilisation categories: Food (including spices); fodder; medical plants; raw material for industrial processing; fuel; decoration (ornamental plants, fragrance); soil conditioner and horticultural growing media. The current potential for paludiculture of about 250 plant species is considered to be good because a market already exists for products made of the above-ground parts. The DPPP serves as an important source of information to combine innovative ideas with existing knowledge and to develop new ways of production for paludiculture.
>
>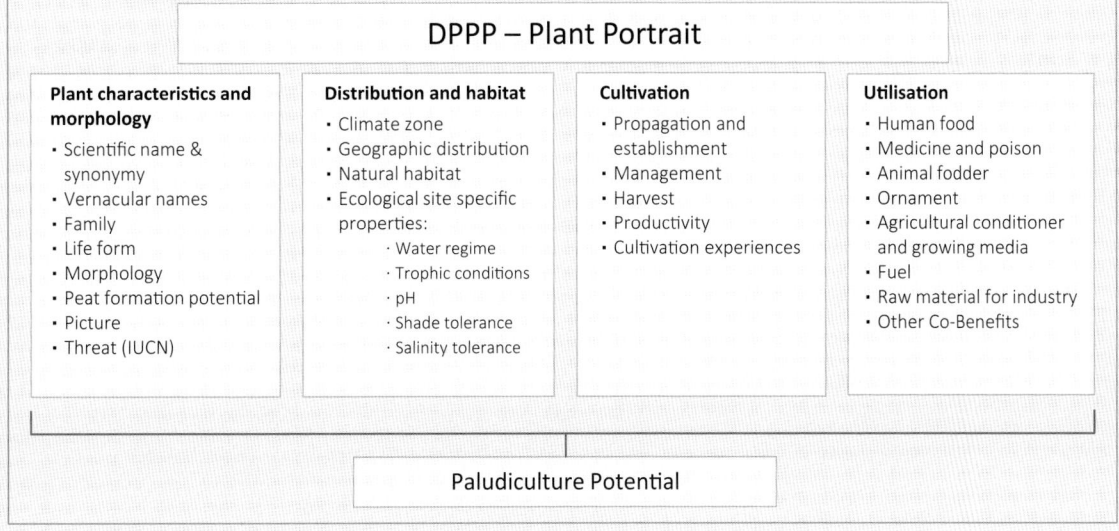
>
> Figure 3.1: Available information about plant species listed in the Database of Potential Paludiculture plants (DPPP) (Abel et al. 2014).

3.1 Promising plants for paludiculture

Claudia Oehmke & Susanne Abel

Paludiculture plants are wetland plants that produce useful biomass in sufficient quantity and quality and simultaneously contribute to preserving the peat. During harvest, the below-ground biomass generally remains untouched to avoid disturbing the peat body. The following chapters present the characteristics, cultivation methods and utilisation options for selected plant species that are already widely used – or would be suitable, for large-scale cultivation in Central Europe. For fen peatlands, the focus is on Common Reed, Reed Canary Grass, Cattail, Saw Sedge, Black Alder and several species of sedge (Colour pictures 13–17). For bog soils, peatmosses (*Sphagnum*) are presented as an example. A more comprehensive overview of plant species suitable for paludiculture is given in the 'Database of Potential Paludiculture Plants' (DPPP; Box 3.1; Figure. 3.1).

3.1.1 Common Reed (*Phragmites australis*)

Common Reed is a reedbed forming grass species of the Poaceae family. The species occurs all over the world along lakes, water courses and ponds, as well

---

Box 3.2: Harvest times

Wendelin Wichtmann

Vegetation composition is largely determined by harvest time; for instance, a lasting use of reedbeds is only possible if reed is harvested in winter (Chapter 3.1.1). The harvest time is on the other hand determined by how the biomass will be utilised (Table 3.1). The utilisation of herbaceous biomass for biogas production or fodder, for example, requires biomass harvested in summer, when the content of soluble carbohydrates and protein is high (Chapter 3.4.7). For using reed as a construction material or as a fuel, the plants for biomass should preferably be harvested in late winter. Many nutrients that may frustrate biomass use for the above mentioned purposes are relocated to the rhizomes or leached out by precipitation during autumn and winter.

The appropriate harvest time must be fine-tuned with the requirements of other ecosystem services that paludiculture is expected to provide. If, for example, harvest should not take place before the 15th of July in order to protect breeding meadow birds (Thomas et al. 2009), this would prevent the use of the biomass for biogas production via fermentation (DLG 2012). In such case, biomass can be harvested later and alternatively used for pellet or briquette production.

Table 3.1: Utilisation options for biomass from paludiculture (after Wichtmann & Schäfer 2007).

Utilisation		Vegetation	Harvest time
Livestock	Fodder (hay, silage)	Wet meadows	Early summer
	Pasture	Wet grasslands	Whole year
	Litter	Wet meadows, reed beds	Summer/autumn
Raw material	Thatch	Reed	Winter
	Mouldings	Reed	Autumn/winter
	Construction/insulation	Reed and Cattail	Winter
	Paper (cellulose)	Reed and Canary Reed Grass	Winter
	Baskets	Willow	Autumn
	Wood for building and furniture	Black Alder	During frost
Energy	Heat by direct combustion	Wet meadows, reed beds	Autumn/winter
	Biogas/Alcohol by fermentation	Wet meadows, reed beds	Early summer
	Liquid fuels	Wet meadows, reed beds	Whole year
Other	Medical herbs	Pristine and near natural mires/speciality crops	Early summer
	Food	Pristine and near natural mires/speciality crops	Summer/autumn
	Growing media	Peatmoss (*Sphagnum*)	Whole year

as in flood plains, swamps and mires. Thus, reed is a major peat forming plant, which is usable in many ways as a building material (Chapter 3.4) and as an energy source (Chapter 3.5).

Depending on nutrient availability and genetic characteristics, reed can grow up to 4 m high. Its dense below-ground rhizome system serves to store nutrients and supports the strong competitiveness of the species. Water is the most important factor for the growth of reed, and its long-term cultivation requires a water level between 0.5 m below ground and 2 m above ground (Ostendorp 1993, Tschoeltsch 2008). Reeds and reedbeds are highly varied because of their genetic diversity. Diploid (2n=24) and tetraploid (4n=48) genotypes dominate in Europe, whereas in Asia and North Africa, octoploid (8n=96) types are widely distributed (Haslam 2010).

The successful cultivation of reed depends on water level, competition with other plants and the method of establishment. In nature reed spreads in a vegetative way or by seeds (Ostendorp 1993, Haslam 2003). After rewetting, reed may within a period of 3 to 5 years spontaneously spread over large areas by vegetative distribution, provided the species has persisted

---

Box 3.3: Hiss Reet Schilfrohrhandel GmbH

Jan Felix Köbbing & Sabine Wichmann

The Hiss Reet Schilfrohrhandel company from Bad Oldesloe (near Hamburg) is the largest reed retailer in Germany (Jedack 2010). The company was founded in 1833 on the isle of Fehmarn and since the 1920s it is involved in national and international reed trade.

Originally, the reed that was traded by the company came from local sources such as the islands of Fehmarn and Usedom, as well as other coastal areas of the North and Baltic Seas. In the 1970s, the demand could no longer be covered from native sources and reed was then imported from Hungary and Austria (Lake Neusiedl). In the 1980s, reed was sourced from Turkey, where a subsidiary company was founded in 1984 (Hiss Kamis Sanayi in Mersin). This event marked the turning of the company from being a wholesaler to become a reed producer. In 1991 a second subsidiary was founded in Hungary (Kecskemet) and in 1998 an additional subsidiary (Stuf Delta Production in Tulcea) acquired the Romanian state-owned reed exploitation structures in the Danube Delta (Kösling 2000; Schwalm 1999, 2005). Additionally, the company buys reed from Ukraine, Poland, Austria and since 2008 also from China (Table 3.2).

In countries with high labour costs such as Germany, the Netherlands and Denmark, combined reed harvesters are used, which automatically pre-clean the reed bundles (Chapter 4.2). In Romania and Turkey, the trend towards mechanisation of harvest did not proceed beyond the use of sickle bar mowers and Seiga machines. In Turkey, reed is still harvested by hand with a sickle, whereas the removal of the reed bundles takes place with amphibian tractors and trailers (Hiss 2013).

In Germany the circumference of reed bundles that are ready for sale is 60 cm. The length of the bundles varies depending on the harvested stock, with different reed lengths being used for different purposes.

The core business of the Hiss company (80% of the total business volume) is reed for thatching, whereas its use for garden products and natural construction materials for interior architectural works has been recently growing in importance (Hiss 2013).

About 10 people work in the headquarters of the company in Hamburg. The subsidiary company in Romania employs 70 workers but in harvesting time their number increases to 160 (Schwalm 1999). The company in Turkey employs 15 workers (Sobottke & Strunk 2007). Hiss Reet also distributes reed in Europe; the main importing countries are Denmark, the Netherlands, the United Kingdom and Ireland (Sobottke & Strunk 2007).

Table 3.2: Reed export to Germany in 2013 (Wichmann & Köbbing 2015).

		Austria	China	Hungary	Poland	Romania	Turkey	Ukraine	Total
Export volume	Million bundles	0.1	0.2–1	0.6–2	0.1–1	0.4–1	0.3–1	0.3–1	2–7
Price	€ per bundle	n.d.	variable	2.60	2.60	3.00	2.40–3.00	2.80	

Table 3.3: Average yields of paludiculture plants at different harvest times and on different sites in t DM ha^{-1} a^{-1}. (f = fertilised, uf = unfertilised).

Species	Yield t DM ha^{-1} a^{-1}	Harvest time	Site type	Country	References
Carex acuta	3.8	Summer	River valley mire, natural vegetation	Poland	Grzelak et al. 2011
	8	July	Fen, natural vegetation	Netherlands	Olde Ventering et al. 2002
Carex acutiformis	6.1	June–July	Rewetted fen, spontaneous succession	Germany	Steffenhagen et al. 2008
	4.2–7.6	Summer	Valley mire	Poland	Grzelak et al. 2011,
Carex riparia	5.3–11.1	May–Sep.	Rewetted fen	Germany	Steffenhagen et al. 2008, Schulz et al. 2011
Phalaris arundinacea	4.5	July	Cutover bog, annual cultivation	Estonia	Heinsoo et. al 2011
	7.3	October			
	5.5	April			
	8	September	Cutover bog, 3-year cultivation, unfertilised	Estonia	Mander et al. 2012
	7.9	April			
	13.9	September	Cutover bog, 3-year cultivation, fertilised	Estonia	Mander et. al. 2012
	12.7	April			
	9.6	Winter	Rewetted fen, spontaneous succession	Belarus	Wichtmann et al. 2013
	4.7–9.3	May–September	Rewetted fen, spontaneous succession	Germany	Schulz et al. 2011
Phragmites australis	8.7	August	Terrestrialisation mires, natural vegetation	Germany	Kühl & Kohl 1992
	10.7	Winter	Rewetted fen peatland, mowing regime	Belarus	Wichtmann et al. 2013
	11	January–March	Fen peatland, spontaneous succession	Germany	Timmermann 2009
	6.2–16.3	August/September	Brackish and freshwater peatland, cultivation	Romania	Hanganu et al. 1999
	6.5–23.8	May–September	Fen peatland, spontaneous succession	Germany	Steffenhagen et al. 2008, Schulz et al. 2011
Typha spp.	~15	March–May	Gley and peat soil (salt marsh), natural vegetation	Canada	Grosshans et al. 2011, Grosshans pers. comm., 04/2013
T. angustifolia	6.9	September/October	Cutover fen peatland, 2-year cultivation	USA	Pratt et al. 1984
	11.5	May–October	Natural vegetation	Canada	Bonneville et al. 2008
	12.5	September/October	Valley mire, 3 year cultivation	USA	Pratt et al. 1984
	6.5–14	Winter	Rewetted fen peatland, cultivation	Germany	Heinz 2011
T. glauca	8.1	September/October	Valley mire, 2-year cultivation	USA	Pratt et al. 1984
	10.3	August	Natural vegetation, fertilised	USA	Woo & Zedler 2002
T. latifolia	4.6	September/October	Cutover fen peatland, 2-year cultivation	USA	Pratt et al. 1984
	6.2	September/October	Valley mire, 2-year cultivation	USA	Pratt et al. 1984
	7.8–12.1	May–September	Rewetted fen peatland, succession	Germany	Steffenhagen et al. 2008 Schulz et al. 2011
T. x glauca and T. angustifolia	4.7–10.5	September/October	Different peat soils, cultivation	USA	Pratt et al. 1988

Colour picture 1: Subsidence of 100 cm in a period of 50 years apparent from drainage pipes installed in a depth of 120 cm in 1958 (Hankhauser Moor, Lower Saxony, Germany; Sabine Wichmann, April 2011).

Colour picture 2: Peatland cultivation by deep ploughing (Rhinluch, Germany; Michael Succow, 1988).

Colour picture 3: Peatland subsidence apparent near Malchin (Mecklenburg-West Pomerania, Germany; Michael Succow, 1978).

Colour picture 4: Flooding and waterlogging after heavy rain as a result of soil compaction (Niederlausitz, Germany; Paul Schulze, June 2013).

Colour picture 5: Peat fires cause immense emissions worldwide and might smoulder within the peat body for several months. Effective re-wetting can prevent peat fires (Central Kalimantan, Indonesia; Hans Joosten, April 2007).

Color picture 6: The use of adapted conventional agricultural machinery equipped with low-pressure tyres in rewetted peatlands is often restricted by the traficability of the site (Relzower Wiesen, Germany; Mike Stegemann, 2010).

Colour pictures 1–20

Colour picture 7: Pricked out reed plants in April after having been sown in January (Botanical Garden Greifswald University; Wendelin Wichtmann, May 2007).

Colour picture 8: When 40–60 cm tall in June, the reed plants are ready for planting (Neukalen, Mecklenburg-West Pomerania, Germany; Wendelin Wichtmann, 2007).

Colour picture 9: Planting of reed in June with a planting machine normally used in forestry (Neukalen, Mecklenburg-West Pomerania, Germany; Wendelin Wichtmann, 2007).

Colour picture 10: Common Reed planted in a 1 x 1 m raster (Neukalen, Mecklenburg-West Pomerania, Germany; Wendelin Wichtmann, 2007).

Colour picture 11: Plantation of Slender Tufted Sedge (*Carex acuta*) (Biesenbrow, Brandenburg, Germany; Wendelin Wichtmann 1996).

Colour picture 12: Already two years after planting a dense reedbed has established (Biesenbrow, Brandenburg, Germany; Wendelin Wichtmann, summer 1998).

Colour picture 13: Saw Sedge (*Cladium mariscus*) forms dominant, highly productive stands in calcareous fens (Przirody Brzezna, Poland; Christian Schröder, October 2012).

Colour picture 14: Reed Canary Grass (*Phalaris arundinacea*) forms highly productive dominant stands in rewetted, nutrient rich peatlands (Zarnekow, Mecklenburg-West Pomerania, Germany; Wendelin Wichtmann, 2013).

Colour picture 15: Plantation of Cattail in the 'Donaumoos' within a research project of Munich Technical University (Donaumoos, Bavaria, Germany; Wendelin Wichtmann, October 1998).

Colour picture 16: After rewetting sedges (*Carex* spp.) often develop dominant stands (Relzower Wiesen, Mecklenburg-Vorpommern, Germany; Christian Schröder, 2011).

Colour picture 17: Plantation of Black Alder (*Alnus glutinosa*) in a rewetted fen in the Trebel river valley (Brudersdorf, Mecklenburg-West Pomerania, Germany; Achim Schäfer, 2006).

Colour picture 18: Water Buffalos have low requirements on forage quality and are well-adapted to grazing rewetted peatlands (Rietzer See, Brandenburg, Germany; Malte Wenzel, 2010).

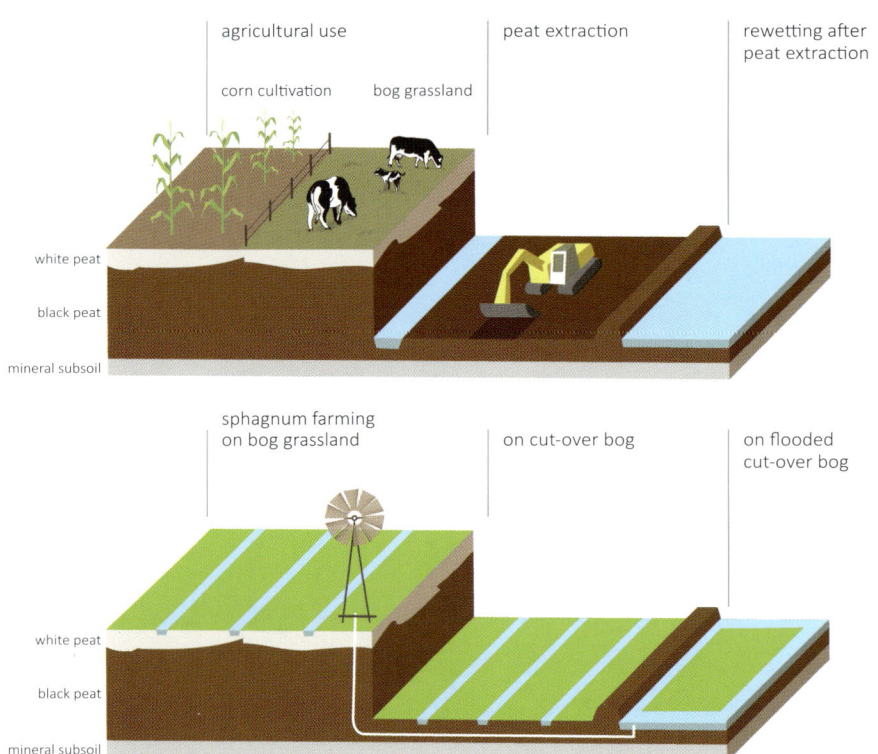

Colour picture 19: Conventional land use on drained raised bogs (top) versus *Sphagnum* farming (bottom) (after Gaudig et al. 2014).

Colour picture 20: Cultivation of Peatmoss (*Sphagnum* spp.) on former raised bog grassland after removal of topsoil. *Sphagnum papillosum* and *Sphagnum palustre* three years after establishment (Hankhauser Moor, Lower Saxony, Germany; Sabine Wichmann, March 2014).

3.1 Promising plants for paludiculture

> **Box 3.4: Fireproof board made of reed**
>
> Anne Wollert
>
> It is important to note that reed is suitable as the sole filling and fibre raw material for manufacturing building and insulation boards that meet the highest standards of fire safety (Colour picture 32). In contrast to other commercially available products, reed boards (thickness 22 mm, density 1,260 kg/m³) are easier to process because of their low specific weight. The boards have good insulating properties (Table 3.4) and a high permeability to water vapour, which improves the indoors climate. The material hardly expands after submergence in water for 24 hours and no irreversible changes are found afterwards, which implies that the board is, in fact, water resistant. Its great compressive strength and resulting good load-bearing capacity makes the board suitable for interior construction. The homogeneous distribution of the particle size of the utilised biomass, which allows attaining high material density of the board, is decisive for the load-bearing capacity.
>
> Table 3.4: Technical data of the fire safety board.
>
> | Property | Data |
> |---|---|
> | Depth (b) | 22 mm |
> | Density (ϱ) | 1259.63 kg m^{-3} ± 10% |
> | Thermal conductivity (λ) | 0.43 W mK^{-1} |
> | Water vapour resistance (μ) | 5.55 |
> | Expansion in depth (b_q) | 0.53 mm |
> | Compressive strength (P_R) | 623.95 N mm^{-2} |
> | Building material class (DIN13501) | A2, not flammable |
>
> The high silica content makes reed resistant against mould, which improves the durability of the building material. Mineral glue is used that does not require high temperatures nor pressure during manufacturing, but needs more time to harden. After use, the building material can be environmentally friendly recycled. All the aforementioned properties can be contrasted to those of conventional chip or fibre boards.

in nearby ditches and drains (Hawke & José 1996). A well-proven way to establish reed is by planting seedlings grown from seeds that have been collected in winter from donor plants on ecologically similar sites in the near surroundings (Hawke & José 1996, Timmermann 1999). Seed germination increases by frost exposure and reaches about 80% when seeds of good quality are used (Tschoeltsch 2008). The seedlings are raised in a greenhouse and planted into the open field from June up to August at the latest (Kersten et al. 1999, Timmermann 1999, Lemm 2005; Colour pictures 7–10). The density of planted seedlings should lie between 0.25–4 plants per m² depending on competition. Tilling prior to planting inhibits competing plant species. During the first two years after planting, the water level should ideally be kept slightly below the ground surface (Kersten et al. 1999). Establishing the reedbed via plants derived vegetatively from rhizomes implies more work and higher costs (Roth 2000). Direct on-site sowing or planting of reed stems is likewise not recommended because such plants are sensitive to flooding, frost and competition (Timmermann 1999, Roth et al. 2001).

About three years after planting, the reedbed can be harvested for the first time (Colour picture 12). In Germany, reed harvesting is regulated by the Federal Nature Conservation Act (*Bundesnaturschutzgesetz, § 39 (5) No. 3*) as well as by supporting guidelines of the individual federal states that regulate harvesting times (Box 3.2, Chapter 7.2). The harvest time depends on how the biomass will be used. Reed for thatch is usually harvested between January and March after the first frost, when the reed stalks are dry and have shed most of the leaves. For thermal utilisation, the reed is harvested in late winter because the content of combustion critical substances decreases during that season (Chapter 3.5.3). If stubbles from last year's cut become flooded in spring, the reed plants become weakened by oxygen deficiency. This means that if regulation of the water level is not possible, the cut should be made high enough to prevent subsequent inundation of the cut stalks (Haslam 2010). Annual mowing during summer restricts growth and photosynthesis during the second half of the growing season (Haslam 1969). As a result, the reed plants become weakened and may be outcompeted entirely because fewer nutrients are stored in the rhizomes to support the sprout in the following growing season. However, summer mowing of reed does foster biodiversity and is often applied in nature conservation to support animal and plant species such as the Aquatic Warbler (Chapter 5.2).

Annual yields of reed depend on site conditions and reed genotype. When harvested between August and September, reed from peatland sites yield 6.5–23.8 t DM ha^{-1}a^{-1} (Steffenhagen et al. 2008, Schulz

Box 3.5: Lower and upper heating value

Christian Schröder & Christian Jantzen

The lower heating value Hi (net calorific value) is the heat that is released with entire oxidation of a fuel. It is the maximum usable amount of heat produced by combustion when no condensation of steam takes place. In contrast, the gross calorific value (or upper heat value Ho) is the heat that is released by combustion including the condensation and cooling down of the combustion gases to 25 °C.

The heat value is determined by quantitative combustion in a calorimeter under an excess of oxygen and high pressure. The heat value of a substance cannot be determined directly. It is deducted by subtracting the vaporization enthalpy of the water from the net calorific value. The mass of vapour water in the exhausting gas is depending on the physical bound water (water content) and the relatively stable content of hydrogen (approximately 6 % DM) in the biomass. This leads to net calorific values that are approximately 10 % lower than the gross calorific values. In practice, the temperature of the exhaust gases can only be cooled down to approximately 60°C, thus the net calorific value is the decisive quantity to assess the useable energy (Kaltschmitt et al. 2009).

With a typical moisture content of 10 % Common Reed and Reed Canary Grass reach net calorific values of 15.1 MJ kg^{-1} and 15.5 MJ kg^{-1}, respectively, i.e. values that lie within the range of wood reference samples (14.5–16.8 MJ kg^{-1}) and above those of reference samples of other gramineous biomass (14.7 MJ kg^{-1}). These data illustrate that positive results are to be expected from combustion of Reed in small scale commercial boiler systems.

Box 3.6: Reed Canary Grass from rewetted peatland as fodder for horses

Luisa Zielke

Horses need fodder that is rich in fibre. Fodder from conventional pastures often results in obesity and 'prosperity diseases' such as laminitis, equine metabolic syndrome and equine cushing syndrome, in particular with ponies and small sized horse breeds. Against this background, the composition of late harvested Reed Canary Grass (*Phalaris arundinacea*) from two typical rewetted peatland sites in Mecklenburg-West Pomerania was determined. The average fructan content was below 5% and therefore unproblematic regarding risk of laminitis. Hay from Reed Canary Grass can be fed in larger amounts than hay from conventional meadows without exceeding the daily requirement of digestible protein (Figure 3.2). The larger amount of fodder keeps the horse busy whilst it eats, which contributes to species appropriate animal husbandry. Late harvested Reed Canary Grass from humid fen peatlands is generally suitable as feed for horses if fodder hygiene and fast drying can be ensured.

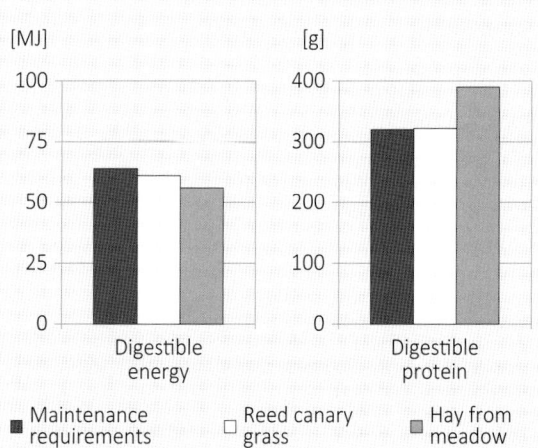

Figure 3.2: Fulfilment of nutritional maintenance requirements of a horse (500 kg body weight) fed with 7.5 kg hay from Reed Canary Grass or with 7.5 kg hay from common meadow per day.

et al. 2011, Table 3.3). Winter harvest of reed yields 4.6–15 t DM ha^{-1} a^{-1} (Knoll 1986, Timmermann 2009, Wichtmann et al. 2014). During winter, the dry biomass available for harvesting is reduced by about 25 % (Knoll 1986).

Reed is a traditional material for thatching roofs and for insulating buildings (Box 3.3; Box 3.4, Chapter 3.3.5). If harvested during winter, reed is also suitable as a fuel that can be burned in heating plants whether in a pure form or mixed with other biomass fuels. The average lower heating value LHV (= net calorific value) of Common Reed is 17 MJ/kg free from water (Kitzler et al. 2012, Wulf 2009, Kask et al. 2007; Box 3.5).

3.1.2 Reed Canary Grass (*Phalaris arundinacea*)

Reed Canary Grass is a perennial robust grass of the Poaceae family that forms naturally dominant stands in floodplains and along rivers (El Bassam 2010). Until the mid 19th century, Reed Canary Grass from floodplain meadows was important for its use as hay for horses (Box 3.6). In East Germany, productive varieties of Reed Canary Grass were cultivated on drained fen peatlands as intensively used sown grassland (Kreil et al. 1982). Nowadays, the species is cultivated in Northern Europe and the USA as an energy crop and as a raw material for paper production (Pahkala & Pihala 2000, Lewandowski et al. 2003, Prochnow et al. 2009).

Reed Canary Grass looks very similar to Common Reed but does not grow as tall. Like reed, the species emerges from a deep and extended stoloniferous rhizome, which forms a dense turf and increases the load-bearing capacity of the ground (Arny et al. 1929). The culms grow 0.5–2 m tall and are rather sturdy. Reed Canary Grass prefers periodically wet and nutrient rich soils with pronounced flooding phases, and a mean water level of 0–20 cm below ground (water level class 4+, Box 5.3). On such sites, the species is very competitive (Colour picture 14). However, Reed Canary Grass does not tolerate constant flooding and is not a peat-forming species.

Reed Canary Grass is suitable for cultivation on moderately wet sites (Wichtmann & Succow 2001). In Northern Europe, the species has been cultivated for several decades, including on cutover bogs (Shurpali et al. 2009, Heinsoo et al. 2011). However, these sites are mostly too dry to prevent further peat degradation and mineralisation. For establishment, 15–25 kg of seeds per hectare are sown during spring until late summer at a depth of 1–2 cm and with a row distance of 12.5 cm (Lewandowski et al. 2003, Kaltschmitt et al. 2009). Already existing stands of Reed Canary Grass may expand spontaneously after rewetting of fen peatlands, but increased flooding causes the plant to be outcompeted by *Carex* species, *Phragmites australis* and *Glyceria maxima* within a few years (Schulz 2005, Timmermann et al. 2006).

For being used as a fuel, Reed Canary Grass is preferably harvested during winter because a late harvest reduces the occurrence of substances that may be harmful during combustion, thus enhancing its suitability as a fuel (Burvall and Heman 1998). In colder climates – like in Northern Europe, Reed Canary Grass is harvested in spring, shortly before the sprouts emerge (Landström et al. 1996). On peatland sites in Northern Germany, the stands of Reed Canary Grass usually collapse much earlier (Timmermann et al. 2006) and for this reason harvest should take place in November or December. It is possible to make two to three cuts when harvesting in summer (Kreil et al. 1982) but it may lead to nutrient impoverishment and a substantial reduction in productivity. Whether fertilisation will become necessary or not will depend on the prevailing trophic conditions and the re-supply of nutrients by atmospheric deposition and flooding.

Reported yields of Reed Canary Grass amount to 1.6–12.2 t DM ha^{-1} a^{-1} on cultivated fields in North America and 7–13 t DM ha^{-1} a^{-1} on various soils in Europe (El Bassam 2010). The yields of Reed Canary Grass cultivated on cutover bog in Estonia were 7 t DM ha^{-1} a^{-1} when harvested in autumn and 5,5 t when harvested in spring (Heinsoo et al. 2011). Spontaneously developed stands on rewetted fen peatlands in Northeastern Germany yielded 5–10 t DM ha^{-1} a^{-1} in summer and 3–5 t in winter (Timmerman 2009; Table 3.3).

Winter harvested Reed Canary Grass is suitable as an uncompacted solid fuel, but also for the production of pellets and briquettes. Nowadays, the advanced technology of boilers even enables the utilisation of hay from summer harvested Reed Canary Grass in heating plants (Chapter 3.5). The lower heating value LHV is on average 16.7 MJ/kg (free from water; Heinsoo et al. 2011, Wulf 2009). If Reed Canary Grass is harvested before blooming, the hay can be utilised as fodder (Korthals 1928, Kreil et al. 1982; Box 3.6) or as a substrate in biogas fermentation plants (Baserga & Egger 1997, Geber 2002). In the past, summer harvested biomass was used as silage or hay to feed dairy cattle and as litter for farm animals. Reed Canary Grass can also be used for paper production (Pahkala & Pihala 2000).

3.1.3 Cattail (Reedmace, *Typha*)

Cattail is a genus of globally distributed perennial water and swamp plants that can form dense stands in nutrient-rich water bodies and ditches. The species native to Central Europe are Narrowleaf Cattail or Lesser Reedmace (*Typha angustifolia*), Broadleaf Cattail or Great Reedmace (*Typha latifolia*) as well as their hybrid (*Typha x glauca*). All species are highly productive

> **Box 3.7: Blow-in insulation material from Cattail**
>
> Rainer Nowotny
>
> Blow-in insulation material from Cattail consists of 100% renewable material. Cattail material is very well suited for insulation against noise, cold and hot weather. It can be applied in steep and flat roofs, in roofs with air circulation, and in roof frame constructions. The potential of Cattail as a regional, environmentally friendly insulation material is significant. The biomass of one hectare (Colour picture 34) is sufficient to insulate the roofs of four houses. An alternative marketing concept was developed in Northeast Germany for the cultivation of Cattail as an agricultural crop. This concept implies a double role for the farmer, as a producer and marketer of insulation materials from renewable resources. The Hanffaser Uckermark – a company originally producing hemp fibre based construction materials, provides advice and technical support to the farmers. Cattail is harvested in winter and baled. Bale pressing during periods of freezing weather is well known in hemp production and does not pose any technical problem. The harvested crop is transported to the company plant and processed for cavity wall insulation. The Hanffaser Uckermark company guarantees a high standard of processing of the raw material. After processing, the ready-to-use product is transported directly to the construction site. Once there, the insulating cattail material is blown into the cavities by a specialised enterprise, which also warrants a professional work. The producer subsequently transfers the ownership of the build-in insulation material to the client, who receives a truly sustainable insulation material of high quality, whereas the money stays in the region.

and therefore interesting for paludiculture. The above ground biomass can be used as a building material (e.g. for insulation, building) or for energy generation. Furthermore, the plant is (similarly to reed) used for wastewater treatment in constructed wetlands (Wild et al. 2001).

Cattails are rhizomatous grass-like, tall emergent herbaceous plants. The stiff leaves can reach heights of 4 m and have a distinct aerenchymatic tissue that allows the plant to grow in up to 1.5 m deep water. Characteristic are the female flowers which form a dense, thick spike (piston) on the stem below the less conspicuous male spike. At the end of the vegetation period, nutrients are translocated from the above ground biomass and stored in the rhizome (Dubbe et al. 1988). As only the dead above ground biomass is harvested in winter, the rhizome facilitates a rapid sprout in the following spring. Peat formation by native species of Cattail is not documented. Thus, paludiculture of Cattail may be peat preserving but the formation of new peat is unclear. More research on this matter is necessary.

Reproduction of Cattail takes place generatively via airborne seeds – which are able to germinate under water, and vegetatively via the rhizome. Stands of Cattail can develop via natural succession or can be established by planting or sowing (Koppisch et al. 2001). Establishment of Cattail stands by planting is fast and successful but more expensive than sowing (Colour picture 15). As the vegetative spread occurs very fast, it is enough to plant 1 or 2 plants per m². June and July are the most suitable months for establishing Cattail paludicultures by sowing because temperatures are favourable. After establishment, the stands require wet soil conditions or shallow inundation to deliver high yields, whereas an ample supply of nutrients is also important (Dubbe et al. 1988, Heinz 2011). Degraded and rewetted fen peatlands with a high nutrient availability are therefore appropriate sites for Cattail cultivation.

Cattail should be harvested in winter over frozen ground to avoid damage to the soil and the rhizomes (Heinz 2011). A cutting height of 10–20 cm preserves the already formed young shoots, which facilitates a rapid sprout of the stand in spring (Dubbe et al. 1988). The biomass yield of native Cattail species depends on harvest time, water level and nutrient availability. The above ground biomass may provide yields of 4.3–22.1 t DM ha^{-1} a^{-1} (Dubbe et al. 1988, Timmermann 2003, Cicek et al. 2006, Leffler 2007, Heinz 2011; Table 3.3) with constant yields over time in case of annual winter harvest (Heinz 2011).

The distinct aerenchymatic tissue makes Cattail biomass ideally suitable for producing blow-in insulation material (Box 3.7) or insulation boards (http://www.naporo.com/, http://www.typhatechnik.com/). The biomass can also be used for energy generation (direct combustion, biogas production). The lower heating value (LHV) of Cattail briquettes, pellets or bales is on average 18.2 MJ kg^{-1} (free from water; Cicek et al. 2006), while the ash content is 3.7–6.7% (Dubbe et al. 1988).

3.1.4 Sedges (*Carex*)

The biomass of sedges can be used for energy generation and as litter for livestock. Establishment of sedge vegetation may take place via spontaneous succession after rewetting of fen peatlands or by active planting (Roth et al. 2001, Timmermann et al. 2006). Several sedge species are important peat formers in fens (Succow & Stegmann 2001). The productivity of sedge vegetation varies considerably depending on species and site conditions (Colour picture 16). Greater Pond Sedge (*Carex riparia*) is highly prolific with a productivity of 3.3–12 t DM ha^{-1} a^{-1} (Timmermann 2003). Other species suitable for paludiculture include Lesser Pond Sedge (*Carex acutiformis*) with a yield range of 4.2–7.6 t DM ha^{-1} a^{-1} and Slender Tufted-Sedge (*Carex acuta*; Colour picture 11) with a yield of 3.8–4.9 t DM ha^{-1} a^{-1} (Timmermann 2003, Grzelak et al. 2011; Table 3.3). All aforementioned values refer to summer harvest. Sedge biomass can be processed into briquettes and pellets. The lower heating value LHV for sedge biomass is 18.3 MJ kg^{-1} (free from water; Zeng et al. 2013), whilst the ash content ranges from 5.6 to 7.1 % (Grzelak et al. 2011).

3.1.5 Black Alder (*Alnus glutinosa*)

Black Alder occurs naturally on wet nutrient rich fen sites. The species is peat forming and thus suited for paludiculture. The highest peat accumulation rates are associated with mean annual water levels of 0–20 cm below ground (Barthelmes 2010), whereas persistent inundation is detrimental for the growth of the species.

Alder wood can be used as a fuel and for timber production (veneer, furniture) (Colour picture 17). Until a few decades ago, traditional coppicing of alder was widespread in central Europe. The productivity of a 20–40 year cycle is at least 10 m³ ha^{-1} a^{-1} (about 5.5 t DM) (Röhe & Schröder 2010). The production of timber on humid or wet sites is possible if there is sufficient groundwater movement (Schäfer & Joosten 2005). The useable wood volume of a 60 year old Alder forest (yield class I) is 424 m³ ha^{-1} (Lockow 1996).

From a modern perspective, harvesting Alder is a challenging task because the peat soil has a limited load-bearing capacity and the winters have become increasingly milder. Consequently, the ground is rarely frozen enough to allow harvesting. No effective equipment exists for harvesting wood from unfrozen wet soils for energy use. The use of cableway cranes is recommended for harvesting Alder timber (Röhe & Schröder 2010).

Alder wood is soft, has an even and fine structure, and is light and very durable under water. The wood is traded as logs, timber and veneer. Alder wood is also used as solid wood and veneer carrier in joinery and furniture manufacturing (Kropf 1985, Grosser 2004).

3.1.6 Peatmoss (*Sphagnum* spp.)

Bog sites are suitable for Sphagnum farming as a form of paludiculture. Various peatmoss species (e.g. *Sphagnum papillosum*, *S. palustre*) are currently being established as new agricultural crops in order to produce a renewable raw material for high-quality horticultural growing media. Renewable peatmoss biomass can replace fossil moss peat (white peat) in horticulture without losing quality, because its physical and chemical properties are very similar to those of slightly decomposed moss peat (Emmel 2008, Oberpaur et al. 2010, Blievernicht et al. 2011). Favourable characteristics of peatmoss biomass include a low pH, low nutrient content, excellent rewettability, high storage capacity for air and water, very low bulk density, low nitrogen immobilisation capacity, and bactericidal and fungistatic properties (Grantzau & Gaudig 2005).

Since 2004, the University of Greifswald (Germany) studies Sphagnum farming on rewetted bog sites (www.sphagnumfarming.com). Besides providing a renewable horticultural substrate, Sphagnum farming is a sustainable, peat conserving alternative to current drainage based land use on bogland (e.g. grassland and maize cultivation, as well as peat extraction) which degrades the bogs and impairs the climate.

Peatmosses grow best at a high and constant water table of few centimetres below the moss surface, which can be achieved with controlled irrigation and drainage (Gaudig et al. 2014). Necessary water reservoirs can also be used for Sphagnum farming by covering them with floating mats. Such mosaic of different production systems could represent an optimal arrangement for Sphagnum farming on rewetted bog sites (Colour picture 19). Drained bog grassland currently offers the largest potential area for Sphagnum farming in Germany.

In 2011, a 4 ha experimental area was established on bog grassland near Oldenburg (Lower Saxony; Colour picture 20). After removal of the top 30 cm of degraded peat and levelling of the area (Krebs et al. 2012), 75–100 m³ ha^{-1} of chopped moss was spread as propagules and covered with straw (cf. Quinty & Rochefort 2003). A number of measures – namely the use of an electric pump, small ditches at 10 m intervals, and overflow facilities for excess water, ensure an optimal automatised water supply. After 1.5 years, a 95 % cover of Sphagnum mosses with a medium lawn height of 8.3 cm (max. 22.4 cm) was obtained (Joosten et al. 2013). First moss harvesting is possible after 3–4 years.

Field trials show that both on cut-over bogs and on former bog grassland sites, annual biomass productivity reaches 3 to 6 t DM ha^{-1} on average, with *Sphagnum palustre* growing best (Gaudig et al. 2014). Also, peatmosses were successfully cultivated on floating mats in flooded cut-over bogs, reaching a productivity of 2–4 t DM ha^{-1} a^{-1} (Gaudig et al. 2014).

Further research and trials on Sphagnum farming for horticultural growing media are carried out in Canada, Finland and Chile (Landry et al. 2011).

Greenhouse gas measurements show that Sphagnum culture emits at least 15 t CO_2e per ha less than the former bog grassland, Chapter 5.1). Additionally, Sphagnum cultures harbour rare and protected bog typical plant species, such as Sundews (*Drosera rotundifolia*, *D. intermedia*), White-beak Sedge (*Rhynchospora alba*) and Bell Heather (*Erica tetralix*). Sphagnum cultures also provide habitat for rare spiders (*Pardosa sphagnicola*, *Bathyphantes setiger*, Muster et al. 2015) and host an extremely rare slime mold species (*Badhamia lilacina*).

3.2 Edible and medical plants from paludiculture

Susanne Abel

Some wetland plants can also be used for food provision or the production of medical drugs. Collecting berries or medical plants from natural peatlands is a widespread tradition that is nonetheless increasingly abandoned. Some wetland species have a promising potential for commercial use, and information about them is collected in the 'Database of Potential Paludiculture Plants' (DPPP, Box 3.1). In many cases, new or adapted cultivation methods have yet to be developed and tested. Important peatland plants that produce berries include Cloudberry (*Rubus chamaemorus*, Colour picture 21), Cranberries (*Vaccinium oxycoccos*, *V. macrocarpon*), Cowberry (*Vaccinium vitis-idaea*), Black Chokeberry (*Aronia melanocarpa*) and Low- and Highbush Blueberry (*Vaccinium angustifolium*, *V. corymbosum*). In North America, Cranberry (i.e. *Vaccinium macrocarpon*) is widely cultivated on drained peat and sandy soils. In 2011, 75% of the global cranberry harvest originated in the USA, followed by Canada, Chile and Belarus – in the latter country, largely deriving from the collection of wild *Vaccinium oxycoccos* (FAO 2013, Colour pictures 22 and 23). Also, 15–20% of the cultivated areas are situated on drained peat soils. The mentioned *Vaccinium* species and Cloudberry grow naturally in acidic, nutrient poor humid habitats, and can be cultivated on former peat extraction sites (Peatland Ecology Research Group 2009; Théroux Rancourt et al. 2009). As long-lasting inundation is not well-tolerated or negatively affects quality and quantity of the fruits (Peatland Ecology Research Group 2009), tightly focused water management is required to keep the water level close to the surface over the year.

Further edible plant species that can grow in wet peatlands are:
- Celery (*Apium graveolens*) – as a vegetable;
- Holy Grass (*Hierochloe odorata*) – as a flavour (e.g. for drinks);
- Water Pepper (*Persicaria hydropiper*) – as a spicy leaf vegetable;
- Wild Rice (*Zizania palustris*) – as a cereal.

The cultivation of medical plants in peatlands is a further paludiculture option. Various peatland species are used for medical purposes in commercial drugs (Table 3.5). Examples for this are Bog Myrtle (*Myrica gale*) and Bogbean (*Menyanthes trifoliata*) (Colour pic-

Table 3.5: Selection of medical plants that can be cultivated in paludiculture (abbreviations: L = leaf, F = fruit, B = bark, S = seed, St = stem, R = root) (DPPP).

Name		Plant parts used	Area of application
Angelica archangelica	Garden Angelica	S, R, F	Gastrointestinal illness, bronchitis
Aronia melanocarpa	Black Chokeberry	F	Common cold (antioxidant)
Drosera rotundifolia	Round-leaved Sundew	L, St	Bronchitis, asthma
Filipendula ulmaria	Meadowsweet	L, F, R	Fever, headache, stomach irritation, diarrhoea
Frangula alnus	Alder Buckthorn	B	Laxatives
Lycopus europaeus	Gypsywort	L, St	Hyperthyroidism, nervousness, tachycardia
Menyanthes trifoliata	Bogbean	L	Rheumatism, arthritis, digestive problems, loss of appetite, fever
Oenanthe aquatica	Water Dropwort	F	Cough, digestive problems, stomach irritation

> **Box 3.8: Cultivation of Round-leaved Sundew (*Drosera rotundifolia* L.)**
>
> Balázs Baranyai
>
> Sundew (*Drosera* spp.) is a typical plant of acidic and nutrient poor peatlands in the temperate and boreal zone (Colour picture 26). It is widespread in natural and near-natural raised bogs, dominated by *Sphagnum* spp.. The aboveground parts of the Sundew plants (*Drosera rotundifolia* L., *Drosera intermedia* Hayne und *D. anglica* Huds.) have been used already in the 13th century as medicine against respiratory diseases like bronchitis and whooping cough (Hegi 1923, Seeholzer 1993, Babula et al. 2009). Originally only *Drosera rotundifolia* L. was used for Droserae herba (Hb. Droserae) (Egan & van der Kooy 2013). However, *Drosera rotundifolia* is a strictly protected species, which strongly restricts the availability of "Droserae herba". Hence, various non European and non-threatened *Drosera* species are increasingly traded as a substitute named 'Herba Droserae longifoliae'. The amount of *Drosera* biomass that is collected from the wild each year to satisfy the European pharmaceutical market, is estimated at 6–20 tonnes of dry mass (Galambosi & Jokela 2002). The most important exporting countries for medicinal *Drosera* biomass include Finland (*D. rotundifolia* L.) and Madagascar (*D. madagascariensis* DC).
>
> The current high demand for Drosera herba and the decline of natural collection sites as well as the protection of European *Drosera* species have led to problems in the provision of sufficient amounts of *Drosera* biomass. Hence, propagation and cultivation trials have been conducted for some time now, testing different *Drosera* species. In spite of the successful trials, there is no large scale cultivation of *D. rotundifolia* up to date because recent methods for cultivation are time and therefore cost intensive.
>
> The recent activities to grow peatmosses as a substrate for horticulture (Chapter 3.1.6) offer new perspectives for the cultivation of sundew. The peatmoss cultivation areas that were established in 2011 offer the first large, artificial field site to test sundew cultivation in Central Europe. A research pilot on a rewetted bog meadow (2011–2012) has shown that a joint cultivation of *Sphagnum* spp. and *D. rotundifolia* is a viable option (Baranyai 2013).
>
> The cultivation of sundew on rewetted bog meadows would have the following positive effects:
> - Protection of wild Sundew populations;
> - Controlled sustainable production of Sundew raw materials that meet the high quality demands of the pharmaceutical industry;
> - Through selection more productive strains with higher concentration of desired substances can be grown;
> - The integrated cultivation of *D. rotundifolia* and *Sphagnum* spp. can save jobs and offer alternative modes of income for farmers;
> - The peat soil is protected and GHG emissions are reduced compared with drainage based land use.

tures 24 and 25). The cultivation of several species on wet peatland was successfully tested in the Peene Valley, NW Germany (Kersten et al. 1999). Most medical plants that are used today are collected by hand in semi natural peatlands or are cultivated on mineral soils. For instance, the demand for Sundew (*Drosera*) as a remedy against cough is covered by manual collection in natural mires in Finland and Madagascar (Box 3.8).

3.3 The production of fodder in paludiculture

Jürgen Müller & Weert Sweers

Globally, one of the main objectives for wetland drainage was – and still is, its utilisation for intensive fodder production (Joosten & Couwenberg 2001). Natural wetlands do not offer good conditions for conventional animal husbandry but they are nonetheless used for fodder production worldwide – e.g. the 'Flooding Pampa' in Argentina, where 90,000 km² are grazed by cattle. Also, the world's largest wetland, the 'Pantanal' in Brazil, is used as pasture for about 10 million cattle and buffalo. In Australia, extensive coastal wetlands such as the Gwydir Wetlands or the Macquarie Marshes are being grazed (Holmes et al. 2009). In contrast, in Europe fodder production in wetlands is only of marginal importance and restricted to a few large areas with long regional tradition, such as the pastures for horses and cattle in the Camargue (Duncan 1983).

The above mentioned examples have not only permanently inundated areas but also areas with periodic flooding. As the extent and duration of inundations vary significantly depending on climate and weather, it is sensible to distinguish between those sites and humid grassland. In the following sections, fodder production in paludiculture is considered for cases where the fields cannot longer be used conventionally because they have become too wet (Chapter 2.3). New and adapted methods are required when inundation is frequent and extends beyond the winter period, or is even permanent (water level class 5+ /4+, Box 5.3).

There are many historical examples of the use of wet areas for fodder production – for instance, the use of Slender Tufted-Sedge vegetation as late hay in the floodplains of the river Elbe. However, such use was born mostly out of necessity (i.e. in case of famine or shortage of land) and was characterized by an unfavourable balance between invested labour and harvested yields. Now, current economic conditions pose higher challenges to labour productivity, economic viability and yield per unit area. Therefore, knowledge of traditional land use is only of limited value for modern enterprises. Ecologically and economically sustainable methods of fodder production on very wet sites (that also fulfil standards of animal welfare and consumer protection) have yet to be developed and tested. In the following sections, first results of such efforts are reported.

3.3.1 Land use as meadow

Wet areas were historically used as meadows. Thus, the term 'humid meadow' (*Feuchtwiese*) is common in German language, whereas 'humid pasture' is not. The reason for this preference of mowing over grazing is the protection of the turf, which is sensitive to trampling (Husemann 1947), and the stimulation of the growth of tall graminoids. The biomass was largely used as litter – especially in areas with a scarcity of arable land, and also as hay for horses. Hay for horses from sedge vegetation is of lower fodder quality. The regular autumn mowing of low quality humid meadows leads to the formation of distinct 'litter meadows' (German: *Streuwiesen*), which have a characteristic vegetation (e.g. the *Molinietum caeruleae* W. Koch 1926). For practical reasons, very wet meadows were cut only once a year, after the time of the cereal harvest when also water levels were low. Due to the fact that dominant species were already over their optimum, biomass quality was very low. Since the 1960s, this type of land use has been declining, to the extent that nowadays it is hardly practised.

Modern meadow utilisation requires mechanising both the harvest and transport of the biomass (Chapter 4.2). If equipment adapted to high groundwater levels is available, swathing and biomass removal remain the main technical and logistic challenge (Chapter 4.3). Conventional grassland management activities such as levelling and rolling can not be used when water levels are too high in spring. The same applies to fertilising, whereas the application of easy soluble fertilisers is anyhow forbidden in the vicinity of water bodies.

The most widespread method of hay-making includes drying the biomass on site (Müller & Bauer 2006), but this is not an option on sites with high soil humidity. Consequently, the production of hay from paludiculture requires methods where the drying happens elsewhere – e.g. under a roof cover. In the latter case, the biomass is subjected to initial wilting on site to reduce the water content to 45–60 %; then it is transported to a drying installation where it is dried to reach a water content below 20 % to enable storage. Depending on the installation, drying can happen as untreated biomass or as moist hay bales. Modern circulating air installations dry the biomass very effectively and allow the use of surplus heat from combined biogas powered heat and power (CHP) plants.

The great investment in time and effort in this way of hay production restricts its application to wetland grass species of sufficient fodder quality, such as Reed Canary Grass (*Phalaris arundinacea*), Redtop (*Agrostis gigantea*) and Reed Manna Grass (*Glyceria maxima*) (Sundblad & Wittgren 1989, Table 3.6).

The common plant communities of Creeping Bentgrass (*Agrostis stolonifera*), Marsh Foxtail (*Alopecurus geniculatus*) and Floating Sweetgrass (*Glyceria fluitans*) are not suitable for exploitation by mowing as their stands emerge too low to guarantee fodder hygiene, for which a minimum mowing height of 12–20 cm is required. Furthermore, their vegetation often holds toxic plants such as Celery-leaved Buttercup (*Ranunculus sceleratus*) and Marsh Horsetail (*Equisetum palustre*).

Reed Manna Grass requires an early harvest (before full ear emergence) to be used as fodder (Schrader & Kaltofen 1987). For this reason and also because its thick leaves are difficult to dry, fodder of this species can best be conserved by silage. Despite its strong buffer effect against the formation of lactic acid during ensilage, the fodder contains sufficient water-soluble carbohydrates to facilitate ensilage (Bockholdt 2001). However, ensilage for fodder cannot always be applied because the method requires early mowing, when the wet soil conditions associated with paludiculture may hamper the necessary initial wilting on site. Even in warmer climates, where cultivated fodder plants such as Cup Plant (*Sylphium perforatum*) are grown on hu-

3.3 The production of fodder in paludiculture

Table 3.6: Forage quality value of typical fodder plants of wet sites. The nutritional energy content correlates with the net energy of lactation (NEL).

Scientific name	Common name	Energy in MJ NEL kg^{-1} DM	Indicator for forage quality (German: *Futterwertzahl*)
Agrostis stolonifera	Bentgrass or Redtop	5–7.5 [2]	7 [1]
Alopecurus geniculatus	Marsh Foxtail	5.5–7.2 [1]	4–5 [1,3]
Carex acutiformis	Lesser Pond Sedge	5.1–5.8 [2]	2 [3]
Glyceria maxima	Reed Manna Grass	5.5–7.3 [2]	5 [3]
Lotus pedunculatus	Marsh Bird's-Foot Trefoil	4.2–6.2 [2]	7 [3]
Phalaris arundinacea	Reed Canary Grass	4.5–7.1 [2]	6 [3]
Phragmites australis	Common Reed	3.6–5.7 [1]	3 [3]

[1] Schönfeld-Bockholt 2005, [2] Bockholt & Buske 1997, [3] Klotz et al. 2002, Futterwertzahl, Klapp et al. 1953.

mid sites, ensilage is often problematic because of high substrate humidity and insufficient fermentable carbohydrates (Han et al. 2000).

Reed Canary Grass and Reed Manna Grass can achieve high yields if the water is oxygen-rich (5.0–8.5 t DM ha^{-1} a^{-1}, Table 3.3). Even yields of 10 t DM ha^{-1} a^{-1} are common on eutrophic sites (Westlake 1966, Christian et al. 2006). Such high yields also indicate a high lignin content, which makes the biomass less suitable as fodder for ruminants. In case of late harvesting, winning hay for horses (Box 3.6) could be an alternative to utilising the biomass as a fuel (Chapter 3.5/3.6).

3.3.2 Peatland grazing

In contrast to temperate climate zones, grazing in wetlands is common in subtropical areas. Historically, in Central Europe, grazing in wetlands on peat soils was also taking place. In the foothills of the Alps, wetland commons often used to be grazed in association with the three-field-system of crop rotation but this custom was abandoned when the latter was no longer practised (Luik 2002). Also, in the coastal areas of the North Sea and the Baltic Sea, the use of low lying and periodically inundated areas for grazing was common because the site conditions – namely tidal creeks, pools and drift lines, did not permit mowing.

The constant increasing standards of cattle raising and the failing competitiveness of cattle or sheep fattening on grass led to the disappearance of traditional grazing systems in many places. New attempts to re-introduce grazing management in wetlands are less motivated by economic profitability than by nature conservation. The main objectives, apart from aesthetic aspects, are the conservation of typical wetland vegetation and its biodiversity (Reeves & Champion 2004).

One option for peatland grazing is to let the animals exclusively feed on the wetland vegetation during the entire grazing season. Trials with Heck cattle on very wet sites often failed because of lacking economic viability (Petermann et al. 2008). Water Buffalo (*Bulbalus arnee*) is considerably better adapted to permanently wet site conditions (Wiegleb & Krawczynski 2010, Colour picture 18), but copes better with shallow peatland underlain by sand (such as in coastal flood mires) than with peatlands with deep peat. Coastal marshes are less prone to become muddy by the trampling of the animals. Water Buffalo prefers energy-rich food: it consumes first all available more energy-rich stands (Enge 2009, Box 5.6) but also utilizes vegetation with less energy (Krawczynski et al. 2008). Therefore, such behaviour has to be taken into account when selecting pastures – the less energy the vegetation contains, the larger the selection options should be, which also implies that livestock density has to be lower. The recommended density for Water Buffalo is 0.8–1.4 livestock units per ha depending on site conditions and conservation objectives. Lower livestock densities are possible but only at the expense of less achievement in conservation management. Both permanent grazing and a combination of grazing and mowing can be applied. Rotational grazing is not recommended, as concentrated trampling easily damages the sward on wetter sites. The allocation of pasture area should be generous but should still allow monitoring the herds in difficult terrain.

Whereas on organic soils an extra provision with minerals is necessary, a supplementation with concentrated feed should be avoided to prevent input of nutrients into the wetland and to keep fodder costs low. A critical point in the pastoral management of Water Buffalo is the organisation of reproduction. The use of breeding bulls is obligatory because spacing between calves is long and usually it is very difficult to identify when a cow is in oestrus. If the main rules of pastoral management and animal welfare are followed, grazing with Water Buffalo can be implemented as paludiculture. This method is particularly useful for reed control if this is the conservation aim, and also to utilise wet vegetation that cannot be used by mowing.

Another option is to let the grazing animals feed on both paludiculture vegetation and dry mineral soil. The proportion of the pasture area occupied by rewetted land depends on the landscape lay-out but can also be determined by the farmer. The integration of rewetted areas into pastoral management can vary in space and time.

This option is particularly relevant in areas where peatlands adjoin mineral soils. The large stretches of land in the valley mires of Mecklenburg-West Pomerania are predestined for this type of grazing management because the strongly fluctuating groundwater level (Müller et al. 2007) frustrates year-round grazing of the peatland. Within this grazing system, the animals can move onto higher mineral ground if the peatland part of the pasture is flooded. This flexibility is, in a year-round grazing regime, easily achieved if there is continuous access from peatland to mineral soil. In a paddock, this is more difficult to achieve and close monitoring of the water levels is required (e.g. via gauges) to intervene in time and move the animals to another area.

For the combined grazing of wet peat and dry mineral soils, conventional beef cattle with rather low nutritional demands – such as Angus or Limousin breeds, are suitable. These breeds are less demanding with regard to fodder quality than Charolais or Belgian Blue but are easier to sell, unlike pure landscape conservation breeds like Highland or Galloway cattle. Another advantage is that breeding stock of the aforementioned suitable breeds is easily available. Also, grazing with dry cows and young cattle is being proposed (Holst 2003).

If peatland areas are only part of the total pasture area in a year-round grazing system, it might be expected that the animals will feed only sporadically on the wet parts if they have the free choice to do so; consequently, the mineral areas of the pasture will receive stronger grazing pressure (Chapter 5.2.3). This is also advantageous because the vegetation on mineral soil develops earlier in spring and the wet parts are subject to generally higher spring water levels. However, this behaviour can become problematic over time, especially in dry periods. By temporarily fencing off the mineral part of the pasture (about 40%), food supply is artificially reduced and grazing on the wet part is enforced. The fenced part can be used for making hay for the winter, or the regenerated vegetation can serve as reserve supply. The pasture should be partitioned about mid June. Experience in Northeast Germany has shown that the wet parts of the pasture are grazed only from mid August, when the nutritional value of the sedge-rich wetland vegetation is very low (Table 3.6, Dovel 1996). As a result, the animals do not gain weight but may even loose some. If such conditions can be prevented by pre-emptive pastoral management, a weight gain of about 800g per day and animal can be achieved at a livestock density of 1.5 per hectare, without supplementary feeding.

3.3.3 Animal husbandry in paludiculture. Conclusions and outlook

Within the temperate climate zone, fodder production in paludiculture – be it for storage (haymaking, ensilage) or for grazing, does not play a dominant role. Nonetheless, this form of biomass utilisation has a certain tradition in rural areas with the advantage that still an agrarian infrastructure exists that allows using these practices. The necessary investments for this type of land use practice are smaller in contrast to the utilisation of the biomass as a raw material or as a fuel. Additionally, a market already exists for products like meat or animals for breeding.

Besides, fodder production from paludiculture is challenged by the continuous rising nutritional and hygienic demands to fodder, which result from the ever higher performance of livestock but also from legislation (feed and food law, consumer protection legislation). These demands are especially problematic in areas with a long-lasting inundation, which therefore are at risk of contamination with pathogens and pollutants (Krüger et al. 2005). Furthermore, fodder produced in paludiculture has to compete with other forms of fodder production within the same farm (such as grass from dry mineral soil, fodder from arable land or from intercrops), for which the methods of production are well-established.

The utilisation of wetland biomass as fodder serves the protection of abiotic resources (Chapter 5.1 and Chapter 5.4) and offers manifold conservational services in addition to the main product (Chapter 5.2; Box 5.6). If the special and expensive methods of fodder production on wet sites could be adequately integrated and rewarded by subsidy or funding schemes, the additional effort would be compensated and the production branch would be economically stabilised. The societal payoff would be considerable because of improved soil protection, water quality and climate, besides conservation and socio-economic aspects (Chapter 6).

3.4 Material use of biomass from paludiculture

Denny Wiedow & Jörg Burgstaler

Biomass from paludiculture is suitable for many material purposes. The material characteristics of the biomass can be specifically changed by preparation, which provides raw materials for a variety of products. The material use of biomass includes mechanical use, use as a raw production material and biological utilisation (e.g. composting) (Kaimer & Schade 2000, Fricke et al. 2009). In the following sections, examples from the conditioning of paludiculture biomass and its material use in the construction sector are provided.

3.4.1 Reed and Cattail as raw material

In contrast to the use of biomass as a fuel (Chapter 3.5), the material use of biomass achieves a higher added value. Some traditional ways of material use of biomass include its utilisation as thatch, for paper production and for weaving. Furthermore, many innovative ways have been developed for using Reed and Cattail as building materials, each specific use being determined by the particular properties of the raw material. The quality of the raw biomass depends on the water and nutrient supply of the plants, as well as on the management and the species composition of reedbeds and cattail stands.

Common Reed (*Phragmites australis*) has been used for millennia mostly as thatch (Box 3.3) and also as insulation material (Kujawski 1972, Schillberg 1996, Ritterbusch 2011, Chapter 10.7). For thatching, the reed culms are tightly pressed together and bond – e.g. with wire (Colour pictures 27 and 28). 8,000 years ago, exterior walls of buildings in many parts of the world were already plastered with layers of straw and reed (Dirlich 2011). Today, mats and boards of reed are still used as insulating basis for plaster and for building with loam (Colour pictures 29, 30, 31 and 33). For building purposes, reed can also be processed into fireproof boards, which meet high standards of fire safety (Colour pictures 32 and 35; Box 3.4) because its fibres have a special tensile strength and are resistant to the effects of wind and water. Thus, the physical properties of the building material can be optimised by adding chopped reed to the clay plaster (Box 3.9).

Also Cattail (*Typha*) can be utilised in manifold applications. High quality fibres from the seeds were already used 90 years ago in Berlin for textile manufacturing (Graebner et al. 1919). Kreisner (1919) emphasised the great importance of Cattail in the German textile industry as a substitute fibre for wool. Cattail is also suitable for the production of building materials. Its leaves contain many air-filled cavities (aerenchym), which provide the material with very low heat conduction and excellent insulation properties. Around 1900, there were already patents for manufacturing insulating materials from Cattail and silica (Schwemmer 2010). Additionally, insulation made from Cattail also balances the indoor climate of buildings because of its moisture regulating properties. Further advantages include the diffusing capacity of the material and the extremely energy saving manufacturing process (Luamkanchanaphana et al. 2012). Cattail can be used for blow-in insulation, for which the entire plant is shredded and used without any further additives (Box 3.7). The plant can also be used for manufacturing insulation boards (Colour picture 36, Holzmann & Wangelin 2009).

3.4.2 Preparation of Reed and Cattail biomass

The different biomass composition (specially regarding lignin content, Table 3.7, and tissue structure of Reed and Cattail) decisively determine the application options and the associated preparation of the biomass. The biomass properties can be altered and optimised for further processing by conditioning. Preparation turns the biomass into reproducible homogeneous charges, which enable industrial utilisation. Preparation may involve simple mechanical activities such as squeezing, ripping, chopping and grinding. Biological methods of prepararation include ensilage, fermentation or a combination of both. The aim of processing may be a shortening or reduction of particle length or size, an alteration of the surface or the structure, or the separation of fibres from the rest of the biomass (Box 3.10).

Table 3.7: Composition of Common Reed (Holzmann & Wangelin 2009) and Cattail biomass (Al-Hakkak & Barbooti 1989).

Parameter (% DM)	Common Reed	Cattail
Cellulose	42–45	41
Hemicellulose	24–27	30
Lignin	22–24	24
Ash and minerals	4.7–5.6	5.0
Acid-Detergens-Fibre (ADF)	51.0	64.0
Neutral-Detergens-Fibre (NDF)	70.9	74.5
Acid-Detergens-Lignin (ADL)	14.2	16.1

> **Box 3.9: Eco-plaster based on reed**
>
> Ulrich König, Jörg Burgstaler & Denny Wiedow
>
> The utilisation of insulating plaster is one way to improve the energy performance of a building and to reduce costs for heating and cooling. Certainly, energy expenditure used for manufacturing and transporting the building material has to be taken into account.
>
> Long term experience has shown that heavy loam or even clay are needed for the production of insulating loam-based construction materials. Clay, in particular, easily expands and has a very positive effect on the indoors climate in terms of temperature and moisture. On the other hand clay may shrink considerably, which has to be minimised to avoid cracking of the plaster. The shrinking can be mitigated and drying times of the plaster can be reduced by adding reed fibres.
>
> The Mineralische Rohstoffmanagement GmbH company (Friedland/Mecklenburg-West Pomerania) developed an environmentally friendly insulation plaster with low embedded energy. Locally sourced reed from rewetted fen peatlands and local clay as a binding agent were used to produce the plaster. The reed for the plaster was harvested at various times of the year and underwent several processes, such as the addition of additives to the fibres – either in fresh condition or after ensilage, in order to develop a product with low density and sufficient strength.
>
> The research resulted in the development of clay plaster systems consisting of three layers: a bonding course for the rendering base, an insulating plaster suitable for various insulation depths, and a coloured covering loam coat (Colour picture 31). Such insulating plasters are mainly suitable for interior walls but can also be used on exterior areas protected from rain (such as conservatories and arcades). Besides its insulation properties, the plaster also improves the indoors climate, which makes it very well suited for restoring old buildings, as well as for application in new ones.

The following biomass properties are relevant for the preparation:
- Particle size (length and width of the chopped material);
- Particle size distribution (defined frequency distribution);
- Surface structure;
- Degree of fraying;
- Water content;
- Density and volume.

The material properties can be tested with physical (e.g. sieving, Raussen et al. 2010) or optical methods (e.g. scanning, photographing, Voigt 1987).

3.5 Solid energy from biomass

Bioenergy is the useful energy that can be derived and supplied from biomass. Biomass from paludiculture can be converted into various forms of useful energy (Box 3.11). The type of conversion is determined by a number of conditions such as the utilised biomass, the desired form of energy, economic necessities, and environmental standards. The following sections introduce the various technical ways of energy supply in form of heat, electricity or liquid, as well as gaseous fuels from biomass grown in paludiculture. As the refining process is largely determined by the end use of the energy carriers, the processing methods are discussed in the context of their final use as end- or net energy. The discussed methods are those which, at the current stage of technology, are economically viable for biomass from paludiculture.

Biomass has a higher oxygen content in comparison to fossil fuels; also, the carbon in the biomass is partially oxidised. Therefore, the energy yield of biomass is per unit mass lower than that of fossil fuels (Hartmann 2009). Besides, the transportability of energy carriers is dependent on their energy density – i.e. the energy content per volume. For herbaceous biomass in particular the low energy density impedes its economic utilisation as a fuel (Hering 2012, Lenz 2012).

3.5 Solid energy from biomass

> **Box 3.10: Manufacturing of Reed fibres**
>
> Denny Wiedow & Jörg Burgstaler
>
> For fraying Reed biomass for manufacturing insulation plaster (Box 3.9) the chopped and air-dried biomass (particle length 5 cm, moisture 40–60% DM) is initially moistened and steam heated. Subsequently, the biomass is transferred to an extruder, where the particle size is further reduced at a speed setting of 43 rpm and a feed opening of 30 mm (Wallot et al. 2011). In the extruder, two opposite rotating worm shafts cause further fraying of the material via shearing forces and friction heat (Cong et al. 2006). Then, the produced reed fibre is further shortened in a disk mill (speed setting 2,000 rpm, distance between discs 0.3 mm) where shearing forces and compression cause the fibrillation (the curly or net-like splitting into endless and connected filaments of the fibre material; Wallot et al. 2011). Finally, the separated fibres are dried and stored until further processing.
>
> Depending on the product requirements, the respective raw material is selected. Processed fibres of winter harvested reed yield the largest average fibre length (1.63 cm), thickness (0.14 mm) and highest mass fibre thickness (1.71 mm) (Table 3.8). If ensilaged reed is used, shorter fibre lengths (1.12 mm) and thicknesses (0.09 mm), as well as a reduction of the highest mass fibre thickness (0.47 mm) are achieved.
>
> Table 3.8: Characteristics of biomass prepared from Common Reed harvested in different seasons.
>
> | Characteristics Average values (mm) | Quality Thatch Reed Winter after frost | Common Reed Winter after frost | Common Reed Summer fresh | Common Reed Summer ensilaged |
> |---|---|---|---|---|
> | Particle length* | 1.63 | 1.21 | 1.30 | 1.12 |
> | Fibre thickness * | 0.14 | 0.12 | 0.12 | 0.09 |
> | Most mass rich fibre thickness * | 0.37 | 0.32 | 0.35 | 0.26 |
> | Most mass rich fibre thickness ** | 1.71 | 0.67 | 0.86 | 0.47 |
>
> *optic analysis (number of particles analysed: 9,321), **physical analysis (sieve tower).

Compaction enhances transportability, simplifies handling and – via physical homogeneity – increases control over dosing. Consequently, compaction is often unavoidable due to economic (e.g. transport and storage costs) and technological reasons (e.g. automatic feeding and regulation of the conversion plant) (Hartmann 2009). Compaction does not affect the fuel characteristics and, besides a small reduction of the water content, only the physical and mechanical properties are altered (Box 3.12). However, the process increases the chemical homogeneity of the biomass, which enables a better adaptation of the combustion equipment to the fuel type, and improves the energy efficiency of the biomass fuel.

3.5.1 Combustion of solid fuels

Matthias Ahlhaus & Christian Jantzen

The utilisation of biomass as a fuel involves mostly combustion. Solid biofuels (e.g. wood, herbaceous biomass) contain carbon and hydrogen as main oxidisable elements. Oxygen is present in these fuels with more than 40% by dry mass. Furthermore, during its life cycle, the plant has absorbed inorganic elements that may lead to the emission of pollutants (e.g. nitrogen, chlorine, sulphur) or remain as ash (Chapter 3.5.3).

For combustion, the water content is important because a high water content – as with fresh biomass fuels, may significantly lower the net calorific value. Depending on harvesting time, herbaceous biomass can be harvested in a rather dry state. Additionally, the biomass can dry further during storage before combustion to enable reaching the maximum net calorific value (Box 3.5).

Box 3.11: Principles of conversion

Hannes Wagner & Martin Kaltschmitt

In many cases, biomass from energy crops, residues, by-products and organic waste has to be processed to be used as solid, liquid or gaseous energy carriers with defined characteristics (Figure 3.3). Processing can be carried out mechanically (i.e. pelleting) or via more complex thermo-chemical, physico-chemical or biochemical techniques. By processing the biomass properties are altered to improve its use as a fuel. Changes may involve energy density, ease of handling, storage and transport, environmental compatibility, fuel standards or the suitability for substituting fossil fuels.

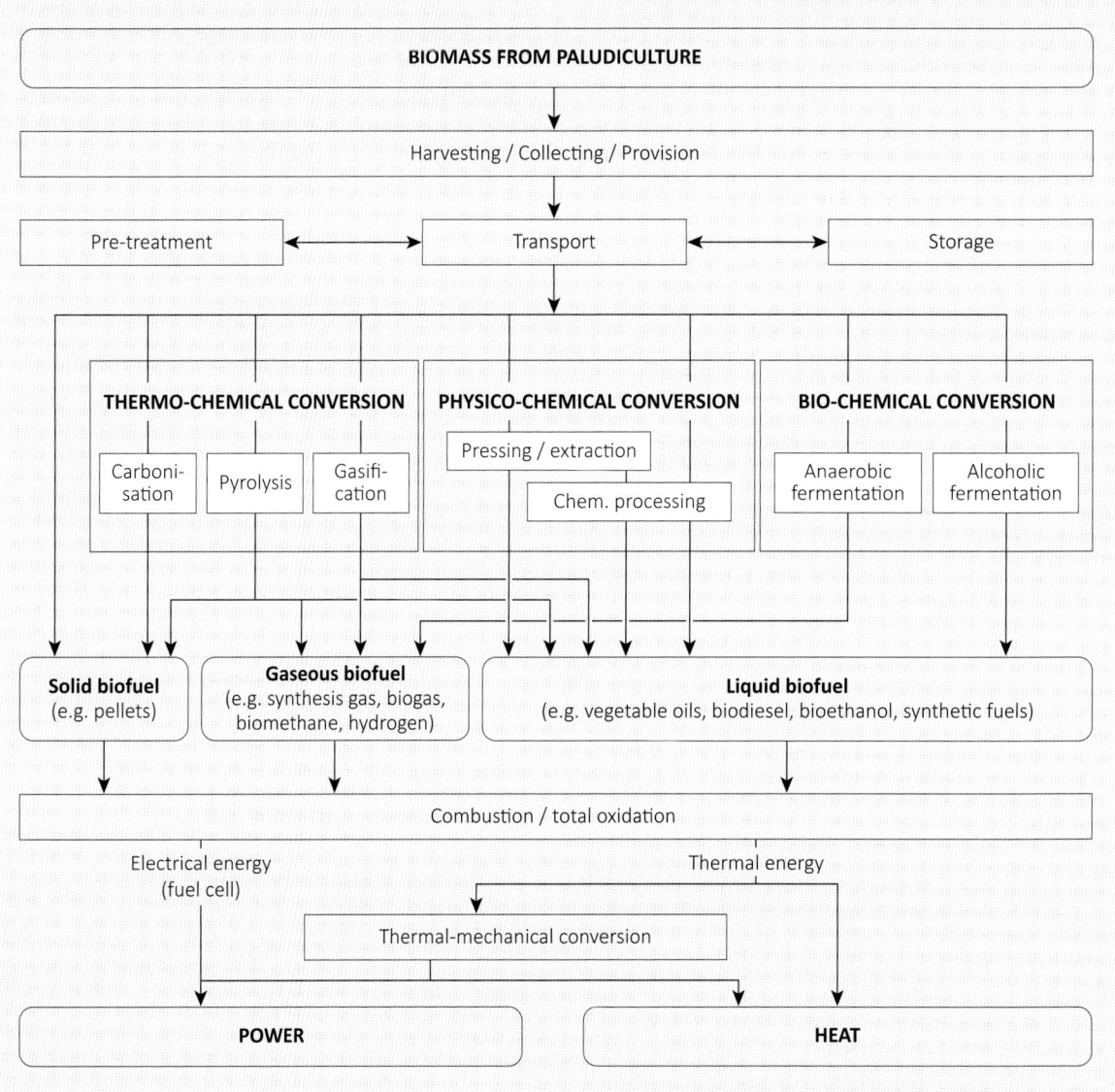

Figure 3.3: Overview of the use of biomass for energy generation (after Kaltschmitt et al. 2009).

3.5 Solid energy from biomass

Box 3.12: Pelletising

Christian Jantzen

Pelleting is the production of pellets – i.e. small cylindrical bodies of compacted material. If the compressed bodies have a diameter larger than 2 cm, they are called briquettes. The energy used for compressing the biomass is solely applied in the form of pressure. In the first phase of processing, the biomass is compressed using little pressure, which minimizes the distance between the particles. At the same time, an initial bonding takes place via mechanical interlocking of the particles. The continuously increasing pressure causes friction between the particles and also between the particles and the equipment and this friction is largely transformed into heat. If the biomass temperature exceeds 75–85°C (the so called glass-liquid transition) the lignin in the biomass changes its phase from solid to viscous. Pressure forces the softened lignin into the gaps between particles. During this process, the bonding between the lignin molecules is dissolved and new bonds are formed. Subsequently, the biomass in the press shaft is displaced by further incoming biomass and leaves the zone of higher temperatures, which causes the lignin to harden again. The reduced pressure causes a reduction of the pellet density, a process called 'relaxation', which can lead to instability. In order to reduce relaxation to a minimum, the cooling of the pellets is accelerated by technical means (Adapa 2011).

Pelleting takes place in a continuous process with flat die or ring die pelleting presses. The biomass is transported via rollers and pressed into press shafts of which the diameters continuously diminish and later widen again. In a flat die pelleting press, the rollers rotate over the horizontally arranged press shafts of the die. In a ring die pelleting press, the vertically fixed ring die rotates (Figure 3.4, Figure 3.5). Furthermore differences exist in how the biomass material is fed into the presses, as well as how the material is distributed over the surface of the die. The conditioned biomass falls onto the surface of the die, is pre-compressed between rollers and die, and then forced into the press shafts of the die. With each rotation of the rollers, the die shafts are filled with new material. During this process, the mechanical energy is transformed by friction into heat; the glass-transition temperature is exceeded causing the separate layers to merge into an endless string, which is subsequently cut mechanically to the desired pellet length (Stelte 2011).

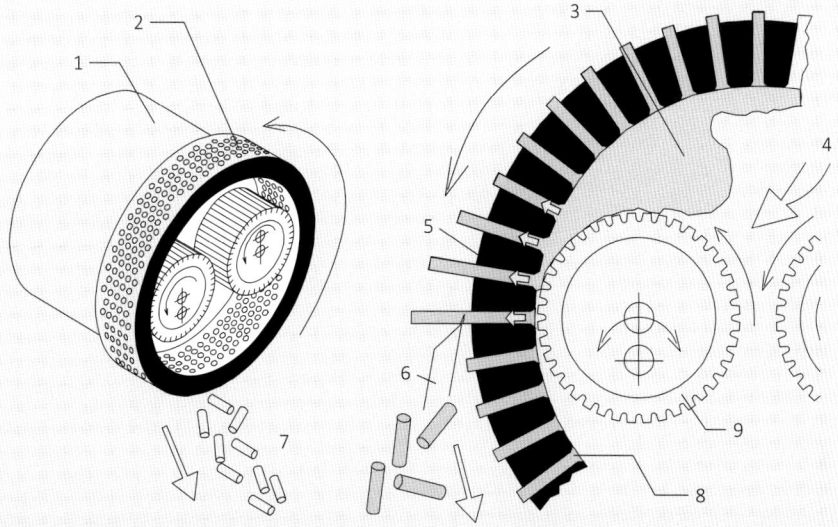

Figure 3.4: Pelleting with a ring die pelleting press 1: Drive, 2: Ring die, 3: Milled biomass, 4: Material supply, 5: Die, 6: Knife ,7: Pellets, 8:Material carpet, 9: Pan grinder (Kaltschmitt et al. 2009).

Since pellets have a high bulk density and very good physico-mechanical properties, pressing into pellets is the most common form of processing wood biomass for fuel. With regard to herbaceous biomass, there is still lack of experience and technology. Pelleting herbaceous biomass quickly reaches technical and economic limits because the conditioning of the biomass (drying, chopping) is technically challenging and expensive (Hartmann & Witt 2009).

The production of high quality reed pellets is a technically sophisticated process and requires a fine-tuning between conditioning of the biomass, the pellet press and its peripheral devices. Pellets produced under ideal conditions have a high true and bulk density (1.18 g cm^{-3}, 665 kg m^{-3}, respectively), a high resistance to abrasion (>98%) and a low moisture content (9.81%). This implies that, apart from the ash content, all requirements of the German standard DIN 51731 for wood pellets are met (Colour pictures 38-40).

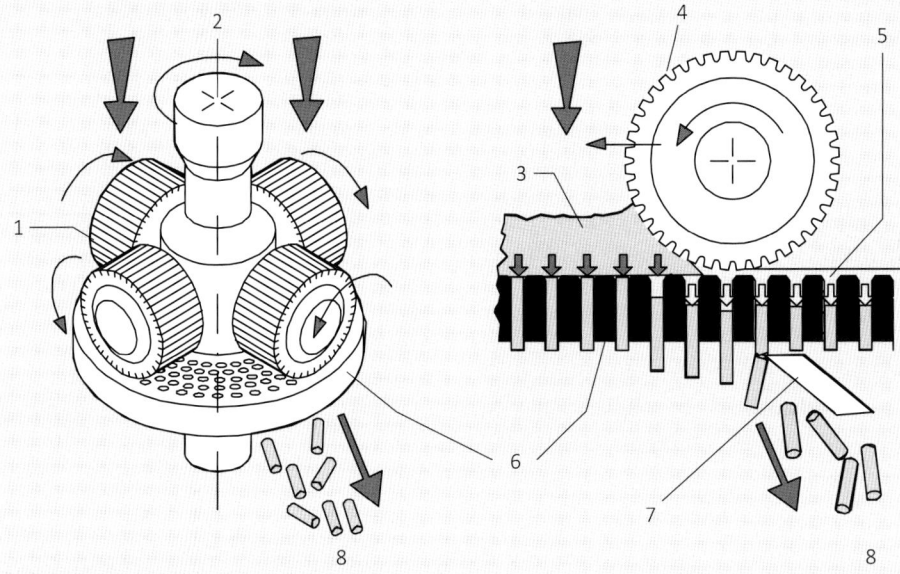

Figure 3.5: Pelleting with a flat die pelleting press. 1: Pan grinder head, 2: Shaft drive, 3: Milled biomass, 4: Pan grinder, 5: Material carpet, 6: Die, 7: Knife, 8: Pellets (Kaltschmitt et al. 2009).

A further important factor is the content of volatile matter that is released pyrolytically during combustion. In contrast to coal, biomass fuels are characterised by a high content of volatile matter (up to 85% of the organic matter). This results in fundamentally different combustion properties of solid biofuels in comparison to coal. The different behaviour requires that the combustion chamber of a biomass furnace has to be adapted to the characteristics of solid biomass fuels, in order to achieve high combustion quality and low emission of pollutants. For that reason, no multi-fuel furnaces can exist that accommodate the properties of different fuels in an optimal way.

Biomass combustion consists of three phases:
I. Drying of the fuel and the release of volatile matter;
II. Main combustion: after ignition of the volatile matter, an exothermic oxidation releases heat, which leads to the further release and oxidation of volatile matter;
III. Burnout: the flameless glowing of the degassed solid biofuel (charcoal).

The air-fuel ratio λ (lambda) describes the ratio of air supply to air required for the complete oxidation of the fuel. Particularly during the burnout phase of the combustion process, it is difficult to transport oxygen from the air to the desired place of oxidation, thereby an oversupply of air is necessary ($\lambda > 1$).

In order to achieve complete oxidation and high efficiency, two separate streams of air supply have to be considered. The 'primary air' supplies the oxygen for the gasification of the solid fuel, whereas the 'secondary air' is required for the complete combustion of the gaseous matter. By supplying air in successive steps, it is possible to regulate the course of combustion to a certain degree. However, the nominal capacity of furnaces with automatic feeding can be regulated more effectively via the fuel feeder. Through the secondary air supply, a high combustion quality is ensured (Kaltschmitt 2012).

3.5.2 Solid biofuels from paludiculture

Matthias Ahlhaus & Christian Jantzen

Winter harvested Reed and Reed Canary Grass from fen peatlands are well suited as a fuel. Numerous practical examples have demonstrated the good fuel characteristics of both plant species, either burned alone or in combination with wood (Paulrud & Nilsson 2001, Paulrud et al. 2001, Kask et al. 2007, Wulf 2009). Kitzler et al. (2012) have successfully performed combustion tests with several combinations of reed and wood chips (30%, 50%, 100% reed) in a 3 MW heating plant. Even with 100% reed no problems occurred. Similar positive experiences were made using Reed Canary Grass from humid meadows in a biomass fuelled combined heat and power plant in Friedland, Mecklenburg-West Pomerania. However, Reed Canary Grass posed some problems during feeding of the biomass and its distribution in the combustion chamber (Barz et al. 2012). Own experiments have shown that combustion of Reed implies higher demands to the combustion plant than burning wood, but in boilers adapted to herbaceous biomass the latter can be burned without facing the typical problems posed by straw (Box 3.13). However, more research is needed in order to improve the combustion of reed briquettes and pellets.

Herbaceous biomass not only has a higher ash content (up to 7%) than wood (< 2%) but also the composition of the ash may differ significantly. High potassium contents result in a lower ash melting temperature with increasing risk of ash sintering on the grate and at other places in the combustion chamber, as well as on the heat exchangers. Sintering reduces boiler efficiency and may even cause combustion to stop completely. However, Reed and Reed Canary Grass from winter harvesting have significantly higher ash melting temperatures than straw, and therefore a lower tendency to ash sintering (Oehmke & Wichtmann 2011).

Besides ash content, the composition of the ash is central in determining the quality of the biomass fuel (Chapter 3.5.4). The ash of Reed has a sintering temperature of 1,450 ± 274.1 °C, which even exceeds that of wood reference samples (1,281.7 ± 219.2 °C) and is significantly higher than that of other herbaceous biomass fuels such as hay and straw (900 ± 23.6 °C). For two out of the seven investigated reed samples, the maximum temperature of the test furnace of 1,500 °C was not high enough to determine the sintering point of the ash. The melting temperature was only for one single reed sample below the maximum analysis temperature. These results suggest that combustion of reed does not pose any problems with regard to ash sintering in the combustion chamber of biomass boilers. The mean sintering temperature of the ash of Reed Canary Grass (1,200 °C) is lower than that of reed but at a comparable level with that of wood.

3.5.3 Critical components of solid biofuels from paludiculture

Claudia Oehmke & Wendelin Wichtmann

The suitability of plant biomass as a fuel is – next to its net calorific value, strongly determined by the content of elements that are critical for combustion (e.g. N, S and Cl). Herbaceous biofuels like straw are well known to contain significantly higher concentrations of elements that are critical for combustion compared to wood. Therefore, boilers adapted to herbaceous biomass and equipped with, for instance, mobile or cooled grates, are more expensive (Hartmann et al. 1996).

Health hazardous NO_X emissions increase with rising nitrogen content of the biomass. High concentrations of sulphur and chlorine in the biomass, for example, may lead to the formation of substances that are harmful to health such as dioxins, furans, hydrogen chlorides and sulphur oxides (Obernberger et al. 2006). At the same time, such nocive substances, together with potassium, are responsible for corrosion of the boiler. Also, potassium and sodium may lower the ash melting temperature – i.e. the temperature at which the ash softens and glues together to form slag on the walls of the furnace (sintering). Calcium and magnesium have the opposite effect: they increase the melting point of the ash. Thus, the chemical composition of the biomass strongly determines the suitability of biomass for combustion.

The concentration of critical elements in herbaceous biomasses can be decreased by a late harvest date in autumn or winter. Since the late 1990s, for example, late harvest of biomass was practised in order to improve the combustion properties of Reed Canary

Table 3.9: Concentrations of elements critical to combustion (% DM), ash and moisture content of winter harvested Reed (*Phragmites australis*) and Reed Canary Grass (*Phalaris arundinacea*) from fen peatlands, in comparison to wood (Hartmann 2009a), miscanthus (Lewandowski et al. 2003) and straw (Hartmann 2009a), *additional values after Wulf (2009).

	Reed	Reed Canary Grass	Wood	Miscanthus	Crop straw
Cl	0.04–0.13	0.02–0.23	0.004–0.006	0.1–0.5	0.19–0.40
K	0.06–0.11	0.08–0.31	0.13–0.35	0.31–1.28	1.01–1.68
S	0.1–0.2*	0.11–0.34*	0.02–0.05	0.04–0.19	0.06–0.09
N	0.38–0.65	0.72–1.4	0.13–0.54	0.19–0.67	0.42–0.55
Ash (550°C)	3.2–5.5	2.0–7.0	0.5–2	1.6– 4	4.8–5.9
Moisture content %	10–18	16–19	15–30	16–62	15

Grass (*Phalaris arundinacea*), (Landström et al. 1996, Burvall 1997). During the longer period until mowing, the above ground parts of the plants die off, and nutrients are relocated to the rhizomes and leached from the biomass by precipitation (Hadders & Olsson 1997, Burvall 1997). Sprinkler irrigation trials with straw have shown that precipitation of 150 mm is sufficient to reduce the concentrations of potassium and chlorine by 60–70 % (Hernández Allica et al. 2001).

Our own studies on Common Reed (*Phragmites australis*) on fen peatlands in Northeast Germany have shown that between October and January/February, the concentrations of chlorine, potassium, sodium, calcium and magnesium were reduced by 43–87 % and of nitrogen by 26 % (Figure 3.6). Chlorine concentrations remained on average considerably below, and concentrations of nitrogen just below, the permitted limits for solid biomass fuels (Obernberger et al. 2006). Also, the potassium concentration in winter harvested reed was very low. In contrast to other herbaceous biomass, the ash melting temperature of reed is significantly higher, which substantially lowers the risk of ash sintering in the combustion chamber of the boiler. Combustion tests running over several days in a 3 MW heating plant and firing pure Common Reed, as well as mixtures with wood chips, showed no problems with regard to combustion (Kitzler et al. 2012). The ash sintering temperature of the used biomass was about 1400°C.

Studies on the relation between site conditions and concentrations of critical elements showed that site conditions are of minor importance in winter compared to autumn (Figure 3.6). Precipitation in Northeast Germany appears to be sufficient to significantly improve fuel properties of reed by late harvesting. Winter harvested Common Reed appeared to be better suited as a fuel than Reed Canary Grass from winter harvest, and than Miscanthus, straw or hay from conservation management (Table 3.9). With respect to the high ash content, a specific ash removal system is required when Common Reed and Reed Canary Grass are used. A reduction of the ash content can be achieved by combining herbaceous biomass with wood.

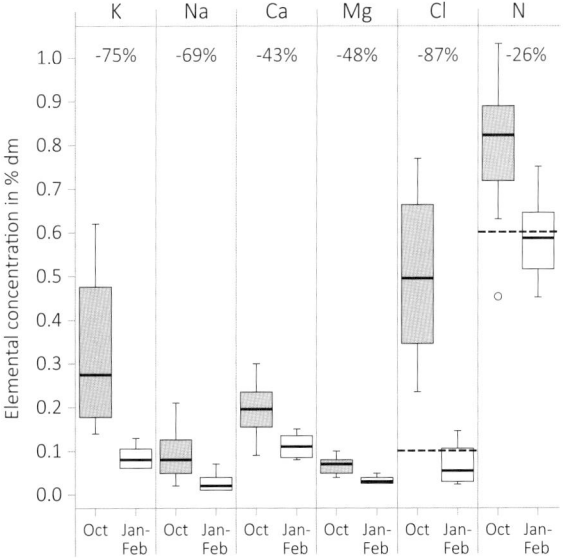

Figure 3.6: Concentrations of elements critical to combustion of five Common Reed stands (*Phragmites australis*) in October and January/February in Northeast Germany (own data). Percentage values show concentration losses. Continuous lines mark the critical values for Cl and N (after Obernberger et al. 2006).

3.5.4 Combustion techniques

Matthias Ahlhaus & Christian Jantzen

Plants for the direct combustion of biomass from paludiculture have to be adapted to the essential characteristics of these solid fuels (Figure 3.7). To achieve complete combustion of the biomass, despite its high content of volatile matter, the gases have to remain at high temperatures long enough and have to be intensively mixed with the supplied air. This requires the regulation of primary and secondary air supply via adjustable fans (Hartmann 2009b). As herbaceous biomass may have a high ash content, the use of a high-capacity automatic ash removal system is generally required. The high silicon content of the biomass may result – depending on the used furnace, in undesirable texturing or structural fixation of the ashes, which hinder the complete combustion of the carbon and reduce the efficiency of energy conversion (Box 3.14; Kask et al. 2007). Both the amount and the composition of ash are equally important (Chapter 3.5.3). Within these boundary conditions, the choice of a furnace is mainly determined by the desired capacity and the type of biomass.

The diverse combustion technology available on the market starts with equipment for domestic use with a capacity of a few kilowatts. This includes home furnaces and central heating boilers with stoker fired or fixed bed furnaces, which are fed manually with wood logs or briquettes (Box 3.15) or automatically with pellets (Box 3.12). Some of these boilers have an automatic and programmable ignition and an automatic ash removal system. Additionally, such boilers provide a high degree of comfort comparable to that of gas or oil boilers. Generally, the furnaces are also able to burn briquettes and pellets from herbaceous biomass but its high ash and silicon content and consequent structure formation might limit its application. Wood chips are typically used in boilers of medium capacity. Boilers in the capacity range of megawatts use moving step grates, stationary/rotating fluidised bed combustion or direct injection of pulverized solid fuels.

Paludiculture can supply both wood (Willow, Alder) and herbaceous biomass (e.g. Reed, Reed Canary Grass). The fuel characteristics of reed are generally between that of wood, which is unproblematic, and that of straw. As straw and herbaceous biomass from paludiculture have higher contents of harmful elements, the risk of increased pollution and corrosion exists (Chapter 3.5.3, Box 3.16).

Baled biomass from paludiculture (Colour picture 41) can be burned in manually fed straw bale gasifiers without further processing, or in big bale cigar burners. The non-continuous feeding of a straw bale gasifier poses, however, the risk of high emissions of pollutants and fine dust. Conversely, cigar burners

---

Box 3.13: Testing reed briquettes in commercial biomass heating plants

Matthias Ahlhaus & Christian Jantzen

Within the project 'Vorpommern Initiative Paludiculture', combustion tests with reed briquettes were conducted in three small-scale boiler systems designed for wood chips, namely:
1.) A biomass boiler suited to burn logs, wood chips and pellets with down-draught combustion and a nominal capacity of 50 kW;
2.) A biomass boiler for wood chips with a flange-mounted tube burner and a nominal capacity of 80 kW;
3.) A biomass boiler for wood chips with reciprocating grate and staged airflow with a nominal capacity of 300 kW.

During the trials it was observed that – in contrast to wood, the ash of the Reed briquettes kept the shape of the original briquette and was unexpectedly not transported out of the combustion chamber. Thus, the ash deposits accumulated in the combustion chamber impaired complete combustion and decreased combustion performance. However, no ash sinter was formed in the combustion chamber or on the heat exchangers during the combustion of Reed briquettes. As long as combustion proceeded without disruption, emissions stayed below the legal limits for small scale biomass boilers. Despite high nitrogen concentrations in Reed in comparison to wood, NO_x emissions also remained below the legal limit. In contrast, legal emission limits are often exceeded during combustion of straw and other herbaceous biomass.

In the biomass boiler with reciprocating grate and staged airflow, complete combustion could be achieved using Reed briquettes. The initial tendency of the ash to form structures and accumulate horizontally on the non-moving grate steps can probably be resolved by modifying the timing of grate movement, and by pneumatically or mechanically rotating the fuel in the combustion chamber. An additional problem – the increased condensation of gaseous inorganic particles, can be prevented or mitigated by a higher return temperature to the boiler. With optimal combustion, process emissions remained significantly below legal limits.

This shows that with an adjusted combustion process, Reed briquettes can be used in biomass boilers designed for woodchips while meeting the legal emission requirements.

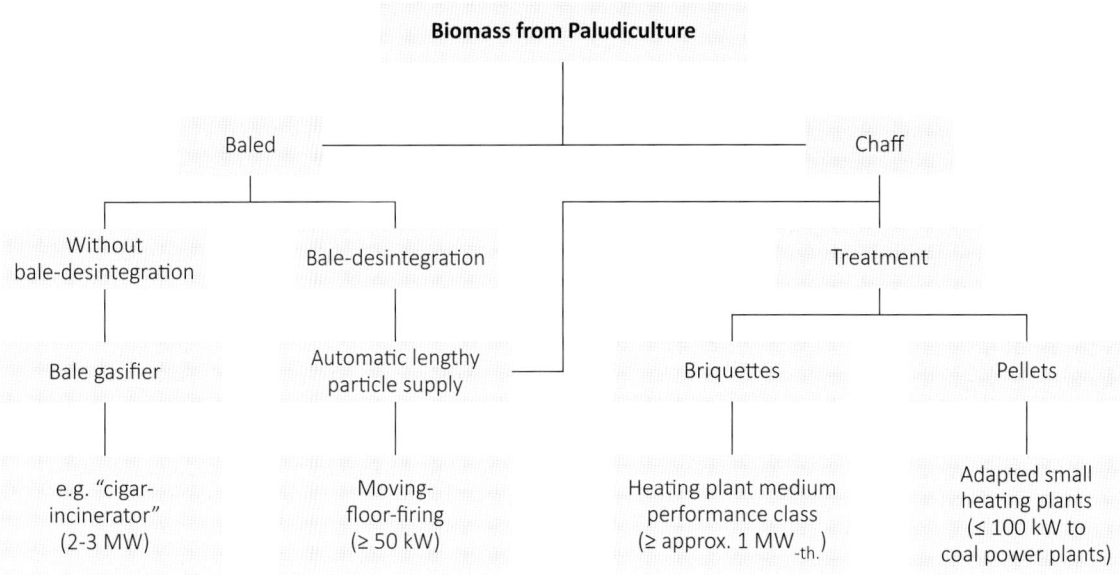

Figure 3.7: Processing biomass from paludiculture as solid fuel (after Kaltschmitt et al. 2009).

Box 3.14: Production of biochar

Jörg Burgstaler & Denny Wiedow

The pyrolysis of plant biomass takes place under the exclusion of air between temperatures of 350–800°C. The end products are biochar, synthesis gas, and heat. Pyrolysis generates gas as an energy carrier and, at the same time, part of the biomass carbon is fixed in the biochar. By further processing and activation, the biochar can attain a large useful surface area. This biochar can be used to improve the fertility and water holding capacity of the soil, and as activated carbon (Brandstaka et al. 2010; Verheijen et al. 2010, Sohi et al. 2009). Biochar not only improves the soil (Burgstaler & Wiedow 2012; Kern et al. 2011) but also serves as a carbon sink because it hardly decomposes.

Another option is hydrothermal carbonisation of biomass. This process imitates coal formation (coalification) by exposing the biomass for a few hours under exclusion of oxygen to a pressure of 80 bar and to temperatures of up to 250°C using steam. Apart from biocoal, by-products such as liquid and gaseous hydrocarbons develop in various concentrations. Possible substrates for producing biocoal are biomass from nature conservation or landscape management, wood chips, and also biomass from rewetted fen peatlands. This coal is of limited value for soil improvement because during hydrothermal carbonisation the cellular structure of the organic material is completely destroyed and its surface area and adsorption capacity are strongly reduced.

---

are fed continuously but are mostly adapted to square bales, which are currently less common.

However, the bale can be shredded before combustion and the loose biomass fed automatically into the boiler via a moving floor or an underfeed stoker (Colour picture 42). Technical risks during this process mostly occur at the bale shredder. The long resistant plant culms, as well as the string used for binding may get entangled in the rotating equipment and cause the stop of the entire plant (Kask et al. 2007, Wulf 2009). The quality of bale shredding determines the flow properties of the biomass and therefore is of crucial importance for the automatic operation of the heating plant. In biomass heating plants in Denmark, bale shredding has reached a high degree of automation (Skøtt 2011). Alternatively, biomass from paludiculture can be supplied chopped, and fed into a biomass boiler (loose biomass chain). In this case, it has to be taken into ac-

3.5 Solid energy from biomass

Box 3.15: Briquetting

Christian Jantzen

In comparison to pelleting, briquetting uses higher working pressures (up to 1,200 bar) and a longer retention time of the biomass in the die shaft. This allows making high quality briquettes also from biomass, with a higher moisture content (maximum 20%) and a larger particle size (maximum 15 mm) of the pretreated biomass. Briquetting is better suited for compacting herbaceous biomass because the demands to conditioning are lower than in case of pellets. During briquetting with lower pressure, the glass-transition temperature is reached only for a short time or not at all, and the biomass is only mechanically solidified. This lowers the energy costs of processing but also the transportability and the bulk density of the end product. Briquettes have various forms and sizes but the most common are those with a cylindrical shape and a diameter of 50–70 mm. The length of the briquettes should be about the same as their diameter to guarantee good mechanical flow.

Among the briquetting methods, string briquetting prevails because it requires less energy and yields briquettes of good quality. During the briquetting process (Figure 3.8, Figure 3.9), the chopped (and if necessary dried) biomass is transported from the storage container into a press shaft via screw conveyors, which pre-compresses the biomass. The rotation speed of the screw conveyor determines how much material is fed into the press shaft. A hydraulic press presses the fed material into the shaft, and condenses it to the diameter of the die. The total pressure is brought about by the reduced diameter of the die at the end of the shaft and the friction between particles, and between particles and equipment. In order to mitigate the relaxation effects, the briquettes are transported through a cooling system, similar to the pelleting process.

The performance of the briquetting press can be significantly improved by reducing the average particle size of Reed choppings and using the ideal moisture content. Briquettes produced in this way even exceed the bulk density of wood reference samples. Reed briquettes are stable and do not disintegrate during transport and storage. However, reducing the particle size results in the formation of more fine particles, which may lead to problems when feeding the boiler (Colour picture 37).

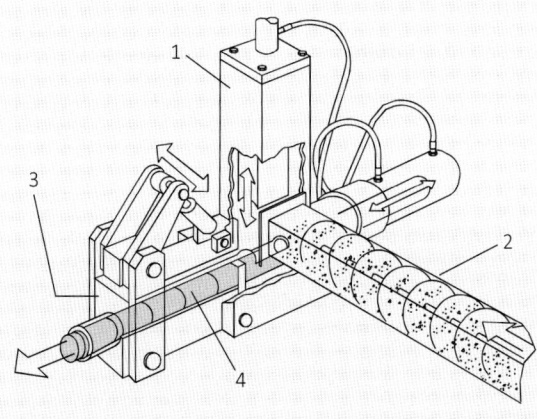

Figure 3.8: Schematic illustration of briquetting with a hydraulic piston propulsion: 1: Hydraulic forced input, 2: Screw precompression (inlet), 3: Pressure control, 4: Press channel (Hartmann & Witt 2009).

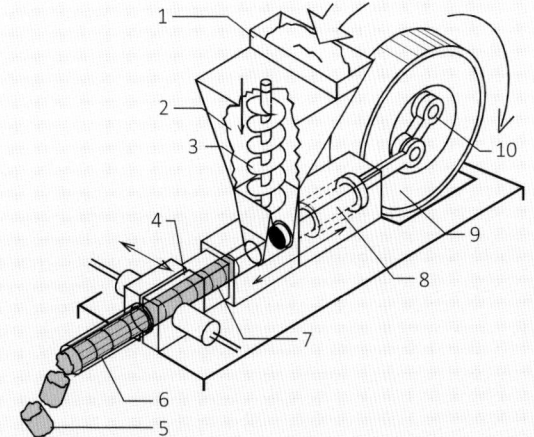

Figure 3.9: Piston briquettes press with a gyrating mass supported power unit 1: Feeding chute, 2: Biomass, 3: Feed screw, 4: Adjustable tube, 5: Briquettes, 6: Cooling zone, 7: Briquette tube, 8: Piston, 9: Fly wheel, 10: Excenter (Kaltschmitt et al. 2009).

count that heterogeneous particle distribution may lead to irregularities during feeding.

Technically, pellets can be applied in biomass boilers of various types and capacity. For the automatic feeding of briquette combustion chambers larger feeding systems are required, which are currently only available for boilers with a capacity over 1 MW (Figure 3.7).

It can be concluded that biomass from paludiculture is well suited for thermal utilisation by direct combustion (Box 3.13). The energy content of the biomass is high and its ash content low compared to other herbaceous biomass and no risk of ash sintering exists. Further research is still required with regard to the emission of dust and fine dust during combustion. This problem can, however, be effectively mitigated by an automatic regulated combustion process with continuous feeding and downstream dust filter systems.

3.6 Liquid and gaseous biofuels

Besides direct combustion, the biomass can be upgraded to liquid and gaseous energy sources to be used for more challenging purposes – e.g. as a fuel for vehicles; for producing electricity, heat, or both, in combined heat and power (CHP) plants (WGBU 2009).

3.6.1 Fermentation to biogas

Denny Wiedow, Jürgen Müller & Jörg Burgstaler

Biogas is a mixture of gases with methane (CH_4) as the main component, which is produced via anaerobic fermentation of organic matter. The gas mixture is usually utilised in CHP plants, but can also be fed into existing natural gas infrastructure after the gas has been cleaned (Weithäuser et al. 2010).

Biogas is produced from a wide range of organic substrates by a variety of production methods. Depending on the dry matter content of the utilised substrate, wet fermentation (< 15% DM) or (more rarely) dry fermentation (fermentation of solids, >15% DM) may be applied. For wet fermentation, generally liquid manure combined with energy rich substances such as maize silage is used and continuously fed into the fermenter. For dry fermentation, the fermenter is fed discontinuously with solid substrates such as solid manure. Fermenters for dry fermentation with continuous feeding are currently being developed.

Although fresh biomass is suitable for the continuous feed of co-ferment, conserved biomass such as silage is currently more commonly used. If harvested late, biomass from wetland vegetation such as sedge

---

Box 3.16: Legal aspects of burning herbaceous biomass in Germany

Thomas Hering

The same legislation regulates combustion of biomass from paludiculture and the combustion of straw, miscanthus and herbaceous biomass from conservation management. The size of the boiler determines which rules and regulations apply.

The legal framework in Germany permits the combustion of straw and similar herbaceous biomass in small boilers with a capacity of less than 4 kilowatt (§ 3, no. 8 of the 1st German Federal Emissions Control Ordinance). Such biomass includes, for example, reed, miscanthus, hay and corn cobs. Since the amendment of the ordinance in 2010, the use of other, unspecified biomass from renewable resources is also permitted, as long as it meets various requirements (§ 3, no. 13 of the 1st German Federal Emissions Control Ordinance).

From a capacity of 100 kW onwards, the boilers are regulated by the 4th German Federal Emissions Control Ordinance and the Technical Guidelines on Air Quality Control (German: *Technische Anleitung Luft*). This implies that for boilers between 100 kW and 50 MW, a planning permission procedure according to paragraph 19 of the German Federal Emissions Control Act is required. This procedure entails that considerably higher requirements have to be met in regard to emission control, which leads to higher costs for admission and inspection of the boiler. This, in turn, results in higher investments – for instance, in feeding, firing and emission reduction technology, as well as in significantly higher operational costs for electricity, maintenance and emission control in comparison to similar wood fired boilers. Depending on fuel quality and conversion concept, similar expenditures are to be expected for flue gas cleaning.

vegetation or reedbeds does not fully meet the requirements – in terms of fermentability and space capacity, to be used as sole substrate in conventional wet fermentation (Figure 3.10). Similarly, this biomass is not well suited for ensilage because it contains few soluble carbohydrates (Box 3.17).

Solid fermentation is better suited for substrates rich in cellulose because of longer fermentation periods and cellulolitic preprocessing. The aerenchymatic tissue of many wetland plants is beneficial for fermentation because it improves gas movement and the uptake of percolation liquid in the solid substrate (Müller et al. 2012). A further advantage of solid fermentation is the high methane content of the yielded biogas, which allows the gas to be easier fed into the existing natural gas grid, because processing to meet the requirements of DIN 51624 (2008) is less complicated. Less favourable is the fact that various wetland plants have rather high concentrations of sulphur, which may lead to high H_2S concentrations in the end product and therefore requires desulfurization. The current state of technology also allows using small amounts of well-chopped biomass from paludiculture in wet fermentation plants. If biomass from paludiculture is to be used as the sole substrate, only solid fermentation is suitable. However, also in this aspect technology still is in progress: The Verbio AG enterprise is currently building installations in Zörbig (Saxony Anhalt) and Schwedt (Brandenburg) for producing biomethane from straw. Till 2019, these plants should process about 20,000 t of straw annually (www.verbio.de). In principle, this technique would also allow the utilisation of late harvested herbaceous biomass (in bales) from paludiculture to produce biomethane.

3.6.2 Fermentation to bioethanol

Denny Wiedow & Jürgen Müller

Bioethanol is produced from vegetable raw materials and is utilised either as an alternative or as additive to liquid fuel (Gray et al. 2006). If the production process costs can be kept low besides keeping producer prices competitive, there is an almost unlimited market for bioethanol.

Ethanol can be synthesised from a wide variety of plant species and several plant parts:
- From plants with high concentrations of water soluble carbohydrates (largely vegetative plant organs, i.e. of sugar cane);
- From starch containing plants (mainly generative plant organs, i.e. from grains, but also from geophytes such as potatoes);
- From cellulose containing raw materials (mainly woody parts of herbaceous plants, shrubs and trees such as wood waste, or material from landscape and nature conservation management).

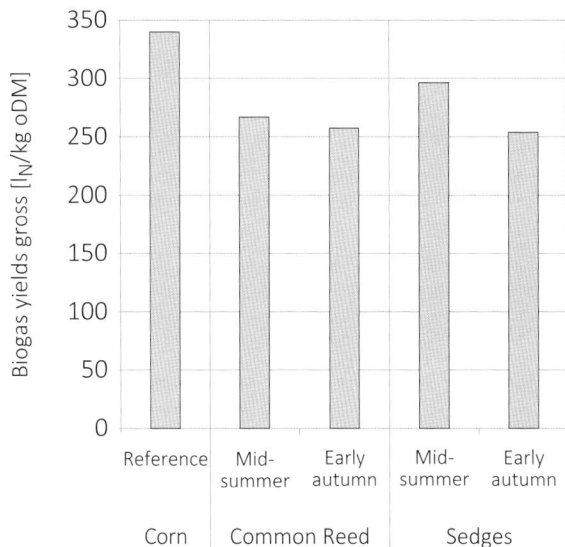

Figure 3.10: Norm-Biogas yields (IN kg^{-1} oDM) and methane yields (m^3 CH4 ha^{-1}) of ensilaged reed and sedges harvested during July and September in comparison to maize silage (Kalzendorf 2011).

To produce bioethanol, the carbohydrates of the plant matter are fermented to alcohol with the aid of enzymes or yeasts. For the production of bioethanol from sugar and starch containing plants, well known techniques are available.

In contrast to conventional bioethanol, ethanol from ligno-cellulose has a more favourable CO_2 balance and does not compete with the production of food or fodder (Alvira et al. 2010). With regard to the physiological characteristics of most wetland plants and the substances they contain, the latter method is the only one that is available for the utilisation of biomass from paludiculture. Although perennial helophytes and hydrophytes with starch containing rhizomes do exist, the harvest of these rhizomes would weaken the productivity of these plants and therefore would not permit a sustainable use.

In contrast, the harvest of biomass from, for example, Common Reed (*Phragmites australis*) or Papyrus (*Cyperus papyrus*) offers an opportunity for the production of bioethanol. However, the techniques to produce ethanol from ligno-cellulose still are at an early stage of development (Balat et al. 2008). The high complexity and costs of the different process stages up to the saccharification are currently still problematic (Sassner et al. 2008). Only when significant progress in this respect is made, the production of bioethanol from paludiculture plants will become a realistic option.

Box 3.17: Ensilage

Denny Wiedow & Jörg Burstaler

Storing and preserving biomass is necessary to secure availability for feeding animals, supplying biogas plants or processing raw material over the entire year. Drying and ensilage are established agricultural techniques for preservation. During ensilage, fermentation is directed to produce organic acids, which preserve the plant matter (Nehring 1972).

To test the suitability of paludiculture biomass for ensilage, Reed Canary Grass, sedges and Reed were harvested in July and September. The biomass was chopped into particles of <4 cm length to increase the availability of carbohydrates by largely destroying the cells, and to compact the material. The biomass was dried to reach a dry matter content of 35% and subsequently ensilaged in 100 litre barrels, as well as in glass preserving jars (1.5 l, 5 samples), for a period of 90 days at 25°C in line with DLG (2000). After filling the containers, the material was compacted and sealed under the exclusion of oxygen. This process suppresses aerobic and non obligatory anaerobic microorganisms like bacteria, yeasts and moulds. During the ensilage, lactobacteria convert the sugar contained in the plant matter into lactic acid, lowering the pH to below 4.0. This environment suppresses the growth of bacteria that impair the fermentation (*E. coli* group, *Listeria* and *Clostridia*).

Table 3.10 shows that Reed silage reaches values similar to high quality grass silage but only regarding dry matter, ash content, crude protein and crude fat concentrations. All other relevant variables are outside the reference range for optimal fermentation of grass silage with 2nd to 4th cut ryegrass (Pries 2007; Spiekers et al. 2007; Dunker 2011). This implies that without silage additives it is impossible to convert Reed into silage that is suitable as animal feed.

Table 3.10: Success of silage and fermentation quality of reed samples harvested during July and September. Assessment was carried out according to DLG (2000) in comparison to silaged grass (Dunker 2011*, Pries 2007** and Spiekers et al. 2007).

Variable	Reed silage		Grass silage
	July	September	
pH	5.5	5.6	4–5**
Gas formation [ml·200 mg^{-1}]	18.3	17.3	> 45*
Dry matter (DM) as percentage of fresh matter [%]	30.5	43.8	30–40
Ash content [% DM]	8.9	9.8	< 10*
Crude protein [% DM]	16.3	14.9	< 17
Crude fibre [% DM]	32.8	34.7	22–25*
Crude fat [% DM]	2.4	2.7	2.5–4.5*
Neutral detergence fibre (NDF) [% DM]	66.4	65.4	40–48
Acid detergence fibre (ADF) [% DM]	36.5	37.0	27.4*
Acid detergence lignin (ADL) [% DM]	7.0	8.9	< 10*
Ammoniac (NH$_3$) [g·kg^{-1} DM]	1.7	1.6	0.3–0.5*
Ethanol [g·kg^{-1} DM]	9.5	12.8	< 2*
2,3-butandiol [g·kg^{-1} DM]	0.5	3.8	< 2*
Acetic acid [g·kg^{-1} DM]	19.3	10.3	20–30**
n-butyric acid [g·kg^{-1} DM]	16.7	2.7	< 3**
Lactic acid [g·kg^{-1} DM]	2.3	12.3	20–30**

3.6 Liquid and gaseous biofuels

3.6.3 Thermo-chemical gasification

Hannes Wagner & Martin Kaltschmitt

During thermo-chemical gasification, solid biomass fuels are converted into combustible gases under high temperatures. This process takes place under oxygen deficiency but an oxygen containing gasification medium is supplied in the form of air or steam, in order to transform the biomass – through several steps, mainly into carbon monoxide (CO) and hydrogen (H_2). The energy required for this process is supplied by the partial combustion of the biomass (Pollex et al. 2012).

The end product of the gasification process is cleaned from dust, tars and other undesired by-products, and can be used as a fuel in engines, turbines or fuel cells, either for electricity or for combined heat and power (CHP) generation. For this purpose, Otto gas engines or spark ignition Diesel engines are commonly used. Thermo-chemical gasification with subsequent utilisation of the gas for CHP and power generation is an interesting option for small and medium capacities – up to 1 MW (Hofbauer et al. 2009).

Many techniques are available for thermo-chemical gasification, such as fixed-bed gasifiers, fluidized bed gasifiers and entrained-flow gasifiers. These techniques are, however, still under development or in the beginning of commercialisation (Kaltschmitt 2012).

The first gasification installations with a thermal capacity of about 10 MW entered in operation in Austria. In Germany the first gasification installations are currently being built. This implies that the technology is available on the market for the operation with wood chips. The gasification of herbaceous biomass such as straw, hay from conservation management as well as Reed briquettes and pellets is possible but needs more research and development because of the described differences to wood. Commercial projects involving the gasification of herbaceous biomass do so far not exist.

3.6.4 Production of synthetic liquid fuels

Hannes Wagner & Martin Kaltschmitt

Alternatively to utilisation in a CHP installation, the gas produced by thermo-chemical gasification can be converted into liquid or gaseous biofuel. During this process, the CO and the H_2 of the producer gas are – most often with application of catalysts, exothermically converted into various hydrocarbons (Kaltschmitt et al. 2009). Such installations have been developed for wood, are currently being introduced to the market (Chapter 3.5.4), and could be a promising opportunity for the utilisation of biomass from paludiculture.

Liquid fuels can be produced via methanol or Fischer-Tropsch synthesis. For both technologies, the synthesis gas needs to be cleaned in order to protect the catalysts from deactivation. Another requirement is that the carbon to hydrogen ratio is close to that of the desired end product. The chemical reactions are usually exothermic, which implies that the released heat should be used in the best possible way. After synthesis, the end products need to be cleaned and processed.

Apart from liquid fuels, also refined gaseous fuels can be produced – e.g. biomethane or Synthetic Natural Gas (Bio-SNG). For gaseous fuels, the same conditions apply with regard to gasification, gas cleaning and synthesis as for liquid fuels. It is generally more efficient to produce gaseous than liquid energy carriers because the catalysers used for gas synthesis work in a more selective way, and the upgrading of the products is considerably easier.

The synthesis of liquid or gaseous fuels from herbaceous biomass is still at the beginning of its technical development, both with respect to optimisation of methods and techniques and reduction of production costs. Successful market introduction is furthermore still hindered by moderate conversion efficiencies (Rönsch et al. 2009, Rönsch & Kaltschmitt 2012).

4 Harvest and logistics

In the light of progressing climate change and the enormous greenhouse gas emissions from agricultural use of drained peatlands, the development of technologies for soil conserving use of wet peatlands becomes increasingly important. The utilisation of wet peatlands requires special adaptations, first and foremost with respect to trafficability. Methods to assess soil trafficability have been developed for drained peatlands, but these have only limited value for degraded and rewetted peatlands. Consequently, new assessment approaches need to be developed to identify suitable management techniques for such areas (Chapter 4.1). Specialised machinery is already being used in landscape conservation management and traditional reed harvesting (Chapter 4.2). Based on this experience, it is also important to develop new high performance harvesting methods and logistic concepts to address the specific site conditions of wet peatlands (Chapter 4.3). These concepts must focus on avoiding repeated movement of vehicles over the same track of peat soil. Departing from the boundary conditions of wet peatland utilisation, recommendations are given for the large scale harvest of biomass from paludiculture (Chapter 4.4).

4.1 Trafficability of wet and rewetted fens

Denny Wiedow, Jörg Burgstaler & Christian Schröder

Trafficability for vehicles, especially harvesting machinery, is of crucial importance for the utilisation of wet fens. Trafficability is determined by the bearing capacity of the sward and the stability of the ground; also, it is controlled by vegetation, groundwater level and soil conditions. The bearing capacity can be defined by two soil variables, namely shear strength and soil penetration resistance. Thus, the use of the aforementioned variables allows assessing the trafficability of wet peatland sites. Furthermore, these variables are useful for monitoring the effects of driving over the peat soil.

4.1.1 Methods to assess trafficability

As it was stated above, the most important variables for assessing trafficability of peatland soils are shear strength and soil penetration resistance.

Shear strength is the ability of a solid material to resist forces that act in opposite directions along parallel faces. Shear strength is the maximum shear stress that a material (in this case the soil or sward) can resist before breaking (Kraschinski et al. 1999; Schmidt 1980; Schreiner 1967). Shearing forces occur when the weight of a vehicle exerts pressure at the point of contact between vehicle and soil/sward. Peat soils usually consist of peat layers with a large pore volume that easily deform. Therefore, trafficability is decisively determined by the root density in the top 10 cm of the soil (Schmidt 1995). As the impact of weight may reach deeper into the soil, it is recommended to access the shear strength of the soil in different depths. Shear strength is usually expressed as pressure in kilopascals ([kPa] = kilonewton per square metre [kN·m^{-2}]) (Lindner 1963).

Soil penetration resistance is the resistance of the soil against the impact of load (Krahmer 1997). This variable depends on soil physical properties such as soil compaction, water content and pore size distribution, as well as on the structure of the organic matter (Horn 1984). To determine soil penetration resistance in wet fen peatlands, a penetrometer is used, which consists of a rod with a conical tip at the bottom. The tip is driven into the soil with uniform speed, whilst soil penetration resistance (in force per unit area = kilopascal) is electronically recorded along the depth profile. Structural changes in the soil (as a result of load) over time can be assessed by repeated measurement (Dürr et al. 1995).

4.1.2 Trafficability after rewetting

For easy assessment of trafficability of drained peatlands, bearing capacity classes have been defined, which use shear strength as a parameter (Table 4.1, Knieper 1999). However, these classes can be applied only to a limited extent to rewetted peatlands. The structure of the soil of rewetted peatlands, which has been subject to secondary pedogenesis following drainage and subsequent rewetting, differs from that of both drained and of pristine wet peatlands. In pristine mires, the top peat layer is usually characterised by a

Table 4.1: Bearing capacity of drained peatlands for the use of conventional agricultural machinery on grassland, based on shear strength of the sward (modified after Knieper 1999).

Bearing capacity	Shear strength [kPa]
Very poor	< 20
Poor	> 20-< 26
Intermediate	> 26-< 31
High	> 31-< 36
Very high	> 36

Table 4.2: Trafficability for conventional agricultural machinery on grassland based on soil penetration resistance (modified after Schmidt 1995).

Trafficability	Penetration resistance [kPa]
Good	> 800
With restrictions	800-500
Not trafficable	< 500

large pore volume and a very low degree of decomposition. Drained peatlands, in contrast, have a significantly lower pore volume and higher density in the top peat layer (Chapter 2.2). Furthermore, the mulmified top layer of degraded and rewetted peatlands consists of an amorphous mass with poor cohesion. In addition, rewetting of degraded peatlands may also have led to swelling of the peat body and floating of the upper peat layer, which greatly reduces trafficability despite the presence of a sward of high shear strength. At such sites, heavy machinery (even with a large contact area) sinks into the ground, which greatly increases shear stress on the soil surface. Thus, trafficability of a site cannot only be assessed by shear strength but also soil penetration resistance needs to be considered (Table 4.2, Box 4.1).

4.1.3 The impact of machinery use on the peat soil

The use of heavy machinery increases performance, but does not necessarily lead to higher ground pressure when the machinery is equipped with wider tyres or tracks. Additionally, wider working widths may result in the proportional reduction of the area crossed by the vehicle (Brandhuber et al. 2008). Therefore, size and weight of the harvesting machinery are less relevant to their suitability for wet peat soils than the ratio of total weight and ground contact area. Besides static ground pressure, weight distribution, vibration, and the kind of pressure transmission onto the ground determine the suitability of vehicles for wet peat soils.

Tracked vehicles and vehicles with wide tyres are used on wet peat soils, and both have their advantages and disadvantages for soil protection (Colour pictures 43–46). The main objective of using tracked vehicles is to avoid damage to the sward by slipping wheels or shear stress. The suitability of vehicles for wet peat soils depends on their use. Small tractors with wide wheels and light weight mowers are particularly suitable for soil conserving management of fen peatlands (Box 4.2), but also larger machines with wide tyres or balloon tyres are tried and true. Alternatively, wide-tracked vehicles with a large ground contact area and a uniform weight distribution can be used. Independently of the machinery used, drive style and experience are of great importance for soil conserving driving in wet peatlands.

4.1 Trafficability of wet and rewetted fens

Box 4.1: Bearing capacities of fen soils

Denny Wiedow, Jörg Burgstaler & Christian Schröder

The influence of groundwater level, vegetation, and peat degradation on the trafficability of wet fen peatlands was investigated on 25 wet and rewetted sites. Soil penetration resistance (with a penetrologger, cone 3.3 cm², angle 60°) and shear strength (with a shear vane tester, vane 20 x 40 mm) were measured in various depths, once in winter and once in summer. The investigated plots showed, on average, a high bearing capacity (after the classification of Knieper (1999; Figure 4.1, Table 4.1). However, the soil penetration resistance values demonstrate that the study plots have to be classified as inaccessible for vehicles according to the criteria of Schmidt (1995; Figure 4.1, Table 4.2). The different outcome of these assessments illustrates that trafficability of rewetted fens cannot be solely derived from either shear strength or soil penetration resistance. Shear strength enables to assess the mechanical strength of the sward, whereas soil penetration resistance allows determining the firmness of the soil.

Furthermore, we analysed which site parameters best explain shear strength and soil penetration resistance. For shear strength in 15 cm depth the vegetation appeared to be the best explanatory variable, for soil penetration resistance the mean water level (soil moisture class). Degradation depth of the peat soil, however, influenced shear strength in a depth of 50 cm (Table 4.3).

Table 4.3: Relevance of the explanatory variables vegetation, peat degradation and soil moisture class to mean shear strength and mean soil penetration resistance at 15 and 50 cm depth. Analysis of variance (ANOVA) for 25 sites; level of significance: *** 0,01; ** 0,05; *0,1.

	Shear strenght				Soil penetration resistance			
	15 cm		50 cm		15 cm		50 cm	
	F-value	Pr (>F)	F- value	Pr (>F)	F- value	Pr (>F)	F- value	Pr (>F)
Vegetation	6.30	0.019 **	0.01	0.938	0.21	0.651	0.00	0.994
Peat degradation	0.96	0.339	7.49	0.012 **	0.00	0.993	0.45	0.508
Soil moisture class	0.01	0.937	0.08	0.781	3.06	0.094*	0.58	0.455

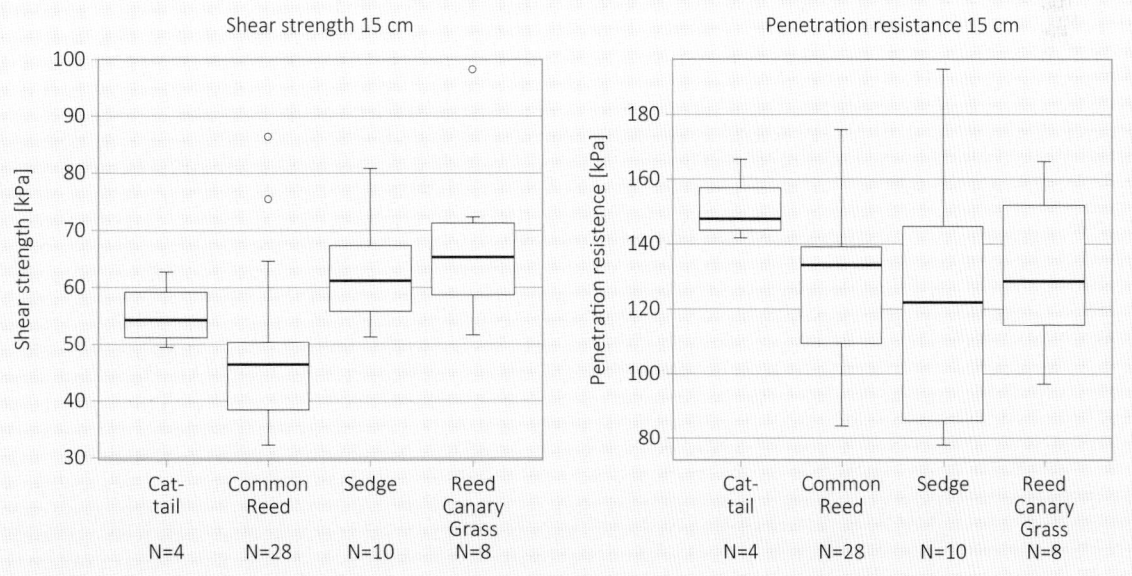

Figure 4.1: Shear strength (left) and soil penetration resistance (right) in different vegetation types.

Box 4.2: Influence of harvesting machinery on peat soils

Denny Wiedow, Jörg Burgstaler & Christian Schröder

The impact of harvesting machinery on wet peatland soils was studied in the Lower Peene Valley (NE Germany) using a) a small tractor with wide tyres (mowing for landscape conservation management); b) a snow groomer (for harvesting energy biomass, Colour pictures 61 and 62); and c) a Seiga reed harvester with balloon tyres (harvest of reed for thatching, Colour picture 45). The studied fen peatland is shallowly drained, with only slightly decomposed peat in the top soil ('Ried', cf. Stegmann & Zeitz 2001). In some parts of the peatland, the upper peat layers had previously been removed. The investigated plots differed with regard to plant species composition, the time of harvest, and the groundwater levels during machinery operation (Table 4.4). Before and after passing the machinery over the peatland, shear strength (at 15 cm depth) and soil penetration resistance (up to 80 cm depth) were measured with a shear vane tester (vane: 20 x 40 mm) and a penetrologger (cone 3.3 cm², angle 60°), respectively (Table 4.4). Using the shear strength scale of Knieper (1999), the summer harvest areas with loose reed had a poor bearing capacity, whereas the bearing capacity of winter harvested areas with reedbeds (which have a dense network of rhizomes and roots) was very high (Table 4.1). However, soil penetration resistance (135–411 kPa) was below the threshold of 500 kPa (Figure 4.2), so that all study plots have to be classified as 'not trafficable' (sensu Schmidt 1995).

Table 4.4: Shear strength of semi-natural fen peatland soil at 15 cm depth after crossing by: a small tractor with wide tyres, a Seiga with balloon tyres (biaxial) or a converted snow groomer (each 160 repetitions), – compared to untraveled control plot (each 80 repetitions).

Type of machinery	Small tractor		Seiga		Snow groomer	
					Tracked vehicle	trailer (25 m³)
Type	Antonio Carraro TTR 4400 HST		Seiga		PB 240	
Power [kW]	28		37		178	
Weight [kg]	1,200		1,500 (unloaded)		7,000	5,00 –9,750
Ground pressure [g· cm^{-2}]	230		83		80	50–100
Harvested area	Ferne Wiesen (West)		Ferne Wiesen (East)		Murchiner Wiesen	
Vegetation	open reedbeds with mixed vegetation		dense monodominant reedbed		open reedbeds with mixed vegetation	
Soil moisture class (Box. 5.3)	4+		5+		4+	
Mean water level during summer [cm]	-5.4		+1.6		-4.3	
Mean water level during winter [cm]	-3.3		+5.7		-2.3	
Water level during harvest	- 20 cm		- 5 cm		+ 10 cm	
Harvest time	Summer		Winter		Summer	
Shear strength [kPa]	control	traversed	control	traversed	control	traversed driven over
15 cm	21.7	20.2	55.7	46.2	26.6	20.9
Bearing capacity after Knieper (1999)	poor		very high		intermediate	poor

4.1 Trafficability of wet and rewetted fens

A comparison between areas that were crossed by vehicles and those that were not did not show any evidence of impact caused by the small tractor. In contrast, the impact of the snow groomer and Seiga reed harvester was very pronounced (Table 4.5). Although these vehicles did not cause any visible damage, the decrease in shear strength showed that crossing the area influenced the sward. Driving across the area with the snow groomer furthermore leads to an increased soil penetration resistance, indicating compaction of the upper peat layer (Figure 4.2).

Whether this compaction is reversible and the sward will recover until the next harvesting season is unclear, but damage to sward and soil must be expected with short time intervals between repeated vehicle crossings. Multiple crossings should therefore be avoided.

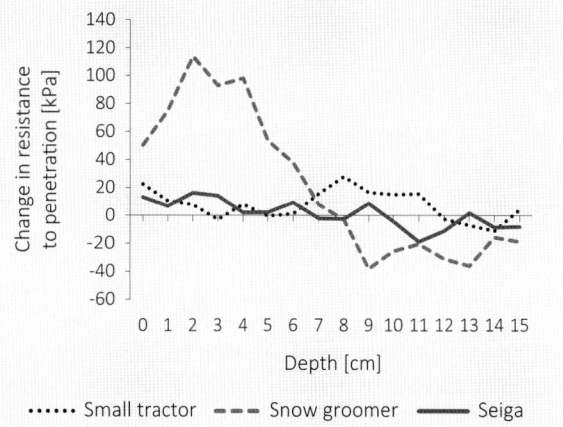

Figure 4.2: Change in soil penetration resistance [kPa] within the top 15 cm of the soil in a semi-natural fen peatland after crossing with a small tractor, a snow groomer and a Seiga (Number of repetitions for each variant: crossed 160, control 80).

Table 4.5: Existing machinery for wet peatland sites.

Type of machinery	Examples	Fields of application and advantages	Limitations and disadvantages
Adapted conventional agricultural machinery (Colour pictures 43, 47 and 48)	Farm tractor with flotation wheels or twin tyres with a light weight baler on a tandem axle; alternatively with bogie- or delta tracks	• Use in moderately wet areas, under dry conditions and with frost • High mowing performance • Biomass removal possible	• Application limited by high water levels and weather conditions • Problems with biomass removal: heavy bales might need to be individually transported to the field margin
Small size machinery (Colour picture 44)	Uniaxial tractor or small tractor with cutter bar	• Application in management of humid meadows to conserve and restore species and habitats • Usually only mowing; occasionally biomass removal (e.g. small HD-bales, loose on a tarpaulin, manual work) (Colour pictures 85 and 86)	• Low performance/ high costs in relation to area → only for small scale applications • Less suitable for large scale biomass removal
Wheeled special machinery (Colour picture 45)	Mostly Seiga machines with two or three axles and balloon tyres	• Used in reed harvest (Europe, China, etc.) • Low ground pressure due to low weight and balloon tyres	• Seiga machines are no longer produced, only old machines or copies are in use • Seigas have limited engine performance • Possible soil damage by slipping
Tracked special machinery (Colour picture 46)	Specialised machinery and adaptation of snow groomers	• Conservation management and biomass harvest (e.g. thatching reed) • Alternative fields of application: peat industry, bunker silo making, landfill remediation • Low ground pressure due to wide tracks, even for heavy machines	• Conversions are mostly individual solutions • Not roadworthy, transport via flat-bed trailers, wide tracks make transport of the machines difficult (transport permission required or tracks need to be removed) • Soil damage may occur during turns

4.2 Agricultural machinery for wet areas

Sabine Wichmann, Sebastian Dettmann & Tobias Dahms

The availability of suitable agricultural machinery is crucial for the practical implementation of paludiculture as a new land use practice. In contrast to conventional land use of drained peatlands, paludiculture requires the adaptation of machinery to wet conditions instead of the adaptation of the land to the machinery. The challenges for agricultural technology are to minimise ground pressure and to reduce the frequency of vehicle crossing, while establishing a high performance production chain. Solutions can be derived from the experiences of reed harvest for thatch and from conservation management. The following Chapter gives an overview of existing technology and discusses its advantages and disadvantages, as well as the need for further technical development.

4.2.1 Adaptation of harvesting machinery

The trafficability of water saturated peat soils is limited (Chapter 4.1). This makes it essential to minimise ground pressure caused by any machinery operation. The upper limit is frequently specified as 100 g per cm². This 'rule of thumb' is likely based on experiences made during the 1950s in the Romanian part of the Danube Delta: 'The reed is collected with […] light tractors with rubber tracks, not exceeding 100 g/sq.cm specific load which is the maximum pressure the Delta soil will stand.' (Chivu 1963). Two strategies to minimise the ground pressure are used:
a) Reduction of the overall vehicle weight (e.g. small tractor, utilisation of light weight vehicle parts);
b) Increase in ground contact surface, for instance, by the use of:
- Twin tyres (Colour picture 44);
- Flotation wheels (wide tyres with adjustable air pressure, (Colour picture 43);
- Balloon tyres (large low pressure tyres without tread (Colour picture 45);
- Bogie tracks (tracks of metal, rubber or plastic which are put over two wheels);
- Delta tracks as wheel replacement for conventional chassis (e.g. Soucy Tracks, Zuidberg Tracks, AmfiBios Tracks made by the Tidue company; Colour pictures 47 and 48);
- Tracked vehicles (e.g. Colour pictures 49–56).

Machinery used on wet sites can – on the basis of these two strategies, be divided into four groups:
1. Modified conventional agricultural machinery;
2. Small size machinery;
3. Wheeled special machinery; and
4. Tracked special machinery.

Examples of such machinery, their field of application and their suitability for paludiculture are presented in Table 4.5 and Colour pictures 43–46.

The operational range of conventional agricultural machinery is limited by weather and groundwater level. Even when equipped with twin tyres, flotation wheels or bogie and delta tracks, the ground pressure of conventional machinery does hardly reach values below 200 g cm^{-2}. This implies that conventional machinery can only be used in areas which are transitional to drier parts or when the soil is frozen. However, in many regions harvest over frozen soil or ice cannot be guaranteed every year because of failing deep frost – as it happens in Northeast Germany (Wichmann & Wichtmann 2009) or due to a too short frost period (e.g. Southern Finland, Hagelberg & Lyytinen 2007). Additionally, an ice layer thick enough to support the machinery takes more time to be formed in reedbeds, and is impaired by fluctuating water levels (Granéli 1984, Komulainen et al. 2008).

Small size machinery is generally used in nature conservation to manage valuable habitats and species. Small or uniaxial tractors (e.g. Brielmaier) are suitable especially for mowing difficult areas (Wippl & Paar 2015), but hardly for biomass removal on a large scale. Machinery with balloon tyres – as applied since the 1950s for reed harvesting (Björk & Granéli 1978), and especially tracked vehicles have shown to be promising and perform well on wet sites. Despite the large size of the machines and the additional load of the harvested biomass, a reduction of ground pressure to less than 100 g cm^{-2} and occasionally even below 50 g cm^{-2} can be achieved (Colour pictures 50 and 71). Overviews of application field and characteristics of individual harvesting machines are presented in Granéli (1984), Schuster (1985), Rechberger (2003), White (2009), Wichmann & Wichtmann (2009), and Wichtmann & Tanneberger (2009). As modified conventional machinery can usually not operate on wet peatlands, and small size equipment is not suitable for large scale harvests, we focus the discussion in the following sections on special machinery with tyres or tracks.

4.2.2 Wheeled machinery

Common examples of wheeled harvesting machinery are Seigas (Colour pictures 45 and 59) – i.e. light weight vehicles with three, four or six wheels with low pressure balloon tyres. Those vehicles were built by the Danish company Seiga (Lumkes 1969) that also developed a device (based on a hemp cutter) for automatically cutting and binding thatch reed (Rodewald-Rudescu 1974). The company does not exist anymore, but the vehicles are still used in many countries (e.g. Germany, Denmark, Poland, Hungary, Romania, France and China). The commercial reed harvesting business currently reproduces and further develops such machines and also

Box 4.3: 'Moortruck' – development of new specialised machinery

Hellmut Hans Kranemann

Within the project 'Vorpommern Initiative Paludiculture', the Kranemann company developed the concept of the 'Moortruck', a vehicle with a soil-conserving chassis to accommodate for the specific requirements of being efficiently operative on large peatland areas (Figure 4.3).

The weight of the vehicle is spread over six large volume low pressure tyres. The machine moves by four wheels, which are separately propelled with individual control of the torque to prevent soil damage. Each wheel is equipped with a hydraulic engine that is supplied with oil by an axial piston pump. Sensors on each wheel measure the rotational speed, the torque, and the steering angle of the wheels. From these data, the electronic steering device computes the oil pressure required for each wheel in order to minimise slipping. The air pressure within the tyre is automatically regulated to adjust the ground pressure to the load of the vehicle. The four propelling wheels can additionally be equipped with tracks. The combination of wheels and tracks has a significant advantage over the use of tracks only: The tyres with a large volume tend to float, which reduces ground pressure and damage to the sward. The two back wheels of the Moortruck have a hydraulic steering mechanism which is also electronically controlled.

The Moortruck consists of a framework of aluminium alloy (AlMgSi). This framework bears the chassis, the driver's cabin, the engine, and a loading area. The framework also supports a three-point interface for additional appliances both at the front and at the back. The floor of the loading area is a conveyor belt to facilitate unloading the harvested biomass. A baler (preferably a round baler) or other appliances can be mounted onto the chassis of the Moortruck instead of the loading area.

The designed vehicle is about 5 m long, 3 m high and 2.95 wide, which makes it roadworthy. The unloaded weight is about 3 metric tons; the loaded vehicle weighs about 6.6 metric tons. The engine capacity of c. 90 kW enables a hydraulically controlled speed of up to 20 km h^{-1}. Modular modifications to the vehicle chassis allow adaptation to specific harvesting requirements.

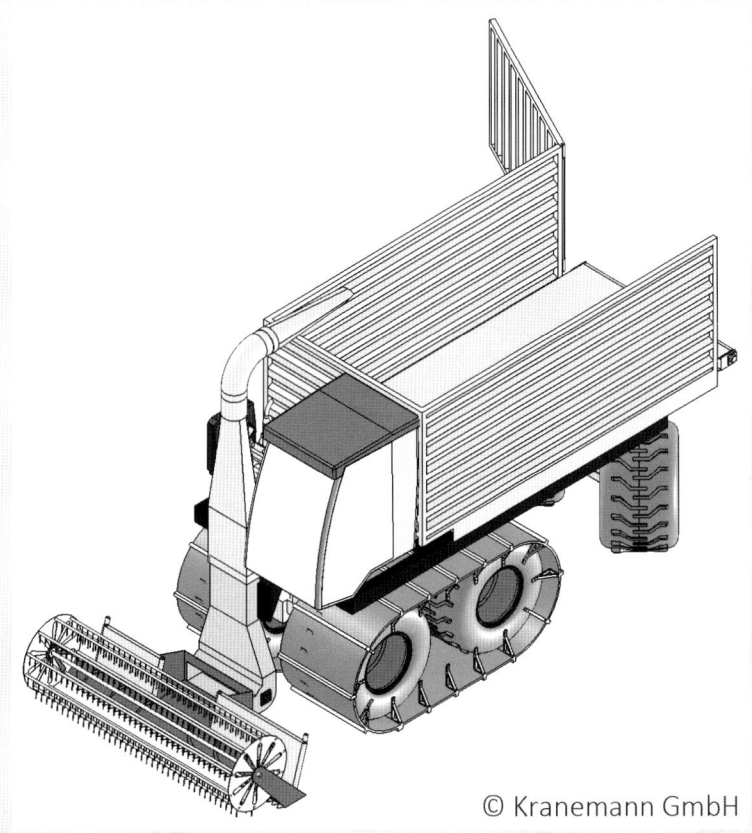

Figure 4.3: The 'Moortruck', a soil conserving vehicle designed for harvesting wet peatlands (Kranemann GmbH).

Box 4.4: The 'Paludi-harvester' in the VIP project

Sabine Wichmann & Christian Schröder

Within the VIP project, a snow groomer was modified and tested in large scale biomass harvesting of wet fen peatlands (Colour pictures 61 and 62). Target of conversion was to achieve a ground pressure of 100 g cm^{-2} or less.

A PistenBully (Kässbohrer PB 240 D, year of construction 1988) with a Cummins engine (177 kW, circa 6,500 h of operation passed) was chosen as a basis machine. The vehicle was given a general overhaul before conversion for biomass harvest. The frame of the vehicle was reinforced and cooling devices of higher performance for engine and hydraulics were installed behind the driver's cabin. The vehicle was equipped with 108 cm wide rubber tracks (made by the company Hans Hall, Colour picture 53) in order to conserve soil and vegetation. For the hydraulic system, biodegradable hydraulic oil was used (AVIA SYNTOFLUID PE-B 30). After conversion, the vehicle had an overall weight of about 7 metric tons, an overall length of 5.85 m, and a height of 2.80 m. The total width including the tracks was 3.35 m (without tracks 2.60 m).

A trailer was developed for transporting the harvested biomass, which is steered via the harvesting vehicle and is likewise equipped with tracks. The steering of the trailer can support the harvesting vehicle in difficult terrain and serves as an extra safeguard in case the harvesting vehicle gets stuck. The trailer has a load volume of about 25 m³ and a moving floor to unload the biomass. The unloaded weight of the trailer is about 5 metric tons. The trailer is 6.50 m long, 4 m high and its overall width – including the tracks, is 3.90 m (without tracks 2.75 m).

The ratio of overall weight and ground contact area (8.6 m² and 9.7m²) results in a ground pressure of about 80 g cm^{-2} for the harvester and 50–100 g cm^{-2} for the unloaded or loaded trailer. However, a much higher ground pressure may partially occur because of the varying weight distribution of the vehicles and the tread pattern of the tracks.

A two-step harvesting method was tested on circa 100 hectares. The first step comprised mowing and swathing the biomass using a front-mounted disc mower (Samasz, working width: 3.65 m). During the second step, the swathes were taken up and chopped using a front-mounted field chopper (made by Mengele, pick-up width: 2 m). Next, the chopped biomass was transferred into the trailer via a pivotable tube (Colour pictures 61 and 62). The trailer was unloaded on the edge of the field using the moving floor. Then, the biomass was reloaded whether onto lorries with trailers for road transport or onto a boat using a lifting grapple mounted to an excavator normally used for the harvest of maize (Colour picture 69).

The harvester was utilised in areas with different biomass yield and trafficability, which is reflected in the varying speed (Table 4.6). Economically relevant are primarily areas with high yields, where the 'paludi-harvester' achieves a performance of up to 1.5 ha h^{-1} for mowing and 1 ha h^{-1} for chopping the biomass. The overall performance (excluding non-productive times for maintenance, transport and dysfunction) is about 0.6 ha h^{-1}.

Table 4.6: Speed of an adapted snow groomer as a function of vegetation, biomass yield and harvesting procedure (n/a = not applicable).

Site	Vegetation	Trafficability*	Biomass yield	Mowing [km h^{-1}]	Chopping [km h^{-1}]	Access road [km h^{-1}]
Teufelsbrücke	Humid meadow (with Molinia caerulea, Carex spp.)	good	low	12	6**	n/a
Relzower Wiesen	Humid meadow (with Carex spp., Juncus spp., Phalaris arundinacea)	with restrictions	intermediate	9–10	6	n/a
Murchiner Wiesen	Reedbed (one-two year old culms)	poor	intermediate-high	6	4–5	n/a
Access way (not reinforced)			n/a	n/a	n/a	10
road***			n/a	n/a	n/a	20

* Subjective assessment, determined by groundwater level, presence of sensitive vegetation, areas with disturbed soil and pits from peat cutting.
** As biomass yields were low, two swaths were added together before biomass chopping.
*** Road trips should be avoided because they cause material stress and damage the tracks.

4.2 Agricultural machinery for wet areas

> The following recommendations for operating on wet areas arose from the tests:
> - As the 'paludi-harvester' is specialized machinery, the operator needs to be specially trained not only in operating the vehicle but also in day-to-day service and maintenance. Training a soil-conserving driving style on wet areas is useful to prevent damage to the sward. Particularly sensitive areas with low trafficability may be indicated by special plant species and need to be recognised by the driver so that he can slow down or avoid these places.
> - Unnecessary turning of the vehicle should be avoided to reduce shear stress on soil and vegetation. At sensitive but frequently accessed sites, the use of mobile road plates has proven to be useful (Colour picture 82).
> - The location of areas inaccessible for vehicles (such as former peat extraction pits and ditches) must be exactly known, particularly in highly productive vegetation, where visibility is strongly reduced. Therefore, it is crucial to have detailed knowledge of the terrain or guidance by a person who is familiar with the site. Satellite navigation technology (GPS) may support orientation.

equips tractors and trailers with large volume tyres to use them for transporting biomass. But, especially if high water levels occur, Seigas can float due to their balloon tires and are then difficult to maneuver. Wheel slipping could also damage the surface. An individual steering of wheels could reduce the risk of wheel slipping (Box 4.3).

4.2.3 Tracked machinery

No machinery from serial production is yet available for the large scale implementation of paludiculture, mainly because the requirements vary considerably, depending on site conditions, biomass processing and logistics, and because demanded quantities are too low. Most diverse experience for wet sites stems from the use of tracked vehicles for reed harvesting and conservation management. The following section and the pictures in the colour plates provide an overview of state-of-the-art harvesting machinery based on tracked vehicles.

The machines currently in use are usually prototypes or modified second hand vehicles. Frequently snow groomers, originally designed and built for use in ski resorts, form the basis of such vehicles. The leading manufacturers of snow groomers are Kässbohrer Geländefahrzeug AG (owner of the companies Formatic and Panaa) and Prinoth GmbH (which belongs to the Leitner group, as does Bombardier). Furthermore, Hägglunds-Bandvagn vehicles developed for the Swedish military are in use. The tracked vehicles are modified by specialised companies or by the users themselves (e.g. Colour picture 46, www.mera-rabeler.de; Mucha 2011). The British company Loglogic designs tracked vehicles for specialised applications such as conservation management of small wetlands – e.g. the Fen Harvester and the Softrak (Colour picture 76; www.loglogic.co.uk). The Netherlands are pioneers in the development of soil conserving chassis, tracks and technical adjustments for various application fields. They have more than 30 years of experience in wetland management with tracked vehicles. Worth mentioning are the Dutch companies Hanze Wetlands (Colour pictures 50 and 71; www.hanzewetlands.com), the successor of the traditional enterprise Wildemann, De Vries Cornjum (Colour pictures 55, 73 and 74; www.devriescornjum.nl) and Van Stipdonk Landschapsinrichting (Colour pictures 60 and 63; www.vanstipdonk.nl).

The tracked vehicles vary in design, materials, weight and width of tracks. Tracks may consist of rubber bands with metal bars (Colour picture 51), rounded metal pipes (Colour picture 52), rubber-coated bars (Colour picture 53), bars made of plastic (e.g. FELASTEC®, Colour picture 54), or have no bars at all (Colour picture 55). Sometimes, the original snow groomer tracks are used (Colour pictures 56 and 58). Sharp edges and inflexible tracks increase the risk of damaging the sward.

The larger the contact area of the tracks is in relation to the overall weight of the vehicle, the lower the ground pressure is. Long tracks, however, cause more shear stress when driving turns. In order to achieve a larger contact area, usually wider tracks are applied. In that way, the machines may reach a transport width of over 3 m (Colour picture 49), which in Germany requires either the removal of the tracks, or a special wide load permission for road transport by a flat-bed trailer. Conversion of the vehicles may also involve the narrowing of the vehicle frame, so that even with wider tracks (i.e. 90 cm) the overall dimensions do not exceed 3 m. A tracked chassis developed by the Dutch company Wildemann has a distance of 25 cm between two 135 cm wide tracks, which provides maximum ground contact (Colour picture 50) without exceeding the permitted overall dimensions for normal road transport.

Next to the large contact area it is crucial to minimise overall weight whilst achieving an even weight distribution. Balancing the vehicle requires considering weight and influence of mounted or attached implements used for cutting, taking up, compacting and transporting biomass. If necessary the use of counterweights is required.

For summer use, snow groomers designed for use in skiing areas need to be equipped with an extra high performance cooling device. Such device must be easy to clean from chaff, seeds and dust by changing the air flow of the fan. Additionally, the frame of the vehicle is often reinforced (Box 4.4) and the chassis may be uplifted to adjust to high water levels.

4.2.4 Application of technology

Depending on the field of application, many practical solutions exist to adapt and combine the basic vehicle with appliances for mowing, uptake, and recovery of biomass. The following elements are relevant for selecting optimal technologies and harvest methodologies:
- Characteristics of the harvested area: trafficability and water level (e.g. inundation, water level fluctuations, edge of water body);
- Envisaged harvest season (summer, autumn, winter; harvest in dry periods or over ice);
- Requirements for biomass processing (e.g. fresh vs. dry; long stems, chopped biomass, round bales, single bundles or bound to large bales);
- Size and shape of the harvested area (i.e. choice of size of harvest machinery, biomass transport with harvester or separate vehicle);
- Location of the harvested area (mobility of harvest machinery: access over paved roads or transport with flat-bed trailer).

Harvest may take place in one working stage or in two or three steps (Table 4.7). If the biomass is deposited on the ground after mowing (methods involving two or three steps) the harvested areas may not be inundated if dry or partially wilted biomass has to be recovered. In case of high water levels, mowing, uptake, and removal of the biomass has to take place as a single stage process (Colour pictures 46, 57–60).

Depending on vegetation and processing requirements, the harvester can be equipped with different types of mowers (working width 1.5–4 m) and biomass collecting devices, including:
- Cutter bars (finger cutter bar mower, double-knife cutter bar): swathing is possible (e.g. centred by using a windrower, or laterally by using a rake conveyor of the Badema company), ability to operate under water, causing less mortality to amphibians and insects compared to rotating devices (Colour pictures 44 and 63);
- Rotary mowers (disc mower, drum mower): swathing is possible, not applicable when water levels are high (Colour pictures 61, 64 and 65);
- Forage harvester or mulcher (flail mower, converted snow blower) with the biomass being directly blown

Table 4.7: Harvest methods for herbaceous biomass from wet peatland.

Method	Product	Method description	Practised in (examples)*:	Harvest time	Biomass-moisture
Harvest in one working step	Chopped biomass	Harvest and transport of biomass in mounted container or in attached trailer on tracks (Colour pictures 46 and 58)	PL, BY, LT, DE, GB, NL	Summer/Autumn	fresh
				Winter	dry
	Bales	Harvester with cutter bar, mounted baler for round bales and bale transport by the harvester or a trailer (Colour pictures 57 and 75)	AT, FIN**, CH	Winter	dry
	Bundles	Harvest of reed or cattail: Harvest of whole culms and tied into bundles for thatch; as building material, transport with harvester (Colour pictures 59 and 70)	NL, DE, DK, PL, AT, HU, RO	Winter	dry
Harvest in two working steps	Chopped biomass	1st step: Mowing and biomass deposition in swaths (Colour picture 61), 2nd step: Uptake, chopping and transfer to container or trailer, biomass removal (Colour picture 62)	DE, PL, NL	Summer/Autumn	partially wilted/ dry
				Winter	dry
	Chopped biomass / long culms	1st step: Mowing and deposition in swaths (Colour pictures 61 and 63), 2nd step: Uptake by trailers and biomass removal (Colour pictures 63 and 64)	DE, NL	Summer/Autumn	partially wilted/ dry
				Winter	dry
Harvest in three working steps	Bales	1st step: Mowing and biomass deposition in swaths (Colour picture 65), 2nd step: Biomass uptake and baling with attached baler on twin tyres or tracks (Colour picture 66), 3rd step: Removal of bales individually or by a tracked trailer with a loading crane (Colour pictures 67, 68 and 72)	DE, PL	Summer/Autumn	partially wilted/ dry
				Winter	dry

* country code: international licence plate code.
** in construction according to Komulainen et al. 2008, no information on realisation and harvesting experience available.

into the mounted container or trailer (Colour pictures 46, 58, 62 and 73);
- Cutting system of a combine harvester with cutter bar: with or without reel and intake auger for the direct uptake of the biomass, possibly transport to a mounted baler (Colour pictures 57, 60 and 75);
- Cutting system for thatch reed with cutter bar with or without rotating brushes (e.g. made by the Italian company BCS): for pre-cleaning the reed bundles, feeding the dry upright reed culms to the automatic binder (via a spindle or a chain with spikes), if necessary with transport of the reed via a conveyor belt to the load platform, with manual uptake and stacking of the bundels (Colour pictures 59 and 70).

For the uptake of chopped material, the biomass is either directly harvested (Colour pictures 46 and 58) or (after swathing) taken up by a field chopper (Colour pictures 62 and 73) or self-loading wagon (Colour pictures 63 and 64). For the compaction of the biomass round balers are used which are either mounted onto the harvester (Colour picture 57 and 75), attached to it, or pulled by a separate tracked vehicle (Colour picture 66). Small high pressure square balers are sometimes used in nature conservation management, but increase the cost and effort of bale removal from the site (Colour picture 86). Large square balers cannot be used on wet peatlands due to their high weight. The loose biomass can also be transported to the field margin and, for higher transport efficiency, compacted into square bales, briquettes or pellets by semi-stationary machinery. Certain raw material (i.e. for thatch reed, reed mats, or typha boards) requires the harvesting of entire culms (see above). The culms are then arranged in a parallel way and – similarly to formerly cereal sheaves with a reaper-binder (Colour picture 59), bound into single bundles, which can be tied into large bundles (Colour pictures 70 and 71).

The transport of the biomass to the edge of the field is either carried out by the harvester itself or by separate transport vehicles that act as a shuttle between the harvester and the edge of the field (Chapter 4.3). Similarly to the harvester, the transport vehicles need to be adapted to the wet site conditions, taking into consideration that the limited load-bearing capacity of peat soil constrains the loading capacity. The biomass is transported in a container or bunker (chopped biomass; Colour pictures 46, 58 and 73) or on a loading platform (bundles or bales; Colour pictures 45 and 68).

For reloading the biomass on the field or at the field margin, the harvesting machinery needs to be equipped with a tipping device, conveyor belt, auger or loading crane. A method developed in Britain involves blowing chopped biomass via pipelines to the edge of the field ('fen blower', Broads Authority 2004). The following methods are used in practice:

- Driving in parallel: the chopped biomass is transported from the harvester to a parallel driving transport vehicle (Colour picture 73);
- Reload of chopped biomass: When the loading capacity of the container or loading trailer is reached, the biomass is reloaded onto a transport vehicle, for example, by tipping the trailer (Colour picture 73), with a conveyor belt (Colour picture 74) or via an auger or pipeline (see 'Fen Blower');
- Reload or uptake of bales: because of the limited load capacity of the harvester, a transport vehicle with a loading crane (Colour picture 72) overtakes the bales or takes the bales up after they were deposited on the field. In reed harvesting, it is common practice to reload large bundles (that consist of several hundred single bundles) onto separate transport vehicles by tipping of the platform or the container, or using a loading crane (Colour pictures 70 and 71).

4.2.5 Performance of harvesting machinery

The economically viable utilisation of biomass as a fuel requires high performance of harvest machinery and logistics, even more than in case of harvesting biomass as an industrial raw material or for nature conservation management. The performance of the machinery depends mainly on:
- The area to be harvested – i.e. its size and form (relevant for the number of turnings), trafficability, distance to the unloading site, productivity, vegetation structure, location of ditches and pits, and so on;
- The basic vehicle and its appliances – i.e. speed, working width, loading volume, capacity of uninterrupted operation;
- The harvest procedure – i.e. one, two or three working steps, time required for biomass recovery or reload onto the transport vehicle.

The specifications on the performance of harvest machinery in technical data sheets are therefore of little value, and the maximum working speed indicated is generally not achievable (Box 4.4). The main emphasis of operation is on minimizing damage to soil and root system, primarily to ensure long-term trafficability and sustainable yields.

Very few long-term data on machinery performance during harvest and biomass recovery are available for large wet peatland sites. On moist meadows in the Netherlands, the Wetlandtruck (Colour pictures 60 and 63) achieved 0.09–0.29 ha h^{-1} in a single stage procedure and 0.19–0.34 ha h^{-1} in a two steps approach (Jong et al. 2003). These values were determined by recording the time required for harvesting plots of 100 × 100 m with an additional transport distance of 50 m to the unloading site. By adding 45% to cover the time required for setup, maintenance, repair and disruption, the overall achievement of the harvester was 0.06–0.24 ha h^{-1} (Jong et al. 2003).

The paludi-harvester in the VIP project (Box 4.4) achieved a productive time performance of 0.6 ha h^{-1} and an overall working time performance of 0.26 ha h^{-1}. The slightly higher performance of the paludi-harvester can be explained by the larger plots, the larger loading volume of the container, and the unloading of the biomass directly at the field margin.

4.2.6 Optimisation and improvement

Whereas a mature and specialised technology and logistic chain exists for the harvest of thatch reed (which has developed over decades), the harvest of energy biomass still has a considerable need of technological development (Komulainen et al. 2008).

Large companies that manufacture harvesting machinery have so far no interest in this specialised market. Yet, snow groomer producers (Kässbohrer, Pinroth) have recognised that these specialised fields of application are an interesting market for second-hand machines from ski resorts; therefore, such companies offer respective conversions in their product portfolio. Multifunctional vehicles for various fields of application (i.e. conservation management, silage preparation in bunker silos, winter road maintenance, and soil conserving operation on vulnerable cropland soils) are interesting both for the provider (sales market) and for the purchaser (capacity utilisation of machinery). The preferred multi-functionality should, however, not be achieved by compromises regarding track width or vehicle weight, which easily conflict with the particularities of paludiculture and go at the expense of soil conservation.

Together with the conversion of machinery originally designed for other fields of application, also the development of new biomass harvesting concepts for wet peatlands should be a target (Box 4.3). The superiority of purposely designed machines is clearly demonstrated by specialised Dutch tracked vehicles for reed harvesting and conservation management. However, also in this aspect the growing demand for biomass for fuel poses new engineering challenges – for example, with respect to the design of harvesters with mounted round balers. Further potential to improve harvest technology lays in weight reduction, increased contact area to the ground, reduction of shear stress and slipping, and the increase in output performance. Starting points for the improvement of harvesting technology in wetlands are:

- The division of tracks into several segments (see delta tracks);
- Adjusted electronic steering such as the steering of individual wheels or delta tracks; separate steering of the tracks of harvester and trailer;
- Use of built-in sensors – e.g. to avoid slipping;
- Use of light weight components to reduce overall weight;
- Improvement of weight distribution and avoidance of leverage effects of attached equipment;
- Change from diesel-hydraulic to diesel-electric drive technology to increase efficiency;
- Precision Farming such as the use of GPS, digital maps and yield registration;
- Reduction of yield losses during harvest;
- Improved coordination of techniques and logistics such as the use of separate transport vehicles;
- Alternative methods of compaction specifically for biomass fuel;
- Technical solutions for collecting biomass from inundated areas;
- Harvest of special crops such as medical plants or peatmoss (Chapter 3.1).

Driving once across a peatland with an attached mower or mulcher is usually unproblematic using existing technology. The challenge lies in removing biomass and avoiding repeated crossings.

4.3 Logistics of biomass production on wet peatlands

Christian Schröder, Sebastian Dettmann & Sabine Wichmann

For the processing, storage, transhipment and transport of biomass from paludiculture, one can revert to the logistical experiences with other herbaceous biomass. However, biomass harvest, removal and pre-processing on site need to be considered separately, as peatlands have limited trafficability and high water levels. Long access distances and restricted access points to neighbouring fields imply a repeated crossing of the peat soil, at least locally. If repeated crossing results in damage to the sward, then harvest and removal of biomass may become a serious problem (Colour pictures 77–80). Therefore, the main challenge to logistics for paludiculture is the reduction of crossings over the same area.

4.3.1 Basics of logistics

In conventional agriculture, the various methods of harvesting plant material have been extensively studied and many economically viable technologies exist. In contrast, with respect to paludicultures biomass harvest, removal and preprocessing in the field there still is great need for technological improvement (Chapter 4.2).

Each type of utilisation requires its own preprocessing of the paludiculture biomass. Preprocessing may take place directly during harvest or afterwards.

4.3 Logistics of biomass production on wet peatlands

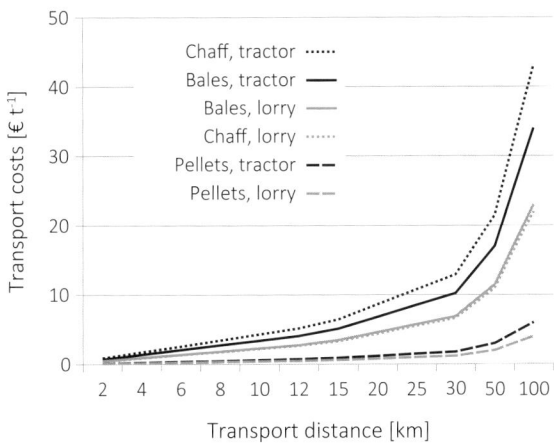

Figure 4.4: Transport costs of chopped biomass, bales and pellets based on diesel consumption and labour costs for a transport by tractor and lorry (after Kowalewsky 2009).

Physical and mechanical preprocessing options include drying, chopping and compacting or a combination of these. A form of chemical-enzymatic processing is ensilage (Box 3.17). Processing increases both storage ability and bulk density of the biomass, which facilitates downstream logistics.

Biomass transport starts at the field margin and ends at the industrial consumer. The goods are either dispatched directly to the consumer or indirectly via intermediate storage facilities. In Germany, the average transport distance for agricultural goods is between 4 and 5 km (Kowalewsky 2009, Janinhoff 2008). Transport over such short distances usually takes place by tractor-pulled trailers. The volume of biomass required by a processing plant determines the area from which the raw material is sourced, and therefore also the transport distances to be covered (Zeller et al. 2011). Meanwhile, the recent establishment of large scale biogas plants and biomass heating power plants has led to transport distances that are significantly larger than 50 km. Figure 4.4 illustrates the transport costs of biomass in differently preprocessed forms and with different transport vehicles. The costs increase with decreasing density, from pellets to bales to loose chopped biomass. From a distance of 15 km onwards, the transport by lorry or Unimog is cheaper than by tractor pulled trailer (Hartmann 2009c, Kowalewsky 2011, Meyer 2012).

4.3.2 Factors influencing biomass removal

The biomass removal from wet sites is challenging. Every additional transport movement increases the risk of damaging soil and sward (Colour pictures 77–79). Thus, before establishing paludiculture on a specific field, the amount of necessary transport runs should be calculated. The effort for biomass removal is determined by:
- Site productivity;
- Transport capacity;
- Volume and weight of the harvested biomass;
- Working width; and
- Harvest frequency.

The number of transport runs for biomass removal increase with the productivity (yield) of the site. In contrast to landscape conservation management, which often operates on sites of low productivity, paludiculture focuses on sites with high nutrient availability as well as on productivity. The productivity of wetland vegetation may lie well above 20 metric tons per hectare and year, which requires higher transport capacities for biomass removal (Table 3.3).

The harvesting approach determines whether biomass harvest and removal take place in one or in several working steps, and what type of machinery should be used (Table 4.7). With regard to transport capacity, the harvesting approaches differ considerably.

The bulk density depends on volume and weight of the biomass and the weight is strongly determined by its water content (Table 4.8). The bulk density of chopped herbaceous biomass is about 36 to 80 kg DM m^{-3}; if chopped very short, also a higher bulk density is possible. Round bales achieve values of about 120–160 kg DM m^{-3} (KAE 2009). More important for the total cargo weight is the water content of the harvested biomass. During summer harvest, the water content of fresh herbaceous biomass is 60–75% (Kalzendorf 2011) and of hay 16–18% (Kaltschmidt et al. 2009). If harvested in winter, the water content of reed is 10–25% (Kask et al. 2007, Kitzler et al. 2012). As trafficability of peat soils is limited, the weight of fresh biomass, in particular, may become the limiting factor for transport. When harvesting dry, loose biomass such as chopped hay or large reed bundles, the volume is the limiting factor for transport.

Working width and transport capacity strongly influence performance. The larger the working width, the less area has to be crossed. Though, working width has no impact on the number of transport runs per hectare needed for biomass removal. For harvesting approaches with only one working step – e.g. a harvester, the distance driven to fill the bunker decreases with increasing working width. Therefore, the working width of a harvester has to be adjusted to the network of roads and the number of reloading points (Chapter 4.4).

The frequency of harvest influences the quantity and quality of the harvested biomass. The frequency of harvest may vary from several times per year to only once in several years. With increasing frequency the volume of biomass per harvest may decrease, which

Table 4.8: Bulk density of gramineous biomass.

Substrat	Dry Matter [%]	Bulk density fresh [kg FM m^{-3}]	Dry Matter [kg DM m^{-3}]
Grass[1]			
Loose	20	330	66
Loose, wilted	33	160–220	53–73
Hay[2]			
Loose	80	80	64
Chopped	80	100	80
Small square bales	80	110	88
Round bales	80	160	128
Straw[2]			
Long material, loose	85	50	43
Chopped (5 cm)	85	60	51
HD-bales	85	80	68
Small square bales	85	115	98
Square bales	85	130	111
Miscanthus[3]			
Chopped	80	118	94
Square bales	80	170	136
Reed (Winter)[4]			
Chopped	80	53	42
Round bales	80	152	122

[1] KTBL 2005.
[2] KTBL 2009.
[3] TFZ 2009.
[4] Wichmann & Wichtmann 2009.

results in less trips for biomass removal per harvest. Simultaneously, the number of trips per year increases, which reduces the time for regeneration of the sward.

4.3.3 Logistic chains

Various harvesting approaches can be conceived for biomass supply from paludiculture (Figure 4.6). Harvest may take place in one, two or three steps, and the biomass may be chopped or processed into bales or bundles (Chapter 4.2). The choice of the method depends on the time of harvest, the site conditions, and the preprocessing required by consumer of the biomass and by the downstream logistics.

Biomass can be compacted into high compaction small square bales or round bales (Figure 4.6a, Colour pictures 43, 66–68). Pressing small square bales requires more time and effort for downstream logistics because of their small size and lower bulk density (Wichmann 2009). Pressers for large square bales are too heavy for peat soils and only suitable for stationary operation at the field margin or at the storage facilities.

In single-stage baling the entire biomass processing takes place in one working step: Harvest, recovery and transport take place without depositing the biomass on the ground. The harvesting machine has a front mower and the harvested biomass is directly fed into a baler mounted on the back of the harvester. The bales are temporarily stored on a platform behind the baler or on a trailer attached to the harvester (Colour picture 57).

Harvest with a towed baling press requires three working steps. The first step includes mowing and depositing the biomass in the field. Next, the biomass is gathered and pressed into bales, if necessary after drying the biomass in the field (Colour pictures 65 and 66). The third step consists of the recovery and transport of the bales using a transport vehicle equipped with grappler and platform (Colour pictures 68 and 72). If the baler is directly attached to the harvester, mowing and baling may take place in one step. The recovery of the bales occurs then as a second step.

Several machines can be used for loading and reloading bales, including tractors with a front mounted loader or grappler, as well as mobile or stationary load-

4.3 Logistics of biomass production on wet peatlands

> **Box 4.5: Performance of a single step harvesting approach used in nature conservation management**
>
> Sebastian Dettmann & Christian Schröder
>
> In the framework of landscape conservation management in the valley of the river Peene (Mecklenburg-West Pomerania, Germany), summer harvest of sedges and reed was performed in a single step approach with a tracked vehicle (Pistenbully, PB 260) specifically adapted for operation in small nature reserves. The vehicle was equipped with a flail mower (working width 2.5 m) and a mounted bunker that can be tipped to dump the contents (loading capacity 9 m³) (Colour picture 46). The time required for labour, transport, unloading, turning and maintenance was recorded (Figure 4.5). The required time for turning includes the interruption of harvest needed to change direction. Maintenance includes the times for breaks, downtime and repairs of the harvester. Two sites were established on opposite margins of the field (circa 1.25 ha) for alternately unloading the vehicle.
>
> Even though only a small area was harvested with short distances between the unloading sites, the largest proportion of total working time was taken by transport (32%). This was due to the limited loading capacity (9 m³) of the bunker mounted to the vehicle. Mulching (with an average working speed of 5.9 km h⁻¹) took 29% of the total working time. Time required for maintenance of the harvester was 23%
>
> (including lubricating and cleaning of the vehicle and the cooling system). Time for turning took 2% of the total working time, biomass unloading 14% (Figure 4.5). The average performance was 0.37 ha h⁻¹, the distance driven per harvested hectare 8.5 km.
>
> This harvesting method is well suited for small areas. When applied to considerably larger areas, the effort required for transport would increase further, reaching the logistic limits of the method.
>
> The recorded values cannot simply be applied to other areas because performance during harvest decisively depends on biomass yields, vehicle accessibility, and transport track infrastructure.
>
>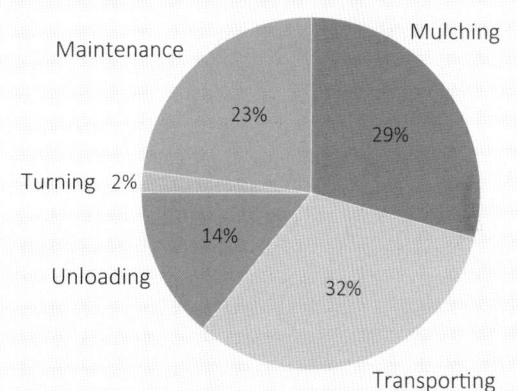
>
> Figure 4.5: Working time required by different working steps of a harvester with tracks used in conservation management (Colour picture 46).

ing cranes. The bales are usually stored in stacks at the field margin or at farm yards, and also in storage facilities of the end user. In order to minimise loss of quality and dry matter by the impact of weather, the stacks are preferably covered with plastic foil or stored in a roofed storage facility.

The harvesting of chopped biomass (Figure 4.6b) differs from that of bales by the type of front-mounted appliances and trailers being used, the temporary storage, as well as the reload of biomass onto roadworthy transport vehicles at the field margin. In a single stage process, the harvesting vehicle is equipped with a field chopper and a mounted container or a trailer (Colour pictures 46, 58 and 62, Box 4.5). Biomass transfer takes place via tipping devices or a drag chain conveyor at the field margin (Colour pictures 73 and 74), or via a reloading conveyor onto a shuttle. Alternatively, an accompanying vehicle may drive parallel to the harvester to gather the biomass (Colour picture 73). If low water tables allow deposition of the swath, loading trailers can be used to pick up the biomass. Thus, mowing,

deposition, and uptake of biomass take place successively in one step (Colour picture 64). Alternatively, the chopped biomass may be collected in a second step after having dried in the field.

In contrast to bales, chopped biomass cannot be temporarily stored at the field margin but has to be transported directly to a storage facility for bulk material. Where appropriate, the chopped biomass may immediately after the harvest be ensilaged in stacks at the field margin. The reload of the chopped biomass may take place via reloading conveyors, wheel loaders, loading cranes, or mobile conveyor belts. Alternatively, the chopped biomass can be compacted into large bales via stationary balers at the field margin, or in a temporary storage facility.

The supply of reed for thatch in bundles (Figure 4.6c) usually takes place as a single-stage process, which may or may not include the removal of weeds. For harvesting, Seiga machines with balloon tyres or tracked vehicles are common; sometimes, several bundles are already tied into large bundles at

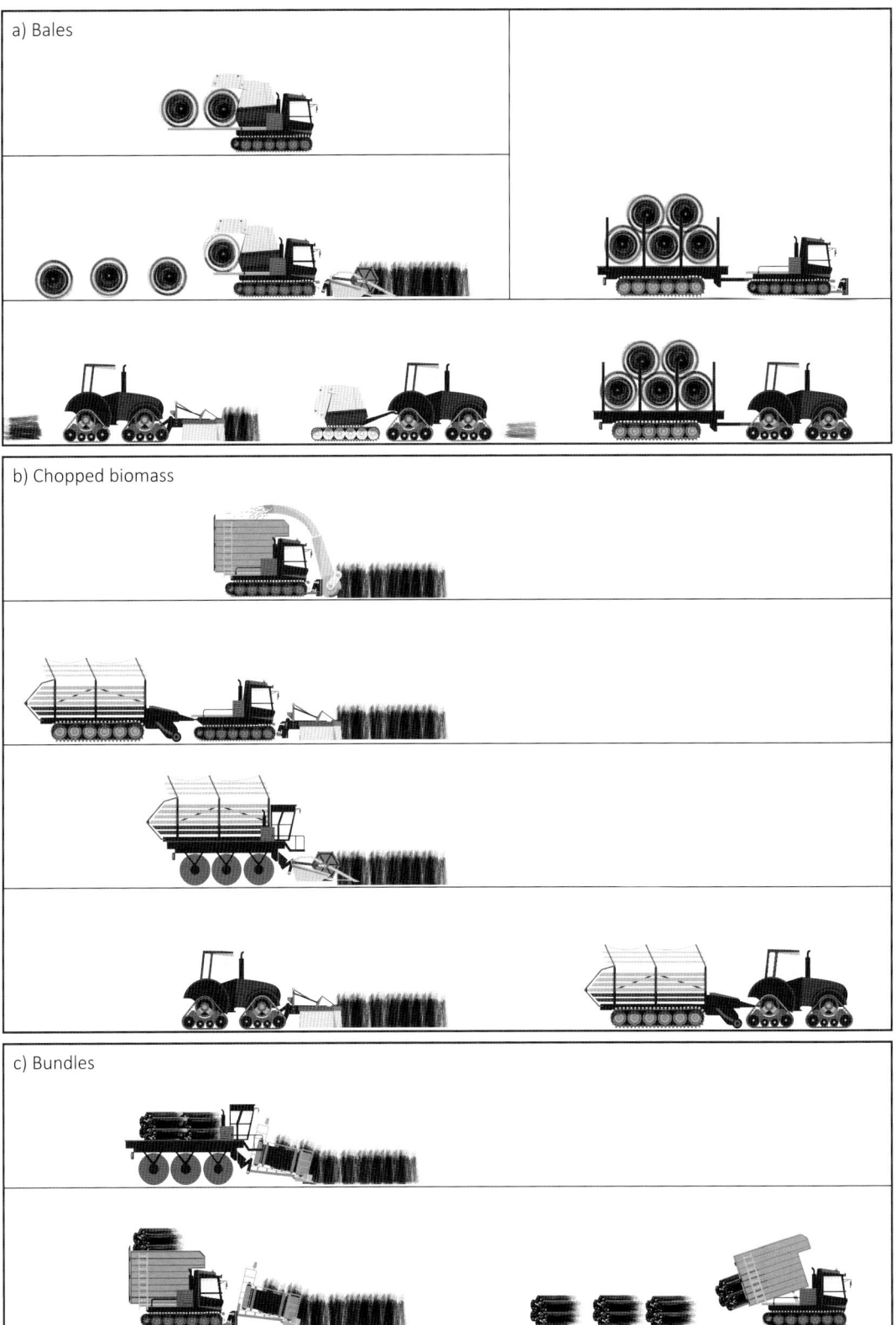

Figure 4.6: Examples for various biomass harvesting approaches in wet peatlands. The biomass is a) baled, b) chopped or cut, or c) bundled. Depending on the water table, the biomass can be temporarily deposited on the ground and harvested in one, two or three working steps.

the field (Colour pictures 58, 70 and 71). Storage of the single and large bundles occurs in sheaves, open air stacks at the field margin, or in roofed storage facilities (Colour pictures 27 and 28). Tractors with front loaders, grapplers and loading cranes are suitable for transhipment of the large bundles.

4.3.4 Biomass removal

In paludiculture, biomass harvest usually takes place only once per year. Biomass recovery is the limiting factor because of limited trafficability of the peat soil. Chapter 4.3.2., Table 4.9 shows exemplarily the effort of biomass removal for the following harvesting methods:

a Chopped biomass collected in a 9 m³ container mounted onto the harvester (Colour picture 46);
b Chopped biomass collected in a separate 25 m³ trailer pulled by the harvester (Colour pictures 61 and 62);
c Bales pressed in a baler mounted onto the harvester with a transport capacity of four round bales (Colour picture 57);
d Bales transported in a separate trailer with a capacity of 14 round bales (Colour picture 68).

The transport of chopped biomass by a harvester with a 9 m³ mounted container (a) requires 6.7–44.4 transport trips per ha (Table 4.9). In order to avoid unproductive trips on the sensitive peat soil, the harvester has to unload every 45–500 m. This method is therefore suitable for small, low productive areas with a dense network of roads.

A 25 m³ trailer (b) requires 2.4–16.0 transport trips per ha and needs to unload after 125–1,389 m. Together with well-planned transport routes and a well-developed network of paths, this concept is applicable for more productive sites as well.

Depending on biomass yield, one hectare can produce between 15.4 and 102.7 bales (bale: 194 kg DM, diameter 1.3, width 1.2). To produce one round bale, the harvester has to cover a distance between 19.4 and 216 m, depending on the yield and the working width of the mower. Thus, the harvesting method with a mounted storage capacity of four bales (c) needs to unload every 78–854 m and is therefore only suitable for sites with low productive vegetation and a dense net of unloading points.

For high productive vegetation stands or large areas the deposition of bales and the removal with a separate trailer or a direct transfer of the bales onto a transport vehicle are recommended. With a transport capacity of 14 bales (d) 1.1–7.4 transport trips per ha are needed to remove the bales from the field.

When comparing scenarios for a yield of 5 t DM per ha, a harvester with a mounted container needs 11.1 transport trips; a harvester with mounted baler and four bales storage capacity, 6.4 transport trips; and a trailer for chopped biomass, 4.0 transport trips. If low water levels permit deposition of bales in the field, the use of a separate transport vehicle is preferable as this

Table 4.9: Effort to remove biomass from wet peatland sites as a function of biomass yield, working width, and transport capacity.

Yield [t dry weight ha-1]	Working width [m]	Chopped biomass*				Bales**			
		Container 9 m³ (~450 kg DM)		Trailer 25 m³ (~1,250 kg DM)		Harvester 4 bales (~777 kg DM)		Trailer 14 bales (~2,720 kg DM)	
		Reload distance [m]	Transport trips per ha [n]	Reload distance [m]	Transport trips per ha [n]	Reload distance [m]	Transport trips per ha [n]	Reload distance [m]	Transport trips per ha [n]
3	3	500	6.7	1,389	2.4	864	3.9	3,023	1.1
	5	300		833		518		1,814	
5	3	300	11.1	833	4.0	518	6.4	1,814	1.8
	5	180		500		311		1,088	
10	3	150	22.2	417	8.0	259	12.9	907	3.7
	5	90		250		155		544	
15	3	100	33.3	278	12.0	173	19.3	605	5.5
	5	60		167		104		363	
20	3	75	44.4	208	16.0	130	25.7	453	7.4
	5	45		125		78		272	

* Bulk density: 50 kg DM m⁻³, own estimation, higher if chopped very short.
** Bulk density: 122 kg DM m⁻³ (Wulf 2009). Bale: 194 DM (diameter: 1.30 m, width: 1.20 m).

needs only 1.8 transport trips per ha. However, also this method quickly reaches its limits. With only one access for a 10 ha field with a yield of 5 t DM per ha, the recovery of the biomass results in 18 transport trips and 72 crossings over the same area nearby the access point (main machine plus trailer, in and out, see Figure 4.7a. Reinforcement of the tracks is required already for such small area to prevent damage to the sward (Chapter 9.4).

Consequently, a soil conserving biomass recovery requires:
- A large transport performance by increasing transport capacity and/or the compaction of the biomass during the harvesting process;
- The separation of harvester and transport vehicle with the aim to use harvesters with larger working width and higher performance, whilst simultaneously deploying light weight transport vehicles;
- The targeted design of infrastructure during the implementation of rewetting by paving the main tracks and removing bottlenecks (Chapter 4.4, Chapter 9.4).

4.4 The feasibility of biomass harvest from paludiculture

Christian Schröder & Sebastian Dettmann

In the previous sections the options and limits of biomass harvest have been discussed. In particular, large areas and high yields pose large technological and logistic challenges. It may be expected that harvest and logistics still can be substantially improved regarding efficiency and soil conservation. In many cases, however, economic use of rewetted peatlands will not be feasible without the development of auxiliary infrastructure. This section illustrates the limits to land use on rewetted peatlands and gives recommendations for improving harvest logistics and accessibility.

4.4.1 Requirements to logistics

When wet peatland sites are frequently passed by harvest and transport vehicles, the risk of damage to the sward and of consequent restricted trafficability increases. Particularly at water levels close to the surface, destruction of the sward may lead to flushing out of material and a deepening of the tracks, which necessitates rerouting to new paths (Colour pictures 77–79). The damaged paths are not usable further and each new path leads to a decrease of the harvestable area. In the Biebrza National Park in Northeast Poland, fascines are used to reduce damage to transport tracks (Chapter 10.5). Wood from clearing shrub-overgrown peatland sites is arranged on the tracks perpendicular to the driving direction (Colour picture 81). Such measures require great effort, but are useful when shrub removal is carried out anyway as a nature conservation measure. Nonetheless, their usefulness is limited in case of frequent and annually reoccurring transport, as the fascines only have a limited life span.

Independently of how well machinery is adapted to wet site conditions, it is quite possible that repeated trips over tracks and transport paths will cause damage. For example, harvesting a 10 ha field with a yield

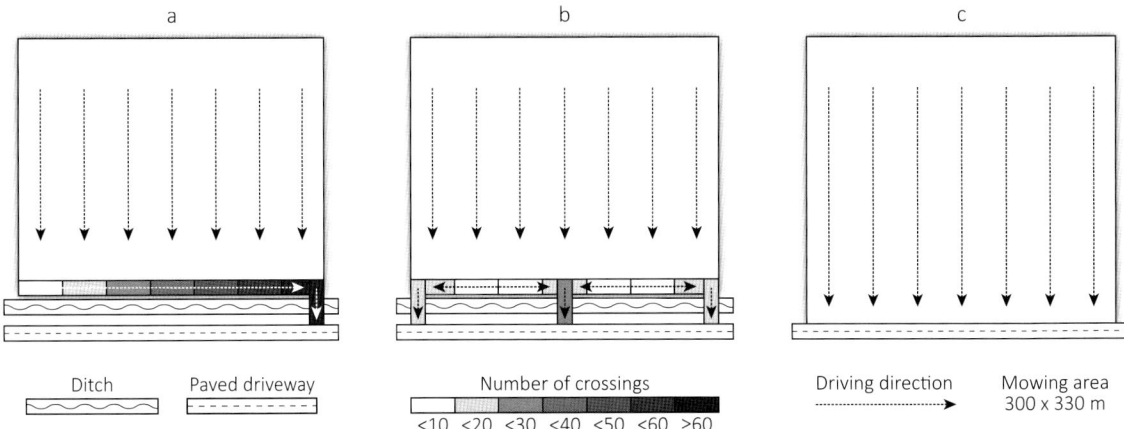

Figure 4.7: The number of crossings to remove the biomass of a 10 hectare field with a) one; b) three access points; and c) no restriction in access, using a productivity of 5 t DM ha^{-1} and a machine equipped with a separate trailer and a maximum loading capacity of 3 t DM.

4.4 The feasibility of biomass harvest from paludiculture

of 5 t DM per ha leads to 72 crossings if only one access point is available (Figure 4.7a). By adding two extra access points still 24 crossing occur, which solves the problem just partially (Figure 4.7b). This example shows how difficult straight course biomass transportation is. It is recommended to spread the transport trips evenly over the entire area to avoid concentration of the crossings at the same places (Figure 4.7c). The following guidance applies to minimise damage to soil and sward during biomass removal:
- Straight course transport over unpaved tracks on wet peat soils needs to be avoided;
- The entire area should be used for biomass removal whilst avoiding concentration of crossings over the same area;
- Bottle necks at driveways or at crossings between fields need to be eliminated (i. e. via filling of ditches) or at least reinforced, and the intensity of their use reduced by establishing additional access points;
- The logistics of harvest and transport have to be adjusted to the characteristics of the peatland and the existing infrastructure

A soil conserving and permanent use of wet peatlands is only guaranteed if all these requirements are fulfilled over the entire area.

4.4.2 Requirements to infrastructure

Biomass removal from wet peatlands should be carried out whilst avoiding repeated crossing of the same site. The transport capacity thus determines the requirements for adjusting the infrastructure. All machine movements that are not directly linked to the uptake of biomass must be avoided. As soon as the load capacity of the harvester has been reached, an unloading or transfer point (e.g. the peatland margin, a causeway, or – in case of water transport, a canal) should allow the harvester to unload and immediately continue with further biomass uptake. With a yield of 5 t DM ha^{-1}, a working width of 3 m and a cargo volume of 9 m^3, a 300 m long strip can be harvested, meaning for a round trip that the turning point is at 150 m distance from the unloading point. With a cargo volume of 25 m^3, a strip can be harvested of 833 m length with the turning point at 416 m. In case of bales, the strip to be harvested is 518 m long (turning point at 259 m) if a harvester with a loading capacity of four bales is used, and 1,814 m (turning point 907 m) if 14 bales are transported via a separate trailer (Table 4.9). Areas located on larger distances from an unloading point require additional crossings for transportation. Such areas can only be harvested during frost periods or if additional causeways are established. Ideally, these causeways could be used by roadworthy vehicles as well. The shorter the strip that can be harvested until full loading capacity is reached, the denser the network of paths needs to be. Regarding the example in Table 4.9, a harvester with a cargo volume of 9 m^3 needs to access a causeway every 300 m, and a trailer for 14 bales needs to do it only every 1,814 m.

In most cases, the existing network of roads in drained peatlands is dense enough to ensure biomass removal after the implementation of paludiculture and with GPS-based planning of transport routes. Precondition is, however, that these roads are reinforced to enable biomass removal during times of high groundwater levels (Chapter 9.4). Nonetheless, in order to permit access along the entire field margin, larger scale infrastructure adaptations have to be planned. Existing drainage ditches, which often constitute bottlenecks for access between neighbouring fields, should – if possible, be filled to allow unrestricted access (Figure 4.7c). For ditches that are essential for further water level regulation or that otherwise cannot be filled, additional crossing points need to be established. The measures to prepare areas for paludiculture are thus considerably more wide-ranging than in case of conventional peatland rewetting.

5 Ecosystem services provided by paludiculture

Wet peatlands provide a number of important ecosystem services. These services are based on societal values attached to distinct ecosystem patterns and processes. For example, wet conditions support strongly adapted organisms that provide biodiversity value. In addition, wet conditions result in incomplete decomposition of organic matter and hence the sequestration of carbon, which provides benefits in terms of climate change mitigation. Many of the services provided by wet peatlands are disrupted when these are drained. Rewetting and subsequent paludiculture aim at restoring peatland ecosystem services – at least partially. This Chapter describes the effects of paludiculture on climate (Chapter 5.1), biodiversity (Chapter 5.2), landscape hydrology (Chapter 5.3), and nutrient retention (Chapter 5.4).

5.1 Greenhouse gas emissions

Gerald Jurasinski, Anke Günther, Vytas Huth, John Couwenberg & Stephan Glatzel

The world's peatlands store significant amounts of carbon and play an important role in the global climate system in terms of greenhouse gas (GHG) exchange. Drainage and land use affect the GHG balance of peatlands and convert them from carbon sinks to carbon sources. However, net GHG emissions can be significantly reduced when the water table is raised again. Thus far, it remains unclear to what extent land use affects the exchange of GHG of wet peatlands. Recent studies indicate that the effect is only minor. Further, simple but detailed approaches are available to estimate GHG fluxes associated with different kinds of wet land use. Thus, rewetting in combination with adapted land use can contribute to the mitigation of climate change.

5.1.1 Greenhouse gas fluxes of natural peatlands

Peatlands occupy only a small part of the World's land surface (3% or 4 million km²), but contain almost 20% of all organic carbon stored in terrestrial ecosystems (Joosten & Couwenberg 2009). Carbon (C) sequestration in peatlands follows from an imbalance between biomass production (sequestration of carbon dioxide – CO_2 – by photosynthesis) on the one side, and decomposition (decay of dead biomass to CO_2) and lateral discharge on the other side, with production exceeding losses (Clymo 1983). High water tables generate anaerobic conditions under which the decomposition of organic matter is inhibited and remains incomplete. At higher latitudes, decomposition of organic matter is further reduced by low temperatures (Whiting & Chanton 2001). Although biomass is produced during the short growing season, it cannot be broken down completely.

Like in other ecosystems, in peatlands a large part of the biomass that is produced during the growing season is quickly decomposed (84–98%), releasing the sequestered carbon again (Päivänen & Vasander 1994). The easily degradable organic matter – including carbohydrates such as cellulose, hemicelluloses and starch – is broken down first. Under oxic conditions, it is directly and completely oxidised to CO_2, whereas in anoxic environments, decomposition is slowed by the absence of oxygen.

Undisturbed, living peatlands (mires) sequester atmospheric carbon. However, their carbon balance needs not be positive each year. Depending on the weather, mires may act as slight carbon sources or have a neutral carbon balance in some years (Roulet et al. 2007, Nielsen et al. 2008, Koehler et al. 2011). Yet, in the long term mires are net carbon sinks. Long term accumulation rates have been determined from dated peat profiles with known carbon content. In northern mires, rates are 100–400 kg C ha^{-1}a^{-1}; in tropical peatlands they can be twice as high (200–1,000 kg C ha^{-1}a^{-1}; Yu et al. 2009, Dommain et al. 2011). Variation in accumulation rates during the Holocene can be in part explained by climate fluctuations (Clymo 1983, Yu et al. 2009, Dommain et al. 2011). It may thus be assumed that peat accumulation rates are also affected by anthropogenic global warming. However, driving factors such as temperature, precipitation and atmospheric CO_2 concentration partially work in opposite directions, and feedbacks between drivers and peatland carbon (GHG) fluxes are complex. As a result, the effects of climate change on peatland biogeochemistry are difficult to predict (Frolking et al. 2011). Whether or not mires will be carbon sinks in the future will depend on atmospheric GHG concentrations and on local climatic conditions (Moore et al. 1998, Charman et al. 2013).

In natural, undisturbed mires, organic matter is only incompletely decomposed. Under anoxic conditions part of the organic matter decays to form methane (CH_4), a much more potent GHG than CO_2. Plants that possess aerenchyma, such as Common Reed (*Phragmites australis*) or Cattail (*Typha* spp.) play an important role in methane emission (Whiting & Chanton 1996, Grünfeld & Brix 1999, Afreen et al. 2007, Lai 2009, Miller 2011, Couwenberg & Fritz 2012). On the one hand, aerenchyma transports oxygen to the anoxic root layer, oxidising CH_4 and limiting its formation. On the other hand, aerenchyma can channel CH_4 directly into the atmosphere, bypassing oxic peat layers. In this way, up to 90% of CH_4 emissions can pass through aerenchymatous plants (Morrissey & Livingston 1992, Grünfeld & Brix 1999, van der Nat & Middelburg 2000). In natural mires, CH_4 emissions vary between 0–650 kg CH_4 ha^{-1} a^{-1}, depending on hydrology, vegetation and temperature (Frolking et al. 2011, Blain et al. 2014).

In general, undisturbed mires do not emit significant amounts of nitrous oxide (N_2O), an extremely potent GHG (Martikainen et al. 1993). The highest N_2O emissions are often associated with fluctuating water tables that result in alternating oxic and anoxic conditions (Davidson et al. 2000). N_2O is produced as a by-product of the soil nitrogen cycle (nitrification and denitrification, Regina et al. 1996). Denitrification appears to be the most important source of N_2O under almost anoxic conditions but N_2O is also formed under aerobic conditions during nitrification (Wolf & Russow 2000). Additionally, denitrification depends on the presence of nitrate. However, the amount of N_2O produced not only depends on available nitrogen, but also on other factors such as moisture, temperature, carbon content and pH (Jungkunst et al. 2006) with soil moisture seeming to be the most important factor.

The climate effect of peatlands can be expressed as its net GHG balance. The most important GHGs CO_2, CH_4 and N_2O have a different global warming potential (GWP, Box 5.1) and different atmospheric lifetimes. When calculated over short time spans (100 years), the climate effect of natural peatlands is neutral or slightly warming (Frolking et al. 2011). In the long term, the importance of CO_2 sequestration dominates, because the emitted CH_4 is oxidised and its atmospheric residence time is only short (12.4 years, Myhre et al. 2013, Moore et al. 1998). In this way, natural, undisturbed peatlands have had a cooling effect on the climate for the past 10,000 years (Frolking et al. 2006, Frolking & Roulet 2007).

5.1.2 Greenhouse gas balance of drained peatlands

Drainage of peatlands leads to a strong change in GHG fluxes, depending on the water table and the subsequent form of land use (Sirin & Laine 2008). The lowering of the water table results in an oxic environment, causing mineralisation of the drained peat layers and the release of large amounts of CO_2. For instance, in Central European drained fens with high intensity grassland management the rates of CO_2 release are between 15 and 35 t $ha^{-1}a^{-1}$ (Couwenberg et al. 2011, Elsgaard et al. 2012, Figure 5.1).

In contrast to natural mires, drained peatlands emit only very small amounts of CH_4 or even sequester it (Joosten & Clarke 2002). However, CH_4 emissions from water filled drainage ditches can be extremely high, up to ~ 0.25 kg CH_4 m^{-2} a^{-1} [2,500 kg CH_4 ha^{-1} a^{-1}], (Chistotin et al. 2006, Minkkinen & Laine 2006, Schrier-

---

Box 5.1: Global Warming Potential

Vytas Huth

The global warming potential (GWP) of a GHG is a measure of the contribution to the greenhouse effect of 1 unit mass of this GHG compared with 1 unit mass of a reference gas (CO_2, hence CO_2-equivalent or CO_2e). As GHGs have different atmospheric residence times, the global warming potential depends on the chosen observation period. As its warming effect changes in a non-linear way, the global warming potential of a specific gas also depends on its actual atmospheric concentration compared with the concentration of CO_2 (Myhre et al. 2013). For instance, integrated over a period of 100 years, the emission of 1 kg CH_4 on average has a 28-fold stronger effect than 1 kg CO_2 (Table 5.1).

Table 5.1: Relative global warming potential (GWP) of the most important greenhouse gases in relation to peatlands integrated over 20, 100 and 500 years (Myhre et al. 2013; 500 year estimate from Forster et al. 2007).

Gas	Residence time (years)	GWP (CO_2e)		
		20	100	500
CO_2	5–200	1	1	1
CH_4	12	84	28	7.6
N_2O	121	264	265	153

Colour pictures 21–56

Colour picture 21: Vitamin bomb Cloudberry (*Rubus chamaemorus*) grows in nordic peatlands (Sweden; Christian Schröder, 2009).

Colour picture 22: Cranberries (*Vaccinium oxycoccos*) are traditionally collected in several countries (Dokudovskoe, Belarus; Wendelin Wichtmann, 2008).

Colour picture 23: Candied cranberries (*Vaccinium oxycoccos*) are a Belarusian delicacy (Christian Schröder, 2014).

Colour picture 24: Because of its essential oils Bog Myrtle (*Myrica gale*) is used since medieval times for brewing beer in some regions of Europe (Sweden; Christian Schröder, 2012).

Colour picture 25: Bogbean (*Menyanthes trifoliata*) is used as remedial plant against loss of appetite and indigestion (Poland; Christian Schröder, 2006).

Colour picture 26: Sundew (*Drosera rotundifolia*) plants on a peatmoss cultivation site (Hankhausen, Germany; Balazs Baranyai, July 2014).

Colour picture 27: Sheafs of reed posted for drying (Lake Neusiedl, Austria, Sabine Wichmann, February 2009).

Colour picture 28: Storage of reed for thatch in a barn (Rozwarowo, Poland; Achim Schäfer, July 2005).

Colour picture 29: Wired plates made of reed for thermal insulation of walls (Lake Neusiedl, Austria, Sabine Wichmann, February 2009).

Colour picture 30: Weaving loom for the production of mats from reed (Zecherin, Mecklenburg-West Pomerania, Germany; Christian Schröder, 2013).

Colour picture 31: Fireproof board made from reed fibres and a mineral binding agent (IBZ Hohen Luckow; Christian Schröder, 2012).

Colour picture 32: Insulation plaster from reed fibres and loam produced by Mineralische Rohstoffmanagement GmbH, Friedland (Nina Körner, 2013).

Colour picture 33: Building panel made from loam and reed. Hanffaser Uckermark eG (Nina Körner, 2013).

Colour picture 34: Blow-in insulation material from 100% Cattail in a display case. Hanffaser Uckermark eG (Philipp Schroeder, 2014.)

Colour picture 35: Compostable form bodies made from reed and waste paper. Left: grass paver, right: carrier for flower arrangement (Wendelin Wichtmann, 1997).

Colour picture 36: Construction panel made of Cattail (typhatechnik.com).

Colour picture 37: Briquettes from reed (Nina Körner, 2013).

Colour picture 38: Pellets from reed (Tobias Dahms, 2013).

Colour picture 39: Loading a mobile pelleting machine with hay from sedges and reed (Neukalen, Mecklenburg-West Pomerania, Germany; Philipp Schroeder, März 2014).

Colour picture 40: Production of pellets from reed in a mobile pelleting plant (performance 1t per hour) (Neukalen, Mecklenburg-West Pomerania, Germany; Philipp Schroeder, March 2014).

Colour picture 41: Temporary storage of round bales at the field margin. The big bales have been harvested from the re-wetted fen peatland in the background (Neukalen, Mecklenburg-West Pormerania, Germany; Christian Schröder, March 2014).

Colour picture 42: Shredder for bales with an automatic feeding system. After shredding the chopped biomass is transported to the combustion unit (Malchin, Mecklenburg-West Pomerania, Germany; Christian Schröder, June 2014).

Colour picture 44: Small tractor equipped with double wheels and bar mower (Murchiner Wiesen, Mecklenburg-West Pomerania, Germany; Wendelin Wichtmann, July 2009).

Colour picture 46: Tracked vehicles: Snow groomer converted for maintenance of wetlands equipped with bunker for transport of biomass (Waschow, Mecklenburg-West Pomerania, Germany; Sabine Wichmann, July 2009).

Colour picture 43: Conventional adapted technique: Harvesting by tractor equipped with 'Terra'-wheels, a disc mower and a baler with tandem axle (Neukalen, Mecklenburg-West Pomerania, Germany; Sabine Wichmann, January 2009).

Colour picture 45: Harvester equipped with wheels: Seiga machine with balloon wheels for harvesting reed for thatch (Sabine Wichmann, February 2012).

Colour pictures 21–56

Colour picture 48: Standard tractor equipped with delta tracks (Germany; Landschaftspflegedienst Hansjörg Fischer, 2008).

Colour picture 50: Special technique with narrow gauge. The space between tracks is reduced to 25 cm. Even with wide tracks (135 cm each), the total width is less than 3m (Hanze Wetlands, Netherlands; Sabine Wichmann, 2013).

Colour picture 47: Standard tractor and round baler equipped with delta tracks (Germany; Landschaftspflegedienst Hansjörg Fischer, 2012).

Colour picture 49: Converted snow groomer with standard track gauge and rubber tracks. Because the total width is over 3m, permission is needed for road transport or tracks have to be removed. Track width: 110 cm; Distance between tracks: 120 cm (Relzower Wiesen, Mecklenburg-West Pomerania, Germany; Sabine Wichmann, 2011).

Colour picture 51: Metal bars (Sabine Wichmann, 2009).

Colour picture 52: Rounded metal tubes (Sabine Wichmann, 2012).

Colour picture 53: Rubber bars (Sabine Wichmann, 2011).

Colour picture 54: Bars made of plastic (Sabine Wichmann, 2009).

Colour picture 55: Plain tracks (De Vries Cornjum, 2009).

Colour picture 56: Standard snow groomer tracks made of aluminium with sharp edged bars (Haberl, 2011).

5.1 Greenhouse gas emissions

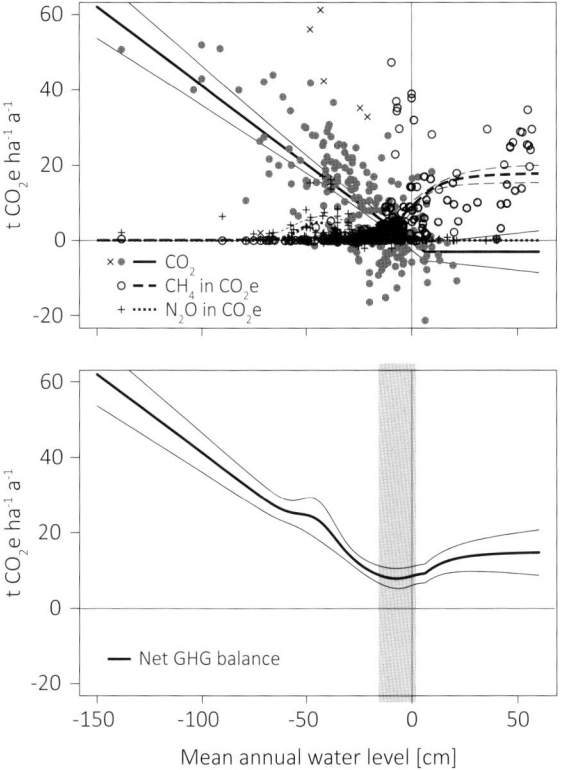

Figure 5.1: GHG emissions and net GHG balance of peatlands depend on the average annual water table. The hairline graphs illustrate the 95% confidence intervals, respectively.

Finland 2014). In Germany, where only 6% of the agricultural land area is peatland, these are responsible for at least 7% of total anthropogenic N_2O emissions (more than 10% of total agricultural N_2O emissions, UBA 2014).

5.1.3 Greenhouse gas balance of rewetted peatlands

Global GHG emissions from drained peatlands are estimated at 2–3 Gt CO_2e a^{-1} (see Joosten & Couwenberg 2009). A large part of these emissions can be avoided by rewetting and restoration (Trumper et al. 2009, Couwenberg et al. 2011, Figure 5.2). Yet, relatively high CO_2 emissions can occur even after rewetting, particularly in case of fluctuating water levels (Yli-Petäys et al. 2007, Höper et al. 2009). Furthermore, it does not appear that large carbon sinks are created by rewetting.

After rewetting, CH_4 emissions will rise to a level comparable to that of natural mires. During the first years after rewetting, CH_4 emissions are typically lower in nutrient poor and higher in nutrient rich peatlands (Blain et al. 2014). Particularly in eutrophic fens, high CH_4 emissions in the first years after rewetting may lead to a higher net GHG balance than during the drained state. However, such very high CH_4 emissions after rewetting have been measured in inundated areas only (several decimetres, e. g. Augustin & Chojnicki 2008, Glatzel et al. 2011, own unpublished data: up to 4,300 kg CH_4 ha^{-1} a^{-1}), whereas CH_4 emissions are

Uijl et al. 2011, Teh et al. 2011). In general, because ditches cover only a small area, the CH_4 emissions from ditches (in CO_2 equivalents) are less significant than CO_2 emissions of the drained areas in between (Drösler et al. 2014; about CO_2 equivalents see also Box 5.1).

The organic nitrogen that is stored in the peat is also broken down under oxic conditions. If denitrification remains incomplete, N_2O is formed and emitted (Martikainen et al. 1993). N_2O emissions may be considerably increased by nitrogen input from mineral and organic fertilisation (Maljanen et al. 2010, Jassal et al. 2011). Annual emissions of N_2O from agriculturally used peatlands are between 6.3 kg ha^{-1} a^{-1} (three annual grass cuts) and 88.6 kg ha^{-1} a^{-1} (crop rotation of potatoes, rye and grass, Flessa et al. 1998, Jungkunst & Fiedler 2007). In peatland rich countries, N_2O emissions from drained peatlands contribute significantly to their total GHG budget. In Finland, for example, only 15% of the agricultural land area is peatland, but it causes 18% of total anthropogenic N_2O emissions (40% of total agricultural N_2O emissions, Statistics

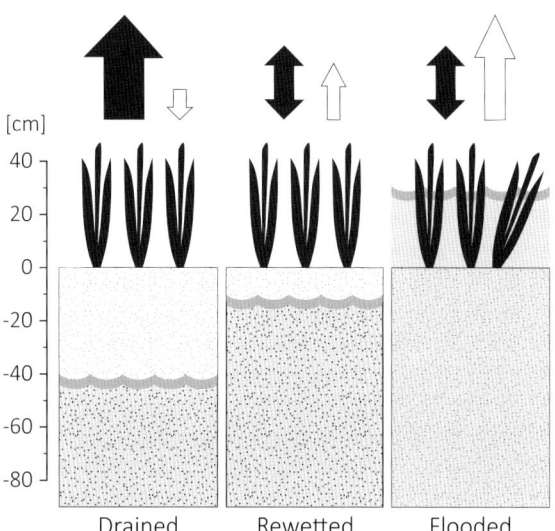

Figure 5.2: Gas exchange in peatlands. Schematic diagram of carbon dioxide (CO_2, black arrows) and methane flows (CH_4, white arrows) related to the water table.

significantly lower when water tables are close to the surface (Hendriks et al. 2007, Huth et al. 2013). A case study in the Trebel Valley in the German federal state of Mecklenburg-West Pomerania revealed CH_4 emissions similar to those of natural mires (190–480 kg ha^{-1} a^{-1}) 15 years after rewetting (Huth et al. 2013). The optimal water table at which emissions are minimised varies across peatlands, depending on various factors. A meta-analysis of published annual emission values shows that the net GHG balance and net climate effect are lowest if the average annual water level is at ground surface (Figure 5.1).

Like in undisturbed peatlands, N_2O emissions are typically small after rewetting. Increasing soil moisture may promote N_2O emissions, but no significant amounts of N_2O are emitted at high water tables (Figure 5.1, Höper et al. 2008, Couwenberg et al. 2011). Overall, N_2O emissions of rewetted peatlands vary between 0 and 50 kg CO_2e ha^{-1} a^{-1}.

5.1.4 Effect of biomass harvesting

If the vegetation is mown and the biomass removed, part of the carbon that was sequestered during the vegetation period is exported. Although a significant part of the plant nutrients may be transferred to the root zone before harvest (depending on the harvest time; Chapter 3.5.3), the question arises, if and how much the export of biomass influences plant growth and thus the carbon exchange of fen peatlands. However, research about the effects of land use management on GHG emissions in rewetted peatlands is generally scarce (Herbst et al. 2011).

Aerenchymatic plants such as Common Reed or Cattail may play an important role in the emission of CH_4 (see above). Mowing these plants during the growing season would likely lead to a change in CH_4 fluxes. Indeed, experiments have shown that CH_4 emissions from aerenchymatic plant stands can decrease directly after cutting (van der Nat & Middelburg 2000, Ding et al. 2005, Duan et al. 2006). A change in the availability of root exudates – that provide substrate for microorganisms – may play a role in this respect (Oates et al. 2008).

Apparently, the effect is not universal and is also strongly dependent on the harvest method, because a reduction in CH_4 emissions was only observed if the plants were cut below the water level probably due to blockage of the internal air spaces by the entering water (Juutinen et al. 2004, Ding et al. 2004, Wang & Han 2005). Additionally, this observation may be explained with the wind-driven Venturi effect, which facilitates convective gas flow from reed culms when cut above the water table. Long-term cutting management could cause a shift in the balance of production and decay in the soil, but so far research into this matter is lacking.

Moreover, it might be expected that renewed biomass growth after cutting supports net carbon sequestration, contributing to a decrease in GHG emissions.

N_2O fluxes from paludiculture fields are essentially determined by the realised water table. Emissions will be insignificant if the water table fluctuates around the soil surface, which is likely to happen when peatlands are used after rewetting (Figure 5.1). In case of lower water tables, higher N_2O emissions can occur, should fertilisation be necessary in order to maintain crop productivity under an intensive cutting regime. If management does not involve fertilisation, like in the case of winter cut reed, N_2O emissions will be considerably lower than under prior, drainage based and often high intensity land use (Hendriks et al. 2007).

A case study in the Trebel Valley (Mecklenburg-West Pomerania, Germany) showed no significant impact of biomass removal on the GHG exchange in rewetted peatlands (Günther et al. 2015; Box. 5.2).

5.1.5 Effects of grazing

To date, there are no measurements of GHG fluxes from grazed rewetted peatlands. However, some data exist on the effects of grazing on the GHG exchange of peatlands that are managed as moist grasslands.

Grazing of rewetted or undrained peatlands has different effects on the GHG balance than mowing. Grazed biomass is hardly exported from the site but largely broken down in the rumen of the grazing animals. Significant amounts of CH_4 are released during this anaerobic predigestion of carbohydrates. It is estimated that 12–17% of global CH_4 emissions derive from ruminant digestion (Lassey et al. 1997, McGinn et al. 2009) and that 23–32% of all anthropogenic CH_4 emissions originate from domestic animals (Kirschke et al. 2013), with emissions varying between species and breeds (Lassey et al. 1997). However, most studies on the impact of grazing on GHG fluxes from peatlands neglect the direct emissions from grazing animals (Langeveld et al. 1997, explicitly e.g. in Teh et al. 2011, see however Herbst et al. 2011), although they clearly exceed the emissions from the peatland itself (Teh et al. 2011, Herbst et al. 2011). The argumentation is that – when land use is changed – the removed grazing unit and its associated emissions are substituted elsewhere, because the demand for animal products remains the same.

The GHG exchange of grazed sites differs from that of unused sites, even when emissions from animals are ignored. Teh et al. (2011) reported N_2O emissions of grazed peatlands in California that are comparable to those of heavily fertilised arable land. The authors explain their findings with the regular application of liquid manure and mineral fertiliser. Also, CH_4 fluxes from grazed drained peatlands can be consid-

Box 5.2: Effects of biomass removal on GHG emissions from rewetted peatlands

Gerald Jurasinski, Anke Günther, Vytas Huth & Stephan Glatzel

Large parts of the fen peatlands in the Trebel valley were drained in the 1960s and managed as high intensity grassland with very low water tables until about 1990. Large scale rewetting took place in two successive steps in 1997 and in 2001. Since then, water tables have stabilised close to the surface.

GHG measurements were carried out between March 2011 and March 2013 at three sites (each with 6 replicates) that were dominated by Common Reed (*Phragmites australis*), Cattail (*Typha latifolia*) and Lesser Pond Sedge (*Carex acutiformis*), respectively. The weather during the first year was exceptionally wet, particularly during summer. The annual precipitation of 821 mm was almost 200 mm above the 30-year average. From June through August 2011, 472 mm of rain caused flooding of the Trebel River valley for a period of several weeks. During this time, water tables were 10 cm above the surface in the reed stand, 20 cm in the cattail stand, and 40 cm in the sedge stand. The second measurement year was much drier; annual precipitation (481 mm) was approximately 150 mm below the 30-year average. Average water tables were close to the surface in the first year (-5 cm in the reed stand and +5 cm in the cattail and sedge stands) and overall conditions represented soil moisture class 5+ (Box 5.3).

In all three stands, harvest was simulated by clipping and removing aboveground biomass. Three of the 6 replicates were clipped, the other 3 served as control. The time of harvest was chosen to mimic land use management aimed at sustaining the stand's species composition. Reed and cattail were harvested during winter and the sedge stands during summer.

The average CO_2 balance of the two measurement years was close to zero for all sites (Figure 5.3), without a systematic difference between the different vegetation types. The sedge stand exhibited the highest CH_4 emissions. Biomass removal did not significantly affect CH_4 emissions for any of the investigated stands. Overall, 14 years after rewetting, CH_4 fluxes resemble those of undisturbed mires of the temperate zone. N_2O fluxes were below the detection limit. Taking the three gases together, all sites had a neutral or slightly warming effect on the climate. The relatively high GHG emissions from the sedge stand may be related to the unusually wet conditions during the first summer. The extended period of inundation must have affected the low-growing vegetation more strongly than the tall reed and cattail stands, suppressing CO_2 uptake and promoting CH_4 production.

In conclusion, the results show that biomass removal has a negligible effect on the net GHG balance of paludiculture areas.

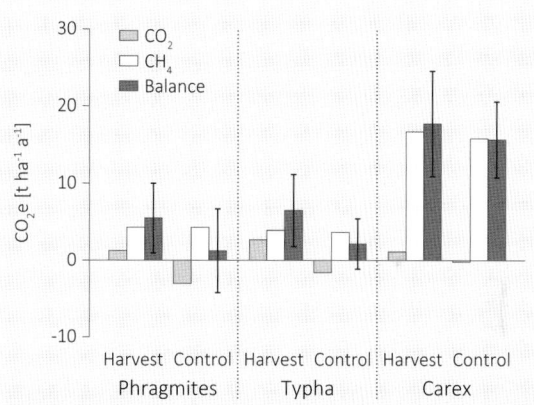

Figure 5.3: Effect of biomass removal on the net GHG balance of different vegetation stands in a rewetted peatland (Trebel Valley). Emissions of carbon dioxide (CO_2) and methane (CH_4), as well as the total balance (including standard errors) are shown in CO_2 equivalents averaged over two years. Positive values indicate sources; negative values indicate sinks.

---

erable if ditches are taken into account. In California, emissions from ditches contributed 84% of the total CH_4 emissions of 125 kg CH_4 ha^{-1} a^{-1}, even though they occupied only 5% of the area (Teh et al. 2011). Excluding emissions from ditches, Langeveld et al. (1997) recorded significantly lower CH_4 emissions on drained and grazed peatlands in the Netherlands (–0.3 ± 0.1 to 0.1 ± 0.1 kg ha^{-1} a^{-1}).

After grazing was stopped in a Californian moist grassland, a decrease in CH_4 fluxes and their spatio-temporal variation was observed. Fluxes were best explained by soil temperature and gravimetric water content. Therefore, Oates et al. (2008) assumed that trampling by animals has a major influence on soil CH_4 emissions, which was confirmed in other studies not concerning moist grasslands (e.g. Boeckx et al. 1997,

> **Box 5.3: Soil moisture classes**
>
> John Couwenberg
>
> Soil moisture classes describe the wetness of the soil or its water table (Table 5.2). The concept follows the approach of Petersen (1952), and was further developed by Koska (2001) as one of the tools to describe site conditions associated with vegetation forms. Soil moisture classes describe a gradient of soil humidity and are a main factor in the description of Greenhouse gas Emission Site Types (Couwenberg et al. 2011; Box 5.4).
>
> Table 5.2: Water supply of a site classified in soil moisture classes Couwenberg et al. 2011, changed after Koska 2001.
>
> | Soil moisture class | | Characteristics | | |
> |---|---|---|---|---|
> | Classes | Description | Median winter/spring [cm] | Median summer/autumn [cm] | Water supply deficit [l/m²] |
> | 7+ | Upper sublitoral | +250 to +140 | | |
> | 6+ | Lower Eulitoral | +150 to +10 | +140 to +0 | |
> | 5+ | Wet (upper Eulitoral) | +10 to -5 | +0 to -10 | |
> | 4+ | Very moist | -5 to -15 | -10 to -20 | |
> | 3+ | Moist | -15 to -35 | -20 to -45 | |
> | 2+ | Moderately moist | -35 to -70 | -45 to -85 | |
> | 2- | Moderately dry | | | < 60 |
> | 3- | Dry | | | 60–100 |
> | 4- | Very dry | | | 100–140 |
> | 5- | Extremely dry | | | > 140 |
> | Seasonally alternating humidity is indicated by a combination of different classes (e.g. 5+/4+). Strongly alternating wetness is indicated by a tilde-sign '~' –e.g. 3~ refers to (winter/autumn 4+ and summer 2+). | | | | |

Yamulki & Jarvis 2002). Here, compaction by trampling tripled N_2O emissions and quadrupled CH_4 emissions (Yamulki & Jarvis 2002). Oates et al. (2008) found that exclusion of grazing did not affect the amount of N_2O emissions, but did lead to less variable fluxes. The above mentioned studies generally did not take into account the direct emissions from the animals' excrements. However, Williams (1993) found that these emissions are negligible compared with the emissions from the rumen.

The damage that grazing causes to the plants may also affect CH_4 fluxes (Oates et al. 2008). Plants must invest more in regrowing biomass, which is presumed to reduce the excretion of root exudates that would otherwise be available as a substrate for methanogenesis. Indeed, cutting and biomass removal experiments (see above) have revealed significantly lower CH_4 emissions in various ecosystems (Whalen 2005). Besides reduced root exudate production, reduced convective gas transport is generally assumed to drive the decrease in CH_4 emissions. However, this attenuating effect is obviously less significant than the strengthening effect of compaction by trampling. In balance, grazing results in rather strongly increased CH_4 emissions and slightly increased N_2O emissions.

Thus, if rewetted sites were grazed, net GHG emissions would slightly increase compared with the ungrazed situation. Compared with the high intensity land use before rewetting, emissions will – depending on the water table – be considerably lower.

5.1.6 Quantifiability

Emission reductions can be monetised and may offer an opportunity to generate additional funds from paludiculture areas. However, such monetisation requires reliable tools for the assessment of emission reductions. Consequently, emission reductions must be measurable, reportable and verifiable.

The spatial variability in GHG fluxes is often not well reproduced because of the limited size of the closed chambers that were used in the derivation of the majority of available emission data. The high spatial variability – particularly in CH_4 and N_2O fluxes (e.g. van den Pol-van Dasselaar et al. 1998, Glatzel et al. 2008, Moore et al. 2011) – is usually ascribed to small-scale heterogeneity in environmental conditions (e.g. van Huissteden et al. 2009, Hendriks et al. 2010). Temporal variability in GHG fluxes can be high as well. Apart from a pronounced seasonality (Kim et al. 1998, Regina et al. 2004, Askaer et al. 2011), interannual variability can also be considerable (Beetz et al. 2013, Günther et al. 2014). As a result, annual emission estimates that are based on seasonal or on single-year measurements are unlikely to reflect multi-year averages. Anyhow, multi-year (> 5 years) measurement series of GHG exchange rates are hardly available.

Plants not only regulate the GHG exchange in peatland ecosystems (Bubier 1995, Whalen 2005, Dias et al. 2010) but they also reflect average site conditions. Therefore, the use of vegetation as a proxy may significantly improve GHG flux assessments (Couwenberg et al. 2008, Couwenberg et al. 2011). The Greenhouse gas Emission Site Type (GEST) approach offers such a vegetation based flux assessment (Box 5.4). The GEST model is open for new data and new delineation of site types. For an improved and more reliable assessment, additional data are needed and additional proxy parameters should be included, such as vegetation fitness, nutrient availability or pH. These variables could then also be used in extrapolation and regionalisation of GESTs.

Sharpening GEST values is feasible through targeted measurements. Some research is still needed in this respect. Available multi-year measurement series from different peatland sites may allow development of more efficient measurement strategies for reliable assessment of mean annual fluxes. Concentrated measurements during short periods of high or low fluxes could considerably reduce measuring effort and still provide reliable assessments of annual emissions. For any future measurement, water tables and basic climatic parameters should be recorded automatically on site.

Box 5.4: Greenhouse gas Emissions Site Types

John Couwenberg

In order to make large-scale assessments of GHG fluxes from a peatland site without comprehensive measurements of gas fluxes, Greenhouse gas Emission Site Types (GESTs) were developed at the University of Greifswald (Couwenberg et al. 2008, 2011). GESTs are based on vegetation forms, which are vegetation types that are defined on the basis of both floristic and ecological site variables (Koska 2001). The concept departs from the idea that certain plant species occur together along an ecological gradient (e.g. from dry to wet), whereas others exclude each other. Undoubtedly, a combination of such species groups is a more precise indicator of plant-relevant ecological site conditions than a single plant species. The ecological variables that most significantly influence vegetation composition (water table/moisture, nutrient availability, acidity, and land use), also affect GHG fluxes and thus vegetation forms are a suitable proxy for the assessment of GHG emissions. The mean annual water table appears to have the strongest effect on GHG fluxes (Couwenberg et al. 2008, 2011, Figure 5.1). In addition, the presence of aerenchymatous plants strongly determines CH_4 emissions from wet sites, because they act as shortcuts (or shunts) between the anoxic root zone where CH_4 is formed, and the atmosphere. Examples of such 'shunt species' are *Phragmites australis*, *Eriophorum angustifolium*, *Phalaris arundinacea*, *Cladium mariscus*, *Carex* spp., *Scirpus* spp., and *Juncus* spp.

Based on these insights, CH_4 and CO_2 emission factors are assigned to specific vegetation forms. Direct flux measurements are not available for all of them, but a matrix of all possible vegetation forms allows for extra- and interpolation of gas flux values along various axes of site conditions. A triangular comparison of i) GHG fluxes in relationship to vegetation forms; ii) GHG fluxes in relation to the groundwater table; and iii) vegetation forms in relation to the groundwater table allows for a verification of the assigned values.

Most of the published data on GHG fluxes concerns relatively stable peatland systems. In the first years after rewetting (1–10 years), the transitional vegetation may not (yet) be in equilibrium with site conditions and match a vegetation form. In such cases, other parameters (particularly the water table) can be used to adequately indicate GHG fluxes. For some vegetation forms, there is a shortage of measurement data on GHG fluxes, meaning that substantial spatial and interannual variability may not yet be sufficiently integrated into the associated GEST values (Beetz et al. 2013, Huth et al. 2013). Further research is required to improve these values (see Bärisch & Tanneberger 2011).

5.2 Biodiversity

Natural and near-natural fen peatlands harbour many highly specialised organisms that put different demands on their habitat. Any form of biomass removal from these fens, whether by mowing or grazing, certainly constitutes a significant intervention in these complex systems. On the one hand, such interventions may be necessary for the long term conservation of certain species in semi-natural biotopes. On the other hand, species that are sensitive to the disturbance may decline or disappear. Thus, general statements about the influence of land use on biodiversity of reed beds or wet meadows are not possible. Before implementing any change in land use management, it is necessary to define aims and objectives and also to assess consequences for nature conservation.

At present, most of the fen peatlands of northern Germany are deeply drained and under high intensity use as grassland or even as arable land; consequently, characteristic biodiversity has been lost. Paludiculture aims both to rewet these areas and to use them in a more environment-friendly way. Here we discuss how different forms of paludiculture affect the birds, the arthropods and the vegetation of fen peatlands. A general distinction is made between reed beds and moist or wet meadows, because of distinct differences in vegetation structure and management: Reed beds are usually harvested in winter, against moist and wet meadows mainly during summer. The frequency of mowing influences the presence of litter and broken culms, thus affecting habitat structure.

5.2.1 The effect of mowing on animals

Sebastian Görn & Klaus Fischer

Numerous studies exist on the effect of mowing on bird populations of fen peatlands, covering species such as the Great Bittern (*Botaurus stellaris*), the Aquatic Warbler (*Acrocephalus paludicola*) and various wader species. In Southern France, winter harvest of reed beds had a positive effect on the Great Bittern population, which was attributed to sparser reed stands (Poulin et al. 2005). The population grew even larger when mowing was suspended for one year, for which reason the authors recommend rotational cutting, each year leaving part of the area uncut (Poulin et al. 2009). In Britain, Gilbert et al. (2005) found no evidence of Great Bittern nesting in reed beds that had remained uncut for a longer period of time, nor in reed beds that had been cut every year. They concluded that long-term conservation of the Great Bittern required regular (but not annual) cutting of the reed beds (see also Tyler et al. 1998). Also Bibby & Lung (1982) do not consider winter harvest of reed as a threat to species such as the Great Bittern, Western Marsh Harrier (*Circus aeruginosus*), Bearded Reedling (*Panurus biarmicus*) and Savi's Warbler (*Locustella luscinioides*) so long as the reed bed is not cut in its entirety each year. Contrasting evidence that winter harvest of reed negatively affects songbirds comes from Lake Kolon (Hungary), where significantly fewer songbird species occurred in reed beds that were regularly cut than in those that remained uncut (Vadász et al. 2008). Among characteristic bird species like Savis' Warbler (*Locustella luscinioides*), Moustached Warbler (*Acrocephalus melanopogon*), Sedge Warbler (*A. schoenobaenus*), Reed Warbler (*A. scirpaceus*) and Great Reed Warbler (*A. arundinaceus*), only the latter was more frequently found in the regularly cut areas. A study from the Netherlands showed a similar negative effect of winter harvest of reed on the occurrence of Sedge and Reed Warblers, including a greater risk of predation (Graveland 1999). Overall, winter harvest of reed not necessarily stimulates bird populations, unless reed stands are partly excluded from annual cutting.

The cutting of reed during the summer vegetation period bears more potential for conflict. Summer harvest supports succession to open wet meadows and is therefore conductive for many waders. Moreover, during summer months, mowing may lead to direct loss of birds, their nests and their chicks. Nonetheless, regular mowing may be essential for the conservation of specific birds, like the Aquatic Warbler, that is nearly extinct in Central Europe. In this case, the mowing regime depends on the nutrient status of the habitat. A stable population of Aquatic Warblers has established in mesotrophic fens in Northwestern Poland, where reed is harvested for thatching (Tanneberger et al. 2009). In that area, winter harvest of reed stimulated the development of short reed stands (< 2 m) that are very attractive to both the Aquatic Warbler and the reed harvesters for the high quality thatch they provide (Tegetmeyer et al. 2007). In contrast, nutrient rich sites require an early cut to prevent the development of too dense stands (Tannenberger et al. 2010). To minimize the loss of nests, mowing should be rotational, each year leaving parts of the reed bed uncut.

Besides the Aquatic Warbler (in the more eutrophic habitats) other birds also depend on summer harvest. Many of the inhabitants of moist low intensity grasslands, like Lapwing (*Vanellus vanellus*), Black-tailed Godwit (*Limosa limosa*), Common Snipe (*Gallinago gallinago*) and Ruff (*Philomachus pugnax*) have faced a strong decline over the past decades (Südbeck et al. 2007). The main cause for the decline lies in the intensification of land use, although the abandonment of land use is responsible to some extent. A positive effect of moderate summer harvest on the breeding bird population (of wet sites) could be shown for the Lower Peene Valley nature reserve in Mecklenburg-West Pomerania (Box 5.5). Mown areas showed by far the

5.2 Biodiversity

highest number of threatened bird species (Figure 5.4). The main problem for summer harvest remains determining the best time to minimise negative effects (Berg 1992, Lebedeva 1998, Valkama et al. 1998, Fischer et al. 2013). Although late cutting may be best from a nature conservational perspective, it is hardly compatible with conventional agricultural production cycles.

The reaction of arthropods on mowing varies significantly. Schmidt et al. (2005) showed that winter harvest of reeds had a positive effect on organisms that feed on plant saps, such as cicadas (Auchenorrhyncha) and aphids (Aphidina), whereas butterflies and moths (Lepidoptera), spiders (Araneae) and in particular isopods (Isopoda) were negatively affected. Moreover, significant changes in the species composition of spiders and beetles (Coleoptera) were observed. Therefore, Hardman et al. (2012) emphasise the necessity of establishing reed beds of different age, as each successional stage harbours its specific association of arthropods. Old stands with a thick litter layer are of particular interest in this respect, as they are home to many arthropods of high conservation value. However,

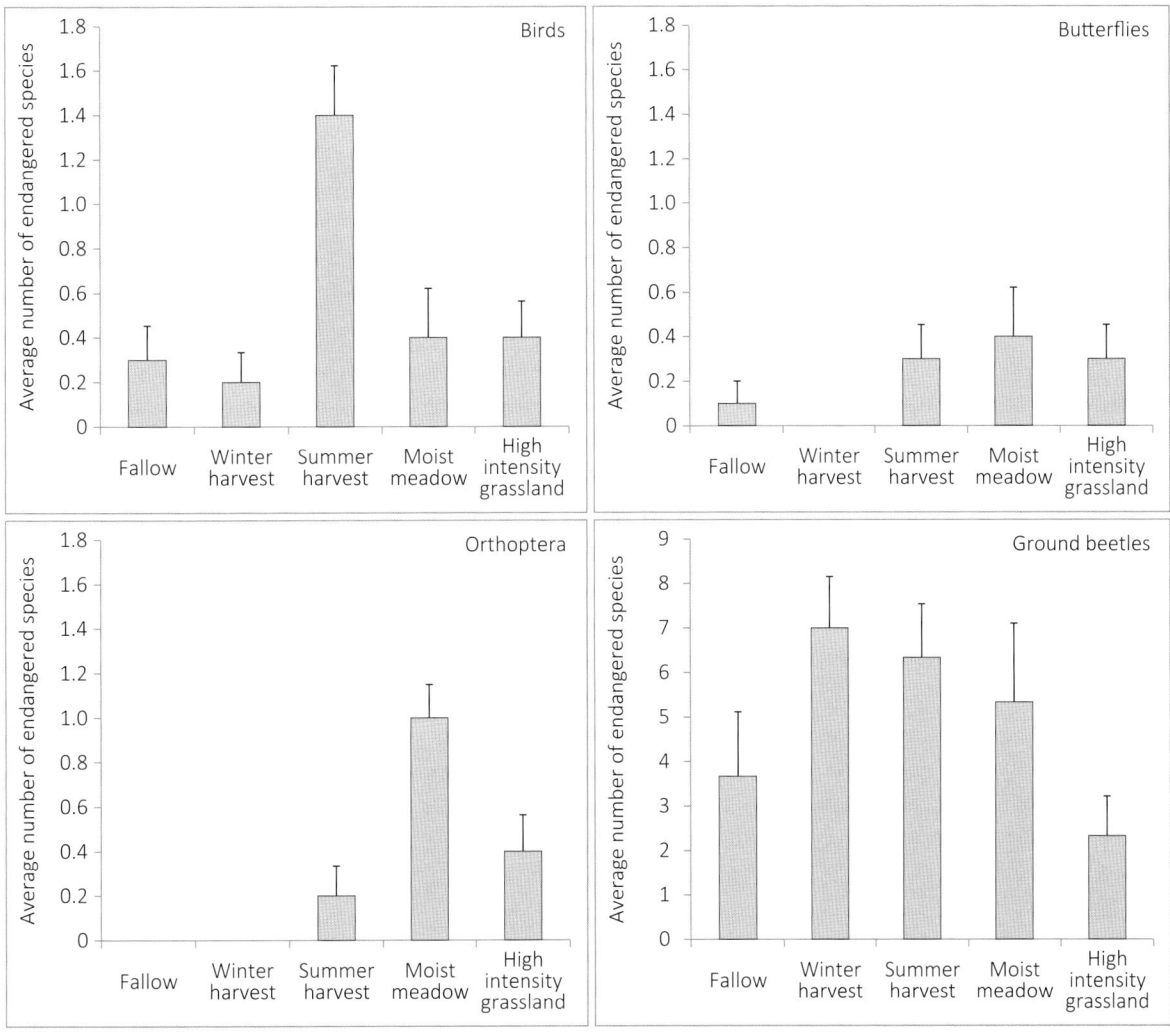

Figure 5.4: Effect of different land use practices on the average number of endangered species of birds, butterflies, grasshoppers and ground beetles in the nature reserve Lower Peene Valley. A species is 'endangered' if it is on the Red List of Mecklenburg-West Pomerania (Wachlin 1993, Wranik et al. 1997, Eichstädt et al. 2003, Müller-Motzfeld & Schmidt 2008). The 'fallow', 'winter harvest' and 'summer harvest' areas are mainly reed beds. Vegetation cover of the 'summer harvest' areas is very sparse, because of mowing during the vegetation period. Graphs show the average number of endangered species per site, including standard error. For each type of land use practice, species numbers of birds, butterflies and grasshoppers were recorded at 10 sites, and those of ground beetles at 3 sites.

as mentioned above, winter harvest had a positive effect on ground beetles (Carabidae) in the Peene Valley (Figure 5.4, Görn et al. 2014). Next to arthropods typical for reed beds, some rare species that require an open habitat structure during spring time were recorded in the above mentioned area. Decleer (1990) found changes in the composition of the spider fauna in a reed bed in reaction to mowing. Compared with unmanaged plots, winter harvest led to a decline in rare specialists and to an increase in ubiquitous spiders. This trend was even stronger with summer harvesting. The effect of mowing on endophagic arthropods remains ambiguous (Kampichler et al. 1994). Only the Twin-spotted Wainscot (*Archanara geminipuncta*) showed a clear decline, because the reed culms of the mown areas proved too thin for the caterpillar of this moth. For owlet moths (Noctuidae), winter harvest is in general problematic because the larvae of many species hibernate inside the culms (Huemer 1996). Also many beetles and spiders spend the winter in reed beds (Neumann & Krüger 1991). Conservation of arthropod species thus demands that reed beds are left, at least partly, uncut for several years (up to four, Hock & Vorbrüggen 1997).

Arthropod communities in moist grasslands are equally affected by mowing (Gerstmeier & Lang 1996). In these semi-natural habitats, permanent management is essential to the entire biocoenosis. Particularly well studied are the effects of mowing on butterflies. Butterflies are severely threatened both by increased land use intensity as well as by abandonment (van Swaay et al. 2006, van Swaay et al. 2010). Highly endangered species that strongly depend on wet meadows – like the Dryad (*Minois dryas*), Dusky Large Blue (*Maculinea nausithous*) and Alcon Blue (*Maculinea alcon*) require a late harvest, preferably in late September (Gerstmeier & Lang 1996, Huemer 1996, Hock & Weidner 1997, Dierks & Fischer 2009). Also in the Peene Valley, moderate mowing in summer had a positive effect on populations of ground beetles, grasshoppers (Orthoptera) and butterflies (Figure 5.4). Again, mowing of wet meadows may negatively affect rare species of spiders (Cattin et al. 2003, Schmidt et al. 2008), grasshoppers and bugs (Hemiptera; Gerstmeier & Lang 1996, Malkus 1997).

In conclusion, any form of management must adapt to the target species of interest. Only a mosaic of differently managed and unmanaged areas can accommodate the requirements of different biocoenoses and taxonomic groups. These requirements should be taken into account and be addressed by specific agri-environmental programmes where novel land use concepts are established on wet peatlands.

5.2.2 The effect of mowing on plants

Stefanie Raabe & Michael Manthey

In order to assess the effect of different mowing regimes on the vegetation, two areas were studied in the nature reserve Lower Peene Valley. Both sites are formerly-drained, high intensity fen grasslands that were rewetted over the past 12 years. At present, site conditions are eutrophic (C/N ratio of 12–14) and medium water tables are around 10 cm below the surface (soil moisture class 5+, Box 5.3). We looked at two different mowing regimes (a single cut during summer versus a single cut during winter), and compared them with an unmanaged control (fallow). In addition, we included drained, high intensity fen grasslands that have comparable trophic status, where management includes multiple cuts per year and fertiliser application. In all sites, vegetation relevées were made, in which all vascular plants and their cover were recorded in 16 m^2 large plots using a 10-part scale (Peet et al. 1998).

When similarity in the species composition of the vegetation relevées is expressed using a non-metric ordination technique (NMDS), the different mowing regimes can be clearly distinguished (Figure 5.7). The most significant difference in species composition is plotted along axis 1, which shows a sequence of mowing regimes from high intensity grassland, via summer harvest and fallow to winter harvest. The two extremes – high intensity grassland and winter harvest, are particularly distinct floristically. The high intensity grasslands are characterised by species poor communities of highly productive grasses, rosette plants and annuals – all with a high tolerance for mowing. In contrast, the winter harvest sites are dominated by Common Reed (*Phragmites australis*).

If wet areas are mown once during summer, Common Reed is suppressed (Gryseels 1989, Güsewell et al. 2000, Ostendorp 1993) and stands will become sparser. The associated increase in the availability of light will promote the occurrence of smaller, less competitive plants (Kotowski et al. 2001). Thus, summer harvested sites are more species rich (Gaisler et al. 2004) and more frequently harbour rare species. We found 21 species that were restricted to the summer harvested sites, 9 of which are on the German Red List of threatened species. Although regular summer mowing reduces available nutrients (Schulz et al. 2011), this reduction will hardly pose a problem for continued land use – at least in the medium term, because excess nutrients are available in these formerly drained and intensively used grasslands. Nutrient removal is desirable from a nature conservation perspective, because mesotrophic sites and the associated plant communities have become increasingly rare (Berg et al. 2004).

5.2 Biodiversity

Box 5.5: Effects of rewetting on the avifauna

Benjamin Herold

Before they were drained, the percolation mires that prevailed in the river valleys of northeastern Germany were characterised by a specific avifauna. The habitat requirements of some 20 bird species were fulfilled only there (Herold 2012). Thus, these species serve as indicator species for percolation mires (see Flade 1994): Garganey (*Anas querquedula*), Montagu's Harrier (*Circus pygargus*), Hen Harrier (*Circus cyaneus*), Black Grouse (*Tetrao tetrix*), Water Rail (*Rallus aquaticus*), Spotted Crake (*Porzana porzana*), Baillon's Crake (*P. pusilla*), Corn Crake (*Crex crex*), Redshank (*Tringa totanus*), Ruff (*Philomachus pugnax*), Black-tailed Godwit (*Limosa limosa*), Curlew (*Numenius arquata*), Great Snipe (*Gallinago media*), Common Snipe (*Gallinago gallinago*), White-winged Tern (*Chlidonias leucopterus*), Bluethroat (*Luscinia svecica*), Sedge Warbler (*Acrocephalus schoenobaenus*), Aquatic Warbler (*Acrocephalus paludicola*), Grashopper Warbler (*Locustella naevia*) and Savi's Warbler (*Locustella luscinioides*) (Herold 2012). During the second half of the 20th century, peatlands were drained for high intensity agriculture, which caused the populations of almost all indicator species to collapse or disappear entirely. Fifteen of the species are listed as (critically) endangered on the Red List of threatened breeding birds of Germany (Südbeck et al. 2007). The Aquatic Warbler is even listed as vulnerable at global level (Tanneberger et al. 2010). Restoration of deeply drained peatlands of Mecklenburg-West Pomerania (which started in the mid 1990s) has raised hopes that lost and endangered bird species will return. Herold (2012) recorded about 60 species of breeding birds in the rewetted peatlands (+ 37 woodland species). Under permanently wet conditions, breeding birds of high conservation value can be found already after a few years. In the meantime, critically endangered and previously lost species like Garganey, Baillon's Crake, Spotted Crake, Corn Crake, Common Snipe, Lapwing, Common Tern and White-winged Tern have returned as breeding birds.

The key factor in the return of the indicator and other endangered species is a permanent rise in the water table to just above the surface, as it is the preferred water table of species of near-natural percolation mires (Figure 5.5, Figure 5.6). Habitat analyses show that the returning indicator species clearly prefer sedge vegetation, corresponding with the original vegetation of the percolation mires (Herold 2012). Although it has hardly been studied thus far, the food supply of the permanently wet sites is probably the main factor responsible for their attractiveness. Areas that have become flooded (about 40 cm) after rewetting provide habitats for equally endangered species of shallow lakes and lake shores, such as the Little Bittern (*Ixobrychus minu-*

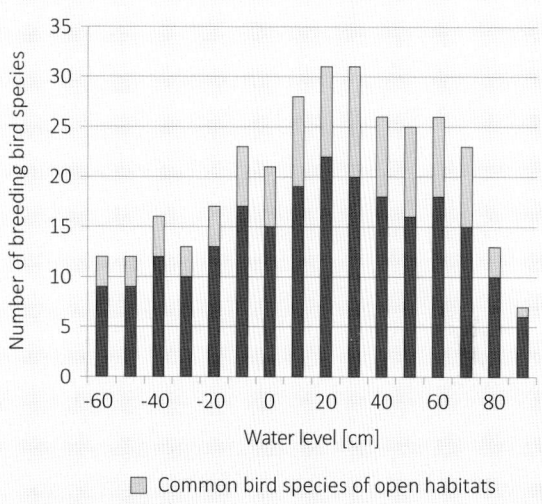

Figure 5.5: Number of breeding bird species of open habitats in relation to the water table (mid of May).

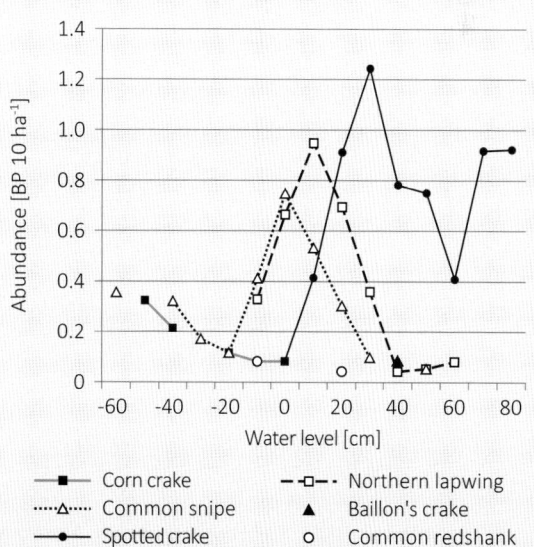

Figure 5.6: Abundance of selected endangered bird species in rewetted peatlands (number of breeding pairs (BP) per 10 ha) in relation to the water table (mid of May).

tus), Great Bittern (*Botaurus stellaris*), Little Crake (*Porzana parva*) and Black Tern (*Chlidonias niger*). In the medium term, a large part of the rewetted areas develop stands of Cattail and particularly of Common Reed (Steffenhagen et al. 2010, Herold 2012). If sites then dry up for longer periods of time, woody plants will establish (Succow & Joosten 2001; Timmermann et al. 2006). These developments lead to habitat deterioration for the indicator species of percolation mires, because these species require extensive open areas. In light of the critical state of several of the indicator species, natural development towards more nutrient poor site conditions would take too long and thus management is required. In this light, paludiculture offers a promising approach.

The species composition of rewetted fallow sites is intermediate between summer and winter harvested sites (Figure 5.7). A lack of mowing disturbance induces denser reed beds than in summer harvested sites, and the accompanying, tall competitive species, characteristic for winter harvested sites occur as well. Overall, the fallow sites showed the highest species density per relevée (Figure 5.8). The number of species that only occur on the fallow sites (22, including 7 Red List species) is comparable to the summer harvested sites. This outcome contradicts studies showing that regular mowing enhances species diversity (Cowie et al. 1992, Humbert et al. 2009). Apparently, the studied fallow plots are at an early stage of succession, in which competitively weak species still survive despite the strong competition for light by taller plants. After an initial increase in the number of species, this number will decline again in the course of succession, ending with the dominance of only a few species (Mälson et al. 2008). In the studied fallow sites, plant species diversity is furthermore increased by rummaging of wild boars, which creates spaces for pioneer species.

Our results show that rewetting – in general and independently of the mowing regime, leads to a significant increase in plant biodiversity compared with drained, high intensity fen grasslands (Figure 5.8), and that endangered species occur exclusively on rewetted sites. Some species that tolerate high water tables occur under all three management regimes in the wet fen areas. Yet, each regime offers optimal conditions to specific groups of plants, promoting biodiversity at the landscape level.

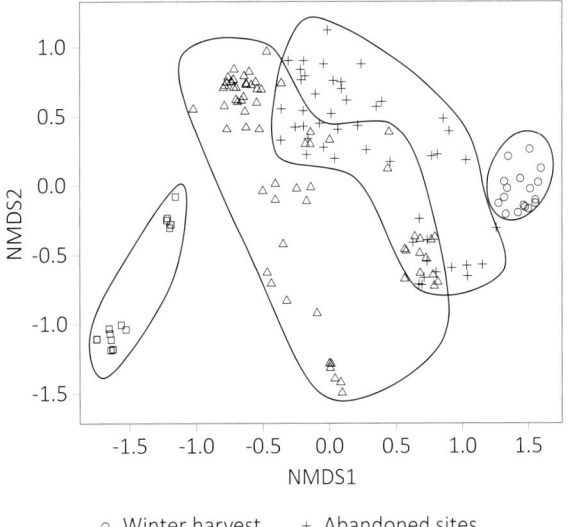

Figure 5.7: Ordination (NMDS) of vegetation relevées from peatlands with different management regimes. Data points that are close together have a similar species composition; distant points, a very different one. The ordination shows clear differences between drained, high intensity grasslands (soil moisture class 2+) and rewetted areas (soil moisture class 4+/5+) with summer harvest, no harvest (fallow), and winter harvest.

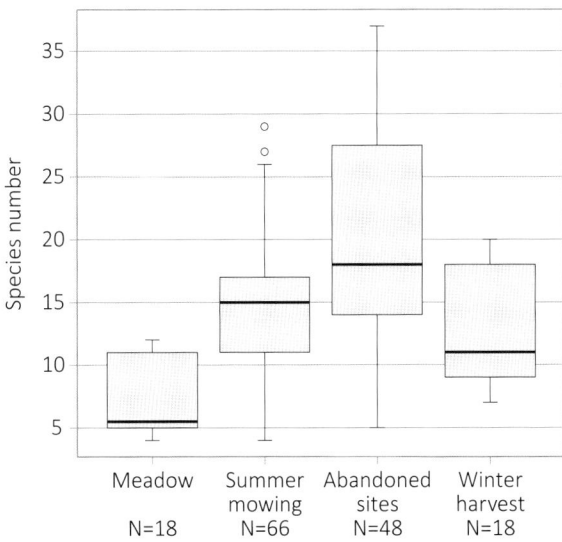

Figure 5.8: Number of plant species for different management regimes on peat soil (high intensity grassland, soil moisture class 2+; summer harvest, fallow winter harvest, soil moisture class 4+/5+; size of vegetation relevées: 16 m²).

5.2.3 Vegetation patterns after multiple years of grazing management

Jürgen Müller & Weert Sweers

Large stretches of the Peene River valley are still in a near-natural condition and rich in plant species (Fischer 1999). The river is accompanied by percolation mires and flows through a landscape of moraines. On the valley edges, transitional areas between organic and mineral soil are ubiquitous (Janke 2002). Where the two soil types meet, low intensity pasturing has established on a large scale. On a 13 ha large area west of the town of Demmin, we investigated the behaviour of young cattle (heifers of commercial crossbreeds) and their influence on grazed vegetation stands (Chapter 3.3.4). The area is very wet and periodically flooded (soil moisture class 4+/5+), and has been grazed for 14 years. Although the mire is accessible to the cattle all year, the animals avoid it during the first half of spring, when they obviously prefer to graze on mineral soil. Starting from the end of May, the cattle occasionally ventures into the peatland, where initially they feed exclusively on grasses (*Glyceria maxima, Festuca pratensis, Agrostis alba*, young tussocks of *Deschampsia cespitosa*) and avoid areas dominated by sedges (*Carex, Scirpus sylvaticus*). Only starting in August, they also feed on the wet mire vegetation (Calthion phytocenosis) because the vegetation on the mineral soils becomes too dry to accommodate the nutritional needs of the animals. In late summer, species like *Holcus lanatus, Carex disticha* and even *Iris pseudacorus* are consumed. A thick layer of litter, washed in by frequent floodings, protects the plants from being grazed short.

The combination of largely undisturbed vegetation development until early summer, and a grazing-imposed shift in the dominant vegetation later on, induces a structural diversity that supports a species

Table 5.3: Selected 'Red List' plant species in a paludiculture meadow in the Peene valley after 14 years of (partial) grazing (field size 13 ha).

Species	Category according to the Red List of M-V[1]	Spatial distribution	Ecological traits			
			MR[2]	MT[3]	GT[4]	TT[5]
Achillea ptarmica	3	Scattered	8	4	4	4
Dactylorhiza majalis	2, § C	Sporadically, concentrated in Groundwater discharge zones	(8)	4	3	3
Helictotrichon pubescens	3	Occasional	indiff.	5	4	4
Lathyrus palustris	3	Widespread	8	3	1	1
Lychnis flos-cuculi	3	Widespread	7	4	–	–
Menyanthes trifoliata	3, § A	Sporadically near water Courses	9	–	–	–
Succisa pratensis	2	Occasional	7	3	3	4
Thalictrum aquilegifolium	R	Scattered	8	3	3	2

[1] Category according to the Red List of threatened plant species in Mecklenburg-West Pomerania after Anonymus 2005.
[2] – Moisture range after Ellenberg 1992.
[3] – Mowing tolerance after Briemle & Ellenberg 1994.
[4] – Grazing tolerance after Dierschke & Briemle 2002.
[5] – Trampling tolerance after Dierschke & Briemle 2002.
– = No information available.

rich community with many redlisted plant species (Table 5.3). The low stocking rate (0.8 livestock units per ha, where 1.5 are allowed) permits the vegetation to retain its moist fen grassland appearance: pastoral indicator species are rare, whereas grazing-sensitive species like *Lathyrus pratensis* have remained common. However, to repress species poor insular stands dominated by *Scirpus sylvaticus* or *Carex acutiformis*, stocking rates would need to be too high, causing damages to the soil (trampling), meadow birds, and the vegetation. Therefore, to improve conditions for more light demanding plants, additional mowing will be necessary. Dense species poor stands could alternatively be opened up by grazing Water Buffaloes (Box 5.6).

Box 5.6: Suppressing reed by grazing with Water Buffalo

Stefan Horn, Weert Sweers & Thomas Frase

The coastal marshes of the southern Baltic region are typically influenced by brackish water from periodical flooding of the Baltic Sea. Brackish reed beds constitute the natural climax vegetation in these coastal marshes (Fukarek 1969), but grazing has transformed this natural vegetation into an anthropo-zoogenic salt marsh (Jeschke 1987). Grazing has kept the habitat open and created a mosaic of tidal creeks, seasonal pools, brackish pioneer vegetation, salt marshes and reed beds (Jeschke 1987). The spatial variety in diverse habitats encountered in these salt marshes likely supports a more species rich plant and animal community than the apparently monotonous brackish water reed beds.

Regular grazing has created a typical salt marsh with many rare species on Schmidt-Bülten, a 28 ha large island in the West Pomeranian lagoon. Abandonment or reduced grazing pressure would cause the vegetation – with its high conservational value, to be lost to succession towards reed beds (Jeschke 1987). In recent years, management of the island focused on grazing by suckler cows with low livestock density (0.6 livestock units per hectare, LU ha^{-1}). However, this low livestock density proved insufficient in conserving the salt marsh. Instead, reed beds developed in the wetter areas (soil moisture class 5+) and regular grazing remained restricted to the drier parts of the island (soil moisture class 3+). Parts of the dry area were even overgrazed, resulting in decreased species richness.

Since June 2010, Water Buffalos have been used for grazing on the island in order to restore the typical salt marsh vegetation and to suppress the encroaching reed beds. A livestock density of 1.0–1.3 LU ha^{-1} and 123–148 grazing days per season proved sufficient to reduce the area of reed beds by 30% (Figure 5.9). Like the suckler cows, the Water Buffalos preferred the drier parts of the island with its high-quality fodder at the beginning of the grazing season. Nonetheless, the animals did disturb the reed beds already early in the season by regular trampling and some grazing on young shoots. As the amount of available fodder dwindles during the course of the grazing season, the buffalos more regularly feed on the young reed culms, allowing understory plants to benefit (mainly *Agrostis stolonifera* ssp. *maritima* and *Juncus gerardii*). Meanwhile, the suppression of reed has led to an increase in grazing area (which, in 2012, required a slight increase in the livestock density) in order to keep grazing pressure at the required level.

5.2 Biodiversity

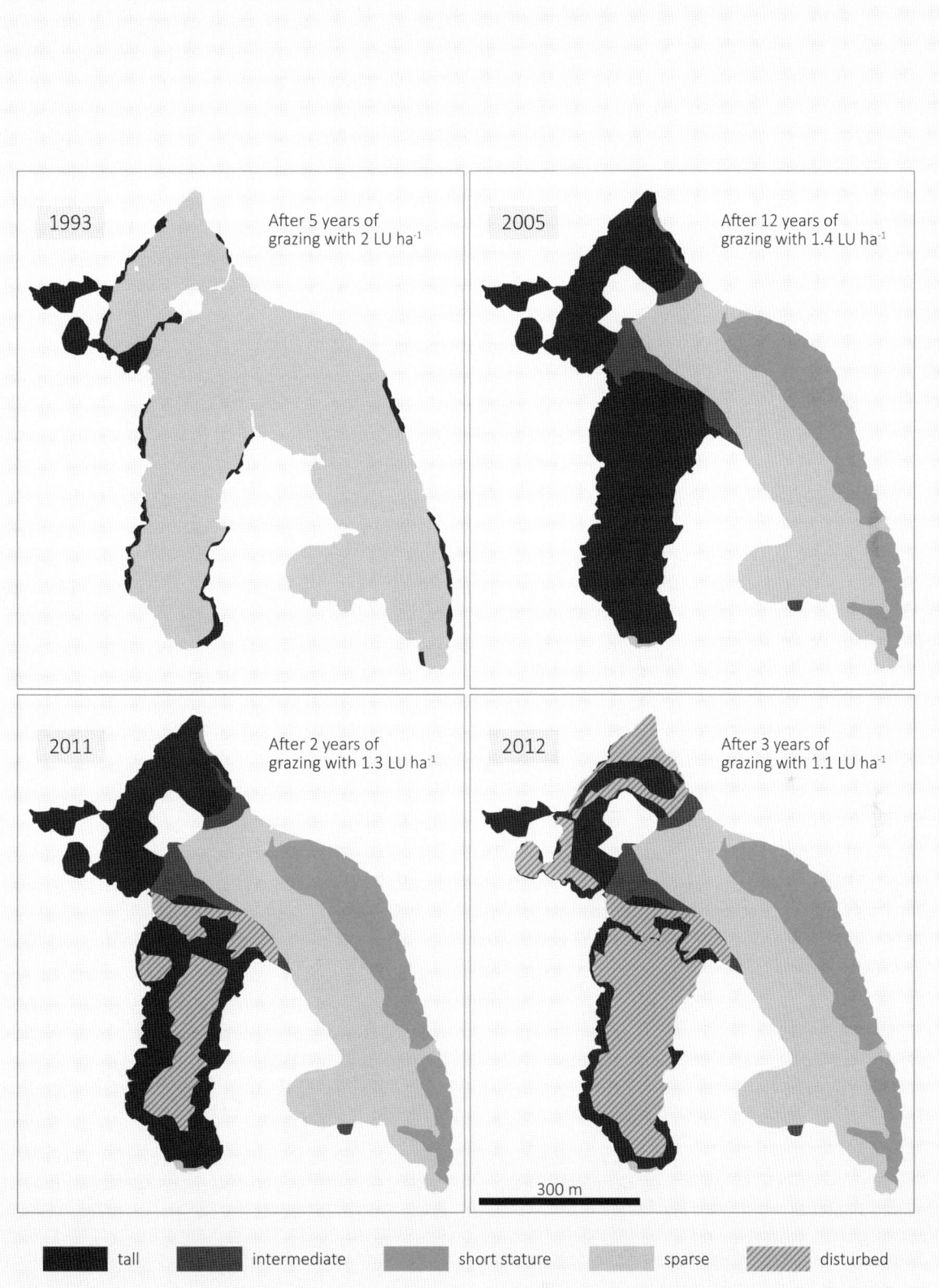

Figure 5.9: Change in reed bed cover in relation to grazing management on the island of Schmidt-Bülten. Before 2010, grazing with suckler cows led to the spread of Common Reed at the cost of salt meadow vegetation (compare 1993 and 2005). Grazing with Water BSuffalos (since 2010) suppressed the reed beds.

5.3 Local climate and hydrology

Andreas Wahren, Kristina Brust,
Ingo Dittrich & Frank Edom

As they are saturated with water for all of the year, mires differ distinctly from terrestrial ecosystems with respect to their climatic and hydrological services. Mires fulfil an important role in the regulation of landscape hydrology and have a marked effect on the local climate. When mires are drained, these functions are largely lost. In spite of irreversible changes within the peat body, re-establishing natural groundwater tables can restore the lost functions to some extent. Paludiculture can help to keep water in the landscape and mitigate weather extremes. The implementation of wet land use can contribute significantly to climate change adaptation.

5.3.1 Water and energy balance of natural mire ecosystems

Following the 1st axiom of peatland hydrology (Edom 2001), the water level must – on average, be close to the surface for peat to be able to accumulate. Mires either receive water from the atmosphere only (ombrotrophic mires) or from the atmosphere and their catchment area (minerotrophic mires). A comprehensive description of the water and energy balance of natural mires is given by Edom (2001).

The water balance of a mire site differs from that of terrestrial sites. The water supply in mires is, in general, sufficiently large during the whole year to cover evaporative losses. This results in high rates of evaporation that are close to the maximum possible rate. Averaged over the year, this high rate of evaporation means that – outside of the polar and boreal climate zones, mires have a cooling effect on the local atmosphere. This cooling effect is generally stronger in a drier and warmer environment. The evaporated water moves into the atmosphere and the moisture content of the air directly above the site increases. If this humid air is not replaced by wind, the lower saturation deficit causes a decrease in actual evaporation.

In particular in raised bogs, there are several feedback mechanisms that limit evaporation and regulate water losses from the bog surface. If water tables fall below a critical level in peatmoss dominated sites, capillary forces break off and evaporation decreases instantly. This self-regulation mechanism temporarily protects the bog surface from strong dessication. Other bog plants – usually with only superficial root systems, receive only little water once the water table drops, because capillary forces are ineffective in the large pores of the soil and the vegetation. This lack of water is one of the reasons why many plants of raised bogs and transition mires have xeromorphic traits (i.e. adaptations to dryness that reduce evapotranspiration by the plants). Thus, in mires dominated by peatmoss, evaporation strongly depends on the water table.

Undisturbed mires typically show a vertical gradient in pore size. Peat formation at the surface results in a high share of large pores in the upper peat layers. A large pore space means that relatively large gains or – particularly, losses of water result in only small changes in the water table. With increasing depth, the effective pore volume deceases rapidly because of advancing decomposition and mechanical collapse of the peat matrix. Fen peats have a larger percentage of small and medium sized pores in the upper peat layers. In general, the physical soil characteristics of fens are closer to those of mineral soils.

Natural mires retain water within the landscape, releasing it only slowly. The main loss of water occurs through evapotranspiration. Many mire types can only exist because of an external water supply (above- or below-ground catchment). They are in hydraulic contact with water catchments, and with their high water table they reduce water discharge from the landscape by creating a low hydraulic gradient. Water that cannot be retained within the peat body is discharged via water courses (direct discharge) or via horizontal and vertical infiltration paths into a groundwater aquifer. Depending on supply, annual infiltration may vary from less than 30 mm (ombrotrophic bog) to more than 60 mm (minerotrophic fen) (Edom 2001).

5.3.2 Water and energy balance of drained peatlands

Conventional land use on peatlands is typically associated with drainage – and thus with increased groundwater table depth, to allow (heavy) agricultural machinery access to the fields (Chapter 2.1). In addition, drainage changes the soil water balance with consequences for the requirements of cultivated plants.

Lowering the groundwater table significantly reduces the amount of water available for evaporation, which in turn results in a loss of the cooling effect that mires exert on their environment. Moreover, the thermal conductivity of the peat is greatly reduced. As the soil heat flux increases, temperature fluctuations within the upper soil layers increase as well, resulting in stronger heating in summer and a larger risk of freezing during the winter (Eggelsmann 1990b). The annual cycle of net radiation is no longer buffered. Although in some cases the radiative input is lower, because of a change in plant cover, this reduction does not compensate the increased heating of near-surface air layers.

Upon drainage some site characteristics are changed irreversibly. Schmidt et al. (1981) mention soil subsidence, peat mineralisation, shrinking and swelling of the peat as examples of processes triggered by drainage (Chapter 2.2). The 2nd axiom of peatland hydrology addresses the water holding capacity and hydraulic conductivity of the peat Edom (2001): 'Due to oxidation and pressure (changes in load) the hydraulic properties of the peat change: pore size and thus porosity, storage coefficient and hydraulic conductivity usually decrease.' Next to a rapid discharge via the drainage system, the water storage capacity is reduced when pore volume and peat are lost. Relatively small losses of water result in considerable lowering of the water table. A lowered water table induces a high wetting resistance in the dessicated peat layers, substantially reducing their ability to absorb water (LfUG 2007).

Drainage affects hydraulic gradients within the peatland and between the peatland and its catchment. The gradient with the surrounding catchment becomes steeper, which results in lowered water tables in the supplying aquifers. Moreover, groundwater recharge (of the aquifers) is reduced, because water is rapidly discharged by the drainage system instead. On the long term, the changes in the pore system of the peat are irreversible and after rewetting, re-swelling is only limited.

5.3.3 The effect of paludiculture on the local climate

In areas with sufficient water supply, most of the available radiative energy is used for evaporation. The evaporation of water requires a lot of energy: To evaporate 1 litre of water requires the same amount of energy as to heat 100 litres of water by 6 Kelvin. As a result, the air over wet areas heats up less than over dry areas and thus wet areas have a local cooling effect on the lower atmosphere. This cooling effect occurs mainly during periods of high solar radiation. In the temperate climate zone, the effect is negligible during winter, when solar radiation is low and vegetation not active, and differences between wet and dry areas are only small. Wet areas can even have a slight warming effect during winter, because they act as heat stores. Overall, temporal fluctuations in air temperature are more moderate over undisturbed peatlands compared with terrestrial ecosystems.

Paludiculture on previously drained areas implies that water tables are brought back to the surface again. As in undisturbed mires, high amounts of water are available for evaporation all through the year. As a result, rewetted peatlands will again have a cooling and

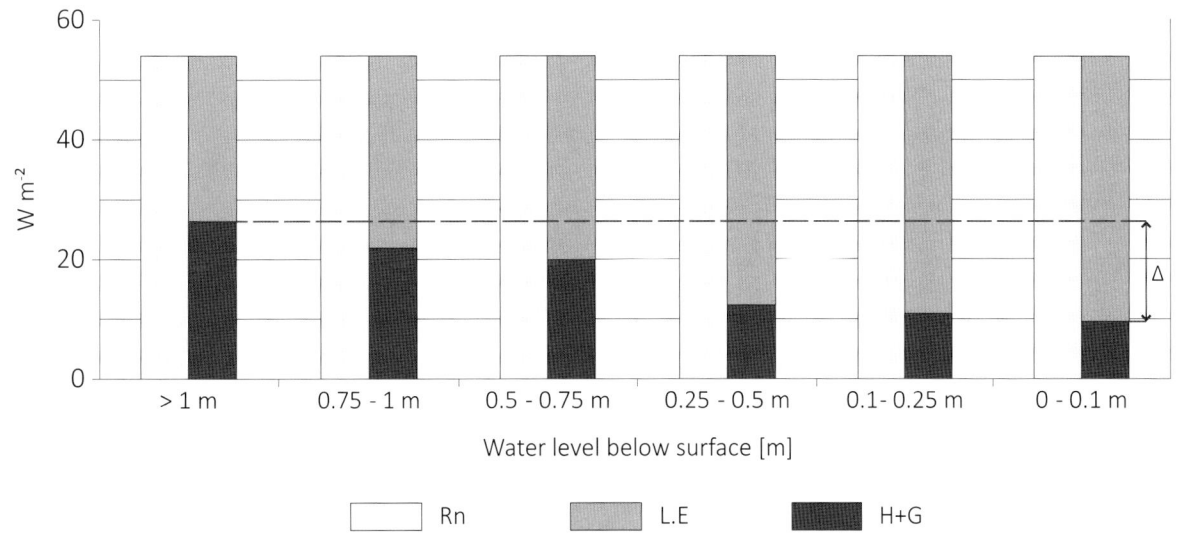

Figure 5.10: Energy balance of grassland on peat soil with different groundwater table depths. Rn: Site specific net-radiation; L.E: Energy used for evaporation; H+G: Energy transformed into air and ground heat. Δ refers to the cooling effect (17 W m^{-2}) of a wet site compared with a drained site. (calculations based on the climate station Marnitz 1997–2011).

humidifying effect and fluctuations in temperature will be moderated. Raising the water table from one metre below to close to the surface reduces the sensitive heat flux of a fen grassland by 17W m^{-2} (calculated for the climate station Marnitz 1997–2011, Figure 5.10). This cooling effect is more than five times larger than the anthropogenic warming effect since preindustrial times (3 W m^{-2}) (cf. IPCC 2013). Note, however, that the effect is geographically very limited and essentially restricted to the peatland itself.

5.3.4 The effect of paludiculture on groundwater stocks

In paludiculture fields retention in the peat body delays water losses from the landscape and high water tables maximize groundwater recharge. The absolute amount of recharge is only small, but in systematically drained regions the groundwater recharge from large scale paludiculture could increase depending on the hydrogenetic mire type.

In most cases, the reduced hydraulic gradient that is achieved by rewetting drained peatlands is more important for the groundwater stock. In minerotrophic peatlands, the reduced gradient delays discharge and a new dynamic equilibrium establishes between water flowing in and out of the aquifer. Thus, after rewetting, groundwater tables in the catchment are higher compared with the drained situation. The delayed loss of water from the landscape enables sites that had fallen dry after drainage to re-connect with the aquifer. The larger groundwater stock will also make the paludiculture fields themselves less vulnerable to droughts. During the planning stage of a paludiculture project, geo-hydraulic investigations should be carried out to determine whether a raised groundwater table could cause damage to neighbouring fields, buildings, infrastructure and possibly other ecosystems of conservation value.

5.3.5 The effect of paludiculture on flood retention

Near-natural percolation mires and raised bogs are temporarily able to retain significant amounts of water via 'bog breathing' (= oscillation: the swelling and shrinking of the peat body). This ability is lost when these peatlands are drained (Chapter 2.2). Peatlands that have steeper slopes (including spring mires) also loose their water holding capacity. Nonetheless, drained peatlands can still contribute to flood mitigation, because they can withstand longer periods of inundation without detriment – particularly if they are situated in riparian zones, water discharge areas, or at lake shores. In case of paludiculture, the peatland ideally is always water-logged and thus unable to take up additional water during floods. Once it has been drained, a peat body is no longer able to swell and shrink. Flood retention is then largely limited to aboveground stores, whose size depends on the relief of the terrain. Flood retention is of particular relevance in rewetted peatlands of riparian zones (e.g. abandoned polders), where the river is reconnected with its natural retention areas when dykes are removed.

On the one hand, paludiculture fields act as water retention areas whose capacity in relation to the river water level is determined by the terrain. If the water course is able to expand, flow velocity is reduced, which in turn reduces peak discharge downstream. On the other hand, no crops are damaged in case of flooding (e.g. crop failure) if the right paludiculture crops are selected. In addition, dykes can be dismantled or abandoned, and funds allocated to their maintenance can be used for other flood control measures. In this way, rewetting has the potential to contribute to the reduction of flood-related damage.

As they have lost their ability to 'breathe' (i.e. swell and shrink), many rewetted peatlands require that winter precipitation is withheld to prevent the groundwater table from falling below the ground surface during the vegetation season (Box 9.4). Such seasonal flooding certainly reduces the potential to mitigate precipitation extremes.

Box 5.7: European Water Framework Directive

Michael Trepel & Bettina Holsten

In the European Union, attitudes towards water resources have changed over the past decades. Before the European Water Framework Directive was introduced, water bodies were mainly regarded as discharge systems, whose efficiency had to be optimised. The Water Framework Directive (Directive 2000/60/EC) was adopted in December 2000 and came into force in 2002. It aims at achieving a good chemical status in all surface water bodies and at restoring a good ecological status in all natural water bodies, as well as a `good ecological potential´ in all heavily modified or artificial water bodies (Mohaupt et al. 2012). Biological evaluation methods were developed to assess the quality of water bodies (rivers, lakes, coastal waters) in terms of fish, macrozoobenthos, macrophytes, phytobenthos and diatoms (Mohaupt et al. 2012). Using these standard procedures, the ecological status of surface waters is assessed throughout Germany with monitoring programmes. At the same time, a federal working group established reference values for nutrients for different water categories (LAW-AO 2015). If the reference value is exceeded, the nutrient status of the water body must be improved. For water courses and lakes phosphorus is particularly problematic, whereas nitrogen is the main problem for the North and Baltic seas (LAW-AO 2015). Recently, there has been a growing recognition that different nutrients may be limiting during different parts of the year (Sterner 2008).

In the north German lowlands, phosphorus and nitrogen loads must be reduced to meet the targets of the Water Framework Directive for water courses, lakes and coastal waters (Holsten et al. 2012). The agreed target for the North Sea is an annual average nitrogen concentration of less than 2.8 mg of total nitrogen per litre, at the transition from freshwater to marine systems (e.g. at the Elbe River gauge Seemanshöft, Gade et al. 2011). For the rivers draining from Germany into the Baltic Sea, it was agreed that their mean annual total nitrogen concentration should not exceed 2.6 mg per litre (LAWA 2014).

Next to avoiding high nutrient loads from agriculture and improving sewage treatment in urban areas, it is also necessary to increase nutrient retention in the landscape, in order to meet the target and protect the marine environment. The EU Water Framework Directive explicitly mentions the restoration of wetlands, including rewetting of peatlands, as a possible measure.

In the decades ahead, the rewetting of peatlands should be a common priority measure for the protection of nature, climate and water resources, as there are considerable synergies to numerous European environmental directives (Table 5.4). Thus, public transfer payments towards fulfilment of the Directive could also be used to establish paludicultures.

Table 5.4: Overview of European Directives that are relevant to peatland management (Trepel 2012).

European Directive	Links to peatland management
European Water Framework Directive (EG 2000/105/EC)	• Nutrient retention • Conservation and development of aquatic biodiversity • Conservation and development of a near-natural hydromorphology
European Habitats Directive (EWG 1992/92/43/EEC)	• Nutrient retention • Conservation and development of aquatic biodiversity • Conservation and development of a near-natural hydromorphology
EU Floods Directive (EG 2007/60/EC)	• Conservation and development of a near-natural hydromorphology • Water retention areas • Preventive flood protection
EU Marine Strategy Framework Directive (EG 2008/56/EG)	• Nutrient retention

5.4 Nutrient balance and water pollution control

Bettina Holsten & Michael Trepel

The role of peatlands in controlling water pollution depends on:
- Their location in the catchment area;
- Their hydrogeological structure;
- Climate and terrain; and
- Land use (water resource management, agriculture or forestry).

Compared with drained peatlands, near-natural, wet sites have a more positive effect on the nutrient and water balance of a landscape. They often act as sinks of nutrients and are referred to as the 'kidneys' of a landscape. In contrast, drained peatlands contribute to increased nutrient loads to surface waters because of short flow paths and mineralisation of the peat. Their natural function as a nutrient sink can be partially restored by rewetting, thus contributing to the implementation of European targets in water pollution control (Box 5.7).

5.4.1 The effect of peatlands on the nutrient balance of the landscape

Peatlands have a high potential for retention and biochemical breakdown of nutrients brought in from their above- and belowground catchments (Howard-Williams 1985; Trepel 2004). This ability is strongly influenced by the hydrological and hydro-geological embedding in the landscape. The biogeochemical process of denitrification drives the retention of nitrogen, whereas sedimentation as a physical process enables the retention of phosphorus (Howard-Williams 1985). However, disturbed peatlands can become sources of nitrogen and phosphorus, and can pollute adjacent surface waters (Kieckbusch et al. 2006).

Denitrification is the conversion of nitrogen bound in nitrate (NO_3^-) into molecular nitrogen (N_2). If this process remains incomplete, nitrous oxide (N_2O, a potent greenhouse gas) may be formed and released into the atmosphere (Chapter 5.1). This chemical transformation removes nitrogen as a plant available nutrient from the soil or water and releases it to the atmosphere. Besides temperature and pH, denitrification is mainly driven by the availability of nitrate and of organic carbon compounds that serve as electron donors. As there is abundant carbon available in organic soils, denitrification can reduce nitrate loads from peatlands into surface waters (Davidsson et al. 2002). The second important process in peatlands is mineralisation, in which organic matter is decomposed and nitrate and ammonium are formed (nitrification and ammonification). The rate of mineralisation increases with drainage depth and both substances can be leached into adjacent water bodies (Grootjans et al. 1985, Schrautzer 2004). Raising the groundwater table reduces the decomposition of organic matter and consequently, the formation of nitrate and ammonium.

Retention of phosphorus (P) mainly takes place via sedimentation. This physical process occurs only in peatlands that are periodically flooded, such as valley mires and – to a lesser degree, terrestrialisation mires. Periodically flooded wetlands that are connected to the river network may retain about 1 kg of phosphorus per hectare and year (Trepel 2009).

Raising the groundwater table in drained peatlands bears the risk of a partial release of phosphorus, which accumulated in the upper peat layers when the land was under high intensity use (Zak et al. 2010). This risk should be evaluated prior to rewetting or introducing paludiculture. A release of phosphorus could induce a (further) decline in nutrient sensitive plant species, and also negatively affect downstream water bodies. Besides a rough model-based assessment of the risks of rewetting using default input parameters (for example with the web-based model 'P-Risiko'), for larger projects the phosphorus content of the topsoil should be measured to allow a more detailed estimate (Zak et al. 2011). Iron plays an important role in chemically binding phosphorus; therefore, the ratio of iron to phosphorus should be determined to assess the risk of phosphorus release. In general, any significant release of phosphorus will be limited to a short period after rewetting – within ten years, stable site conditions will have re-established with only low phosphorus release rates.

5.4.2 Effects of land use on the nutrient balance

Calculating nutrient balances is a common method to assess the role of peatlands and peatland sites in nutrient cycling. Such balances can be compiled at the site level – as is customary in agriculture, or at the landscape level – as is customary in landscape ecology and water resource management. The nitrogen balances of four typical fen peatland sites of the north German lowlands (each with different land use management) are shown in Table 5.5.

Wet, unfertilised and near-natural sites such as sedge reeds are well-balanced in their nitrogen dynamics. Input of nitrogen occurs via atmospheric deposition and mineralisation, via groundwater as well as via surface water if the site is inundated. The available nitrogen is almost entirely absorbed by the vegetation, although some of it is denitrified and even less is leached from the system. The difference between input and output (Table 5.5) – which is mainly bound in the plant biomass, is recycled back onto the site as litter and partly sequestered in peat (Rosenthal 1992).

5.4 Nutrient balance and water pollution control

Table 5.5: Nitrogen fluxes (in kg N ha^{-1} a^{-1}) of typical fen sites of the north German lowlands (after Trepel et al. 2000, Schleuß et al. 2002, Schrautzer 2004).

Vegetation type	Sedge	Moist meadow	Periodically inundated grassland	High intensity grassland
Land use management	None	1 cut	Grazing	3 cuts
Average drainage depth	-10 cm	-25 cm	-25 cm	-50 cm
Input				
Deposition	20	20	20	20
Fertilisation	0	0	60	160
Mineralisation	30	100	100	300
Output				
Biomass removal	0	80	60	200
Denitrification	20	30	50	80
N-Leaching	5	10	15	20
Sum of input	50	120	180	480
Sum of output	25	120	125	300
N-balance (input – output)	25	0	55	180
Evaluation	Peat formation	Peat loss	Peat loss and eutrophication	

Also slightly drained but unfertilised sites such as wet meadows are usually balanced in terms of nitrogen cycling. Drainage causes more mineralisation, which increases the availability of nitrogen and other nutrients for plant growth, compared with (near-) natural, wet sites. The nitrogen balance of wet meadows is usually neutral, because larger amounts of nitrogen are removed with the harvested biomass. However, in the long term, peat decomposition will cause noticeable subsidence and renewed drainage will be required. In terms of peat conservation, this form of land use is not sustainable (Trepel 2013).

As high intensity peat grasslands are usually deeply drained, mineralisation rates are high. In addition, fertilisers are often applied to these sites at the beginning of the growing season to stimulate plant growth. Nitrification, as well as nitrogen emissions via denitrification and leaching, increase with depth of drainage and the amount of fertiliser applied. As a result, the nitrogen balance is no longer neutral in these peatlands. Instead, peat decomposition and eutrophication increase (Kuntze 1988; Zeitz 1991).

Moreover, canals and drainage ditches collect inflowing groundwater and direct it into water courses via the shortest pathway, which drastically reduces denitrification. Furthermore, periodical inundation hardly occurs in drained peatlands. Water and nutrient exchange is disrupted by straightening and deepening of river channels, as well as by dykes and pumping stations. River water no longer enters the peatlands, and neither denitrification of the surface water nor sedimentation of phosphorus takes place anymore.

5.4.3 Effects of paludiculture on the nutrient balance

In terms of water resource protection, the implementation of paludiculture will have a positive effect on the nutrient balance of the site as raising water table will slow down peat mineralisation and in case of ceasing fertilisation by reducing nutrient input. As a consequence, nutrient leaching is reduced.

If peat soils are rewetted for paludiculture, the nutrient retention mechanisms of natural mires can be used. Removing drainage infrastructure and allowing groundwater to percolate or flow over the peat soil will rapidly result in increased rates of nitrate reduction. At sites fed by artesian water, nitrate is almost completely broken down already across short flow paths (Blicher-Mathiesen & Hoffmann 1999). Thus, afforestation of such sites with Alder trees for wood production can effectively contribute to removal of nitrate. Also, restoration of periodically flooded wetlands would enable nitrogen retention. The removal or retention efficiency for nitrate depends on the nitrate load from the catchment, and the duration and size of the flooding. In the agricultural landscape of the north German lowlands, small periodically flooded peatlands retain on average about 100 kg nitrogen per hectare and year (Trepel 2009). Thus, the restoration of hydrological site conditions by removing ditches and canals promotes the retention of nitrate. Reed beds for biomass production can be established in periodically inundated areas, where conditions support denitrification as well as sedimentation with associated phosphorus retention. Additional nutrients are removed by harvest (Chapter 3.5.3). Moreover, such sites with natural flood dynamics retain water and can serve as a flood protection (Chapter 5.3.5).

6 Economics of paludiculture

Large-scale implementation of paludiculture will only be successful if the practice is economically viable from a business point of view, and also benefits the national economy. Land users will quickly adopt paludiculture when they are convinced that the new land use practice is profitable for their enterprise. This Chapter explores the profitability of using biomass from paludiculture for construction purposes and energy generation, as well as for animal husbandry (Chapter 6.1). Furthermore, we discuss whether certification of the land use practice or its products could increase market opportunities and profitability (Chapter 6.2).

Paludiculture may be relevant for the regional and national economy. Particularly, if the biomass is processed nearby the production sites, paludiculture may increase regional added value (Chapter 6.3). With cost-benefit analysis we demonstrate that paludiculture has beneficial effects for social welfare also at the national scale (Chapter 6.4). Rewetting of drained peatland is a comparatively low-cost means of climate change mitigation and adaptation. Combining rewetting with continued productive land use will, in comparison to rewetting without subsequent paludiculture, result in considerably higher value creation and acceptance. The many good reasons for large-scale implementation of paludiculture are substantial arguments for providing incentives to change land use, on drained peatlands, to paludiculture.

6.1 Economic aspects of paludiculture on the farm level

Sabine Wichmann

Paludiculture is a new land use practice on wet peatland and has besides nature conservation management and traditional harvesting of litter for bedding and reed for thatch, so far not been established on a large scale. For acceptance and implementation, it is crucial that paludiculture is economically viable on the level of the individual enterprise.

Similarly to all agriculture, practicability and prospects of paludiculture depend on a variety of factors (Figure 6.1). Farm-specific factors include: actual business organisation, local production factors (acreage, machinery, labour) and their utilised capacity, as well as the nature of corporate management (readiness for innovation or taking risks and management skills). Willingness to change and profitability are strongly influenced by socially determined agri-political and legal conditions, which may exert both supporting and opposing effects (Chapter 7).

This Chapter focuses on the costs and revenues of harvesting gramineous (herbaceous) biomass from paludiculture for material and energy use, as well as for animal feed. The calculations do not take into account additional income, which could substantially influence profitability, such as EU agricultural subsidies, certification or payments for ecosystem services (Chapter 6.2, Chapter 7.3).

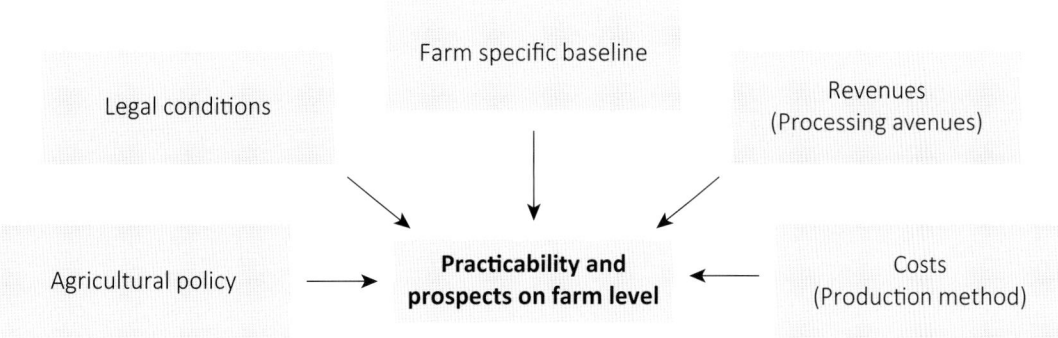

Figure 6.1: Selected factors influencing practicability and prospects of land use on farm level.

6 Economics of paludiculture

6.1.1 About the profitability of paludiculture

Sabine Wichmann

Reliable statements on the profitability of paludiculture require solid underlying data. There are long-term experiences and established markets with stable prices regarding some forms of biomass utilisation – e.g. reed for thatching. Other products and markets, however, are still being developed (e.g. for Water Buffalo, or cattail biomass as insulation and building material). Consequently, prices reflect a willingness to pay rather than a market value. Potential revenues can partially be assessed by comparison with prices of substitutes, such as straw as a substitute for reed as a fuel. Difficult to quantify are business-internal services, for instance, the utilisation of waste material for energy in a company that produces reed for thatch. Also, ecosystem services provided by paludiculture, such as the reduction of greenhouse gas emissions or the conservation of biodiversity, are not reflected.

Compared to conventional drainage-based arable and grassland use of peatlands, paludiculture is often associated with a low land use intensity, including little or no crop management, lower harvest frequency, permanent crops instead of crop rotation, and no – or only single-time, crop establishment. However, harvesting expenses are higher, because the limited bearing capacity of the soil results in high costs for the acquisition and maintenance of specialised machinery, and in lower acreage performance. Hardly any reliable and transferable data exist about acquisition and operation costs, because long-term experiences of large-scale applications are lacking. Most harvesting machines in use are individual solutions (Chapter 4.2) and their performance varies with biomass yield, soil bearing capacity and transport distance to the field margin. The efficiency of harvesting methods and logistics can still be considerably improved.

These facts explain why hardly any data on paludiculture economics are available in literature. The following calculations are therefore largely based on own field research and surveys amongst practitioners.

Setting up paludiculture as a new production branch on the farm level involves much uncertainty: technically and economically consolidated production methods and reliable standard data for cost calculations are lacking, just as a clear political and legal framework. These uncertainties have to be taken into account when considering which costs should be included, and how to deal with the wide range of input variables.

Assessing profitability takes place by various ways of cost accounting, which differ with respect to the cost positions included. In ordinary contribution margin accounting (CM I), for example, only those costs are subtracted from the revenues that vary in dependence of the production volume (i.e. 'variable costs'), such as expenses for fuel or fodder (Table 6.4). However, in practice the revenues must cover further production costs as well as proportional general and administrative costs.

To assess paludiculture as a new production method that requires capital investments, an extended cost accounting approach (multi-level contribution margin – CM II) is preferable. In addition to the variable costs,

Table 6.1: Comparison of the profitability of three ways of utilising reed-dominated vegetation based on A) deterministic calculations using realistic single values, and B) Monte Carlo simulation with information about range and risk of loss (Figure 6.4) (after Wichmann submitted).

		Chopped biomass for biogas (summer)	Round bales for direct combustion (winter)	Bundles for thach (winter)
A) Deterministic calculation				
Revenues from biomass sale	€ ha^{-1}	100	416	1,000
Direct costs*	€ ha^{-1}	0	0	0
Variable machinery costs	€ ha^{-1}	196	156	112
Labour costs	€ ha^{-1}	65	70	280
Ordinary contribution margin I / CM I	€ ha^{-1}	*-161*	*190*	*608*
Fixed machinery costs	€ ha^{-1}	162	199	125
*Multi-level contribution margin ** / CM II*	€ ha^{-1}	*-323*	*-10*	*483*
B) Monte-Carlo-simulation model				
Range of DAKfL**	€ ha^{-1}	-1,036–179	-287–677	-162–1,542
Mode (most frequent value) of CM II	€ ha^{-1}	-195	53	572
Risk of loss	%	*98*	*18*	*<1*

* In case of established reedbeds, no direct costs accrue.
** with attributable fixed costs.

also those fixed costs are taken into account that are attributable to the production method, such as the depreciation of specialised harvesting machines or the costs of direct marketing (Table 6.1, Box. 6.3). To assess long-term profitability, it is necessary to include all expenditures. Full-cost accounting is, however, only meaningful for farm specific cases because general, administrative and area-related costs vary considerably among farms.

The input variables for cost accountings are fixed neither for revenues nor for expenditures. Variation occurs in agriculture every year because of annually fluctuating yields and market prices, as well as differences in operational management.

These fluctuations and uncertainties due to the limited experience with paludiculture can be addressed by stochastic simulation (Chapter. 6.1.2) and sensitivity analysis (Chapter. 6.1.3) (Mußhoff & Hirschauer 2011).

6.1.2 Biomass supply for material and energy use

Sabine Wichmann

Biomass from paludiculture can be used in various ways (Chapter 3.4, Chapter 3.5). We investigated the profitability of material use through the example of the well-established use of reed bundles for thatch. For use as a fuel, we took winter harvested biomass supplied in bales for direct combustion, and summer harvested chopped biomass used in a biogas plant. Both technologies are mature and can be realised in decentralised plants with short transport distances.

All three production methods used reed (*Phragmites australis*) dominated stands and had comparable starting conditions. The deployed machinery was suitable for large-scale harvesting, and fitted out with tracks and the necessary accessory equipment.

The profitability of the three production methods and three ways of utilisation was tested using a Monte Carlo simulation (Wichmann submitted). This computer based stochastic simulation allows varying diverse input variables simultaneously whilst regarding their interdependencies.

The profitability was assessed based on a multi-level contribution margin accounting, so that the output is free of direct costs and operating costs (KTBL 2011, Table 6.1). In our case, direct costs – e.g. for seeds, did not accrue. Operating costs included the variable and fixed costs for machinery and labour. The costs included the biomass supply to the field margin, but excluded the costs for biomass transport and storage. The latter costs differ considerably among individual farms, but can be estimated based on experiences and standard data of conventional biomass logistics (Box 6.1).

For high quality reed for thatching one-year-old straight culms are harvested in parallel position. Traditionally, the harvest starts when frost has caused the shedding of the leaves. The start of harvest varies from year to year and from region to region, but usually begins in December or January. Aim is to harvest storable – i.e. dry (water content << 20%) reed. Ideal harvesting days are sunny, with freezing temperatures and windy weather – conditions that in many regions are rather rare. Hoar frost, snow or (after high water) ice attached to culms, impair harvesting.

This production method has as an advantage that a high quality product is supplied, for which well-established efficient harvesting technologies and logistic chains exist. A disadvantage is the small number of days with suitable harvest weather, which consequently limit yields and machine capacity utilisation. Furthermore, harvesting reed requires licences of nature conservation authorities, and is restricted in time (usually until 1st or 15th of March) and area (see Chapter 7.1).

Cost calculations were based on the use of a harvester on tracks and a separate tracked vehicle for transporting the large bales to the field margin (Colour pictures 70 and 71). The mowing device of the harvester was equipped with brushes to comb out waste (reed leaves, grasses, weeds). This reduces the workload considerably compared to harvest without pre-cleaning, which requires combing and shaking the bundles afterwards. The cost accountings include the bundling of the pre-processed reed for sale. The yield was taken to be 300 to 1,000 bundles per hectare, with revenues of 1.90 to 2.50 EUR per bundle, while 500 bundles and 2 EUR were assumed as being the most common values.

The Monte Carlo simulation for the multi-level contribution margin accounting resulted in a range of -162 to 1,542 EUR per hectare; a risk of less than 1% that the costs of producing saleable reed bundles will not be covered by the revenues (Wichmann submitted, Table 6.1, Figure 6.4).

Regarding thatching reed, regional and international markets exist (Box 3.3, Wichmann & Köbbing 2015). Bundles with a most common standard circumference of 60 cm are traded. To cover one square meter of roof, 10–12 bundles are required. A very high proportion of the thatching reed is imported: 80–90% in Germany (QSR 2008) and about 80% in the Netherlands (DeVries, pers. comm. March 2013). Main reasons for the large import rates are the limited domestic area that can be harvested and the limited time period over which harvest is permitted. Further reasons are the lack of investments in efficient machinery, and the high labour intensity required both for harvesting and processing reed, which intensifies the competition from low-wage exporting countries like Romania, Turkey and China. Nonetheless, there is demand for regional products of

Box 6.1: Supply costs for energy biomass: Comparing reed, straw and miscanthus

Sabine Wichmann

Within the project "Energy biomass from fen peatlands" (ENIM 2007–2009), a deterministic calculation of the supply costs of round bales of reed harvested in winter was carried out (Wichmann & Wichtmann 2009). Compared to the Monte-Carlo-simulation model (Table 6.1) average (e.g. for yields) and optimistic (e.g. for acquisition of machinery) values were assumed for the input variables.

The acreage performance of tracked vehicles is an influential but little known variable, hence a sensitivity analysis was conducted assuming a time requirement of 1, 1.5 and 2 h per ha (Figure 6.2). The costs of biomass transport and storage are shown in addition to harvesting costs. The harvests of existing (column 1 to 3) and newly established reedbeds (column 4) are compared.

In Table 6.2, the supply costs of reed biomass are related to the energy yield by multiplying biomass yield and energy content, and then compared to other gramineous energy sources. In comparison to straw and miscanthus, harvesting natural reedbeds seems to be not only competitive but the most cost-effective alternative (10.5, 12.2 and 13.8 EUR per MWh). This can be explained by the fact that next to the supply costs, the price of straw is also determined by its fertiliser value, whereas mis-

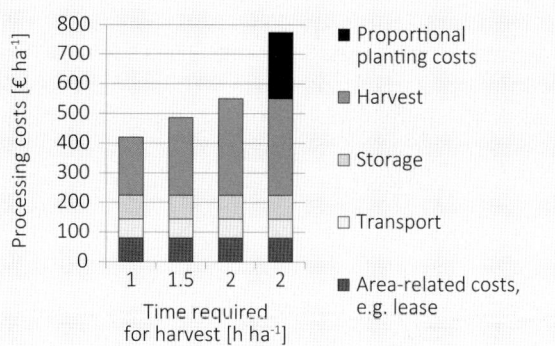

Figure 6.2: Sensitivity analysis of the dependency of supply costs from acreage performance of harvesting operations, and the planting costs (expressed per year), when reedbeds are newly established.

canthus requires crop establishment and management, including soil preparation, planting, fertilisation and weed control. Harvesting planted reedbeds (Box 6.2), under unfavourable conditions (high time requirement of 2 h ha^{-1}, medium yield of 8 t DM ha^{-1}) is, with 19.4 EUR per MWh, comparable to using straw with high market prices or miscanthus cultivated under unfavourable conditions (plot size 2 ha, yield 10 t DM ha^{-1}).

The supply costs between 10 and 20 EUR per MWh (Table 6.2) are rather low in comparison to the prices of other energy sources (Figure 6.3). The comparability is, however, limited. The determining factor for economic efficiency is, ultimately, the cost

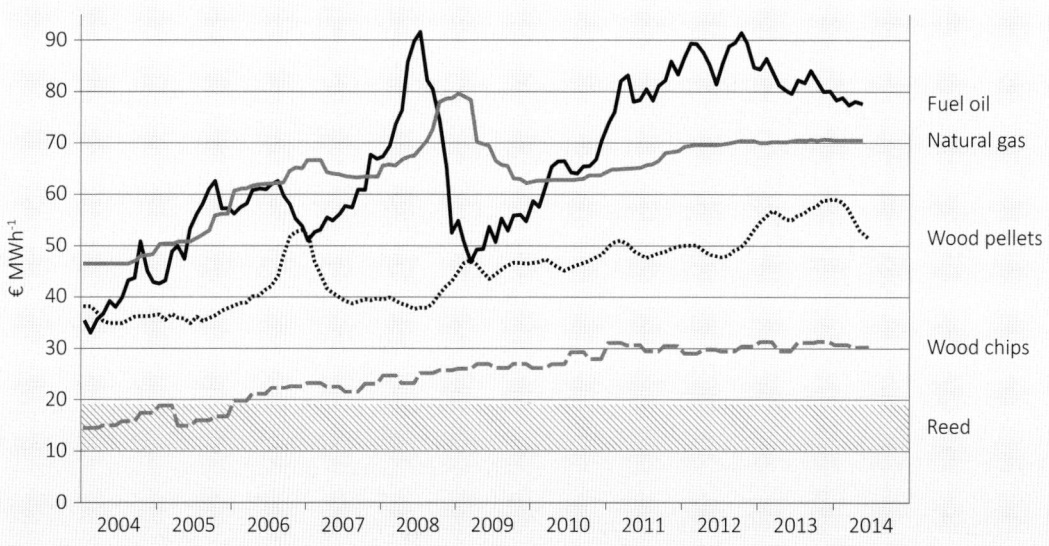

Figure 6.3: Development of fuel price of wood chips, wood pellets, natural gas and heating oil, compared to the price range calculated for reed (grey bar); after http://www.carmen-ev.de/infothek/preisindizes/holzpellets/graphiken.

6.1 Economic aspects of paludiculture on the farm level

per unit produced heat. Especially relevant are – apart from the fuel costs, the individual case specific investment and operating costs, which are considerably higher for biomass combustion plants than for heating oil. In addition, they are yet more expensive for gramineous crops than for wood fired boilers. Because of lower fuel costs, boilers fuelled with gramineous biomass are competitive compared with oil and gas boilers, but only if a high number of full load hours can be achieved by: a) optimising plant size and energy demand, and b) a heat consumption that is evenly distributed over the year.

Table 6.2: Biomass supply costs for reed, straw and miscanthus, delivered to the heating plant without subsidies (after Wichmann & Wichtmann 2009, KTBL 2006).

		Reed natural reedbed			Reed cultivation	Straw by-product		Miscanthus cultivation	
Variations		1 h ha^{-1}	1,5 h ha^{-1}	2 h ha^{-1}	2 h ha^{-1}	65 €	100 €	2 ha	20 ha
Biomass yield	t DM ha^{-1}	8	8	8	8	5	5	10	15
Energy yield*	MWh ha^{-1}	40	40	40	40	24	24	49	73
Costs	€ ha	420	486	550	774	325	500	1,120	1,134
	€ t^{-1} DM	53	61	69	97	65**	100**	112	76
	€ GJ^{-1}	2.9	3.4	3.8	5.4	3.8	5.8	6.4	4.3
	€ MWh^{-1}	10.5	12.2	13.8	19.4	13.5	20.8	22.9	15.5

* Lower heating value: Reed: 18 MJ kg^{-1}DM; Straw: 17.2 MJ kg^{-1} DM; Miscanthus: 17.6 MJ kg^{-1} DM.
** Market prices for straw (moisture content: 15%) calculated for dry mass.

high quality, especially since in recent years low-quality imported reed caused premature decay of thatched roofs (Behrens, personal communication March 2012). Regional tradition and identity (Chapter 6.3), established markets, and the economic superiority of material use compared to energy use, are clear advantages of using regional reed for thatch. Whereas cultivation of reed in rewetted fen peatlands was successfully demonstrated (Chapter 3.1.1, Box 6.2), further research is needed on whether high-quality reed can be produced on nutrient-enriched sites, and how.

For the winter harvest of round bales for combustion, we considered a procedure in which harvest is being carried out by a harvester on tracks, and bale removal, by a separate tracked transport vehicle (cf. Wichmann & Wichtmann (2009) (Colour picture 57). The advantages of this production method are: the opportunity to harvest sufficiently dry biomass (15–25% moisture content), and the good suitability of reed for direct combustion (Box 3.13). Disadvantages are: the limitation of harvest to dry weather conditions and the suboptimal harvesting machinery, which is in need of further development and improvement (Chapter 4.2).

The considered yield ranged from 5–15 t DM ha^{-1}, whereas the revenue – based on the price of straw (Figure 6.5) had a range of 45–110 EUR and a mode (most frequent value) of 65 EUR per t dry matter. The results varied from 287 to 677 EUR ha^{-1} and the probability of the production costs (including harvest and delivery to the field margin) not being covered by revenues was 18 % (Wichmann submitted, Figure 6.4, Table 6.1).

Whereas the Monte Carlo simulation reflects the full range of conditions occurring in practice, deterministic calculations with precise values allow the assessment of specific cases, and the comparison with the supply costs of straw and miscanthus (Box 6.1).

The combustion of bales is suitable to satisfy the heating demand of an individual farm (e.g. for piglet raising) and makes the enterprise less susceptible to rising prices of fossil fuels (Figure 6.3). The combustion of bales may also be used to cover the base load of decentralised heating plants in a district heating network (Box 8.4). Pelleting of biomass is useful in case of larger transport distances, or when biomass is to be used in individual furnaces, or also for co-firing in combined heat and power (CHP) plants. Pelleting is connected with additional costs, varying from 45 EUR (throughput of 1.5 t per hour, two-shift operation) to 72 EUR per t DM (0.7 t per hour, one-shift operation) for miscanthus (KTBL 2012).

Krasuska & Rosenqvist (2012) point out that energy producers in Poland pay lower prices for straw, miscanthus and similar biomass than for wood, because the fuel characteristics of gramineous biomass are considered to be problematic. On the one hand, the lower bulk and energy density and the higher ash content of gramineous fuels (Chapter 3.5.3) cause higher operational costs. On the other hand, the gramineous biomass is used there in unsuitable furnaces. In contrast, straw has been used successfully as a fuel for decades in specifically designed heating plants in Denmark. In this regard, Denmark is a leader in Europe, as 70% of all straw fuelled power and CHP output in the continent, is installed in this country. The technology is now so developed that one quarter of the straw produced in Denmark is used for energy production. For instance, one third of the straw used for energy production is utilised by small farms. This well-established combustion technology is also suitable for the combustion of reed.

To calculate the costs of summer harvested chopped biomass for biogas production, a two-step production method was considered, including mowing and swathing of the biomass, followed by uptake and transport to the field margin. In both steps, a tracked vehicle was used (Colour picture 61 and 62).

The advantages of this method are: the possibility to use a long harvesting season, and the option to combine biomass harvest with nature conservation objectives, such as keeping the landscape open, fostering target species, or nutrient removal. There are some experiences on the use of this method of biomass removal in the Netherlands (Jong et al. 2003), Germany and Poland. One of the disadvantages is that cellulose-rich biomass is not very well suited for fermentation in biogas plants. Conventional wet fermentation plants (Chapter 3.6.1) require additional technical adaptation to avoid the development of floating biomass layers, and also to achieve better digestion with shorter residence time and higher gas output. Currently, most biogas fermentation plants are not ready to process 'difficult' substrates. A 2012 survey of all biogas plant operators in the German federal state of Mecklenburg-West Pomerania (237 plants, response rate 19%) revealed that 53% of the plants were not equipped to process grass-like biomass, and only 29% of the operators were interested to utilise biomass from paludiculture (own unpublished data).

The cost accountings assumed a biomass yield of 3–8 t DM per ha, and a revenue of 0–35 EUR (mode: 10 EUR) per t of fresh biomass with a dry matter content of 30 to 50%. A revenue of zero euros was included

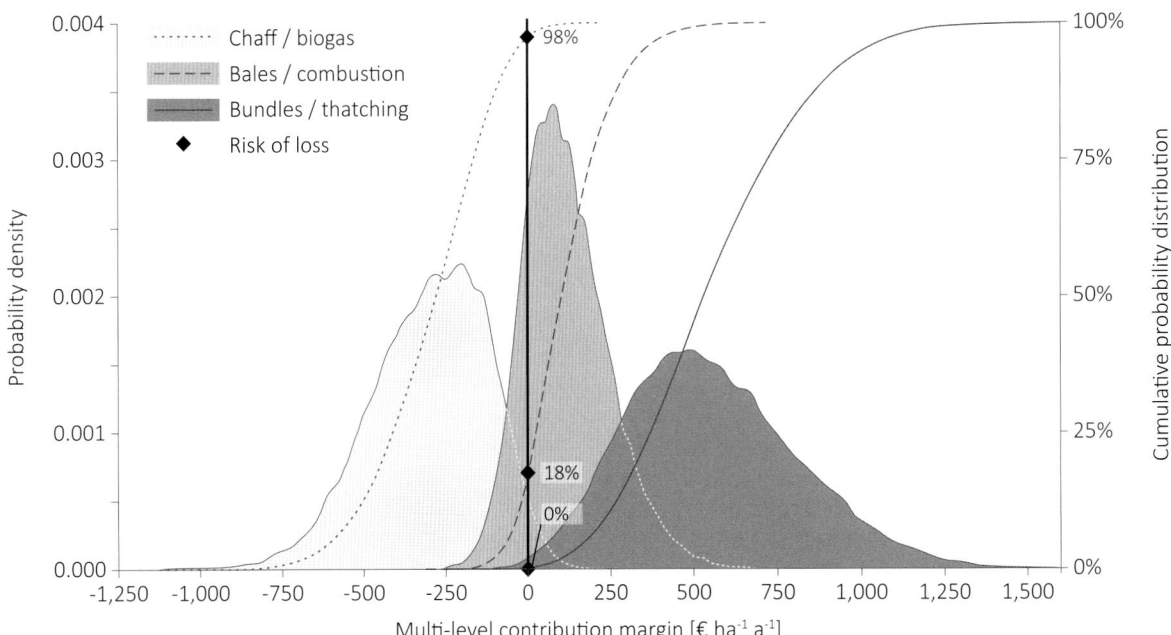

Figure 6.4: Results of a Monte-Carlo-simulation for three variants of reed biomass supply for material and energy use. The grey curves represent the probability distribution of the results of multi-level contribution margin accounting. The cumulative distribution function reflects the probability that the result takes a value of x or less, implying in the case of $x \leq 0$ a risk of loss. Number of iterations: 10,000 (after Wichmann submitted).

Box 6.2: Costs of establishing paludicultures

Sabine Wichmann

Current paludicultures generally consist of naturally established reedbeds. However, it is also possible to purposely create paludiculture fields in order to: a) accelerate the establishment of a harvestable vegetation stand after rewetting; b) introduce target species; or c) control the quantity and quality of the biomass via the choice of cultivars and genetically fixed characteristics.

Whereas in several pilot areas reed or cattail were manually or mechanically planted, in most cases precise cost assessments are not available. Reported costs of planting material range from 0.20 or 0.30 EUR (Tschoeltsch 2008) to 0.40 EUR (Schätzl et al. 2006) or 0.44 EUR (Dahms 2009) per seedling. Apart from propagated seedlings, reedbeds can also be established with rhizomes, which is the prevailing and most cost-effective method for establishing miscanthus plantations (0.16 EUR per piece, KTLB 2012).

Although Schätzl et al. (2006) and Dahms (2009) assumed similar costs for planting cattail and reed (including seedlings, machinery and labour costs), the overall investment and annuity differ clearly (Table 6.3). Crucial differences are in site preparation (e.g. cost-intensive construction of shallow ponds for cattail), planting density, and the expected useful life of the established stands over which the costs have to be depreciated. For the establishment of reedbeds by rhizomes (i.e. digging up, transport and planting, but excluding water management) Daatselaar et al. (2009) estimate overall costs of 8,000 EUR per ha, namely 400 EUR or 300 EUR per ha and year, for a lifespan of 20 or 30 years, respectively.

The targeted establishment of reedbeds makes sense only for winter harvested reed, because reed is weakened when cut during the growing season (Asaeda et al. 2006). For utilisation in summer (e.g. for biogas production), natural succession is preferred or, if necessary, the establishment of crops that also supply sustainable yields when regularly cut during the growing season. When winter harvest is intended, a targeted establishment of reedbeds (or cattail) may be economically sensible, if this enables an utilisation with higher revenue potential (biomass quality), or a faster cover of the investment costs by higher yields (biomass quantity). Such an evaluation can, however, be made only for a specific area under consideration of all alternatives.

Table 6.3: Costs of planting reed (Dahms 2009) and cattail (Schätzl et al. 2006).

		Reed	Cattail	
			Cost reduced version, 20 ha	Normal version, 4.7 ha
Earthworks (removal, filling-up, dam construction, excavation)	€ ha^{-1}	0	922	7,821
Pipes (irrigation)	€ ha^{-1}	0	414	1,079
Planting material	€ piece^{-1}	0.44	0.40	0.40
	Piece ha^{-1}	5,000	10,000	10,000
	€ ha^{-1}	2,200	4,000	4,000
Machinery and labour costs	€ ha^{-1}	580	500	500
Overall investment	€ ha^{-1}	2,780	5,836	13,400
Life span	years	30	10	10
Interest rate	%	4	5	5
Annuity	€ ha^{-1}	224 *	750	1,690

* Including area-related costs (e.g. lease, water board) of 80 € ha^{-1}.

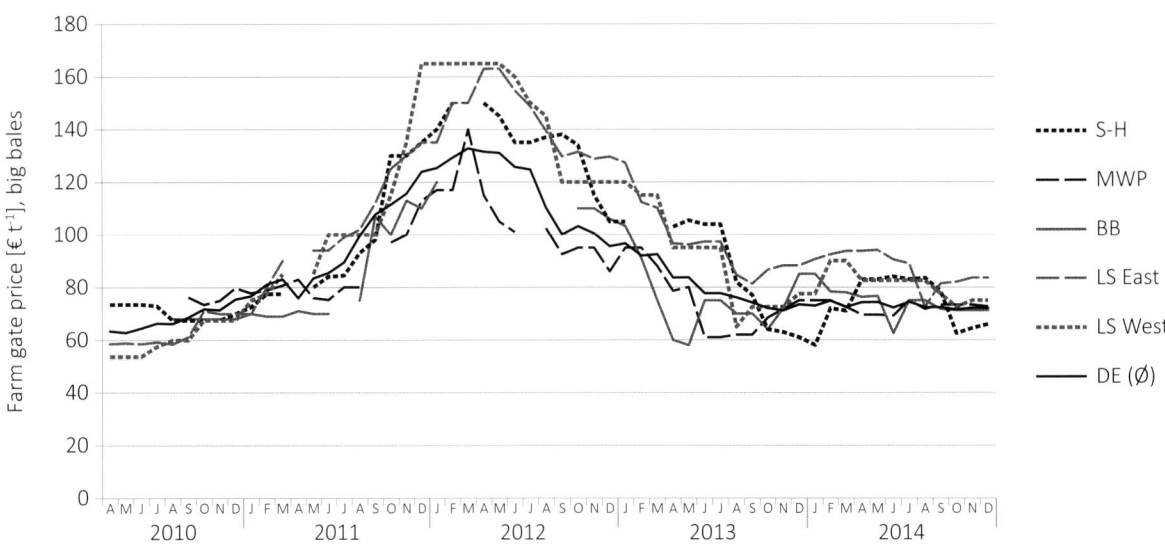

Figure 6.5: Development of prices of straw in Germany (DE) and selected federal states (MWP: Mecklenburg-West Pomerania, BB: Brandenburg, S-H: Schleswig-Holstein, LS: Lower Saxony). Data from BauernZeitung (2010–2014).

as a minimum level because: a) biomass derived from conservation management is not yet established as a substrate for biogas production; and b) currently, biomass from wet meadows is often considered as waste, and its disposal is subject to a fee. The Monte Carlo simulation gave a result between -1,036 and 179 EUR ha^{-1} a^{-1}, with a risk of 98% that the costs of harvesting and biomass removal are not covered by the revenues (Wichmann submitted, Figure 6.4, Table 6.1). The main reasons for the negative result are the high costs of harvesting and the low monetary value of the biomass. Even if the revenue per tonne of fresh biomass is fixed at the assumed maximum value of 35 EUR, the risk of a loss would be 67%. Therefore, using biomass as a substrate for biogas generation can usually be profitable only if agricultural subsidies or payments for nature conservation management can be claimed. If other objectives like habitat conservation or nutrient removal do not require summer mowing, winter harvest of biomass for direct combustion is preferential, because the risk of loss would only be 18%.

6.1.3 Biomass utilisation in animal farming

Weert Sweers & Jürgen Müller

Grazing is another option to utilise biomass of wet peatlands. This section assesses the economic efficiency of conventional cattle grazing (on a combination of wet peat and dry mineral soils), and of summer grazing of Water Buffalos without supplementary feeding (Chapter 3.3). To do so, deterministic calculations with fixed input parameters have been used. A sensitivity analysis (Mußhoff & Hirschauer 2011) was applied to the decisive variables 'market price' and 'fertility management'.

Both low-intensity grazing regimes had a livestock density lower than 1.4 livestock units per ha. The cost-efficiency of low-intensity grazing is – besides the revenues from meat production, mostly determined by subsidies from agri-environmental schemes. These subsidies are usually attached to the size of the agricultural area, whereas the costs and revenues from meat production depend on the number of animals. This implies that, with decreasing livestock density, the importance of landscape conservation subsidies increases. Our cost calculations do not include these subsidies.

The combined grazing of wet peat and dry mineral soils was done with cross-breed suckler cows (Fleckvieh x Limousin x Angus) (Chapter 3.3.2 and 5.2.3). Under normal conditions, with a calf mortality of 5%, 0.95 weaners (calves of 260 kg life weight) are produced as a saleable product per cow and year. The advantage of this farming method is the good marketability of weaners and cows on the existing market, so that no investments for direct marketing are needed.

As demand is currently high (2015), weaners of good quality are sold for 850 EUR (male) and 550 EUR (female), respectively. The grazing scheme thus yields a profit of 38 EUR per cow and year (Table 6.4), despite increased management costs. Apart from variable costs accounted for in the profit margin, further costs

6.1 Economic aspects of paludiculture on the farm level

need to be covered – e.g. for buildings, fodder storage, administration, and so on. It is not feasible to cover these costs without income from agri-environmental or nature conservation schemes, even when the market situation is favourable.

The cost-efficiency of summer grazing (Chapter 3.3.2) was assessed using the example of suckler cow farming with Water Buffalos (Colour picture 18). Similarly to the combined grazing scheme, the objective is to produce as much weaners as possible per cow and year. However, with Water Buffalos, a rate of 0.95 weaners per cow cannot be achieved because the calving interval lasts between 365 and 761 days (Golze 2004, Spindler 2008), which is significantly longer than with conventional cattle. With a calf mortality rate of 2.5%, only 0.47 weaners are produced per cow and year if the calving interval is 730 days. At 'Gut Darß' estate in Mecklenburg-West Pomerania, we achieved a calving interval of 430 days, which resulted in 0.82 weaners per cow and year. Higher reproduction rates cannot be attained at economically viable costs because of physiological limitations (longer pregnancies, longer anoestrous phases, and so on) (Perera 2011). This clearly shows that fertility management is of crucial importance in buffalo husbandry.

In contrast to conventional suckler cow farming, Water Buffalo weaners and cows are difficult to sell as there are no established markets to ensure continuous sales. Besides selling Water Buffalo meat, income may be generated by using Water Buffalo in nature or landscape conservation management, for which the species is well-adapted (Georgoudis et al. 1999, Hoffmann et al. 2010, Wiegleb & Krawczynski 2010; Box 5.6).

We calculated the economic efficiency of Water Buffalo farming for two reproduction variants and two meat prices (Box 6.3, Table 6.5). The outcome confirms the crucial importance of fertility management. Furthermore, it becomes clear that the market price decisively determines the results. Therefore, in planning Water Buffalo farming, prior market research is absolutely necessary. The amount of meat that can be sold – and the prices that realistically can be achieved, vary considerably within Germany, depending on the regional purchasing power. Moreover, it must be considered in advance how the less valuable cuts of meat (minced meat, soup meat) are going to be marketed, because these parts form a significant portion of the carcass. The Gut Darß is an example of how processing and marketing can be successfully organized (Box 6.3).

An alternative marketing option is the sales of breeding animals. During the pioneer phase of Water Buffalo farming in Germany, pregnant heifers were sold for up to 4,000 EUR and older cows for 3,000 EUR (Spindler 2008). Currently, the average price for pregnant heifers is 2,500 EUR and for older cows 2,000 EUR (Möhring 2012, pers. comm.). The market for breeding animals is a temporary one, which will inevitably be saturated as the herds build up. Therefore, the sale of breeding animals should be considered as a welcome additional business but not as a continuous source of income.

If direct marketing is the business objective, further branches of Water Buffalo farming may be developed – e.g. milk production. However, producing milk for diary specialities or cosmetics (Spindler 2008) is almost not feasible in paludiculture due to practical reasons (e.g. labour intensity, long distances, and udder hygiene). Milk production is only an option for parts of the herd on drier sites.

Table 6.4: Contribution margin of low-intensity suckler cow farming on humid meadows (calculations after Dietze & Heilmann 2010, including adapted selling conditions and additional expenditure for wet area husbandry).

Method		Suckler cow farming (grazing with supplemental roughage feeding)
Price level (3.30 € kg^{-1})		
Calving interval		365 days
Revenues from marketing Older cow 700 kg (350kg carcass weight)* Heifer (not yet gestate) Weaner	€ Cow^{-1} a^{-1} € Cow^{-1} a^{-1} € Cow^{-1} a^{-1}	192 21 577
Direct costs Suckler cow farming (feed, labour, litter, cattle tax, machine costs, veterinary costs, membership fees)	€ Cow^{-1} a^{-1}	-752
Contribution margin including labour costs (DB)	€ Cow^{-1} a^{-1}	38

* 1,155 € every six years.

Box 6.3: Water Buffalo farming at Gut Darß

Weert Sweers & Jürgen Müller

The agricultural enterprise Gut Darß has, since 2011, been monitoring Water Buffalo farming for landscape conservation management of coastal marshes, in order to study its economic efficiency. Gut Darß had already established direct marketing structures, enabling the use of existing marketing channels. The main reason to build up the Water Buffalo herd was landscape conservation management in the National Park Vorpommersche Boddenlandschaft.

Since the establishment of a herd of 15 Water Buffalo cows in 2007, offspring has raised the number of Water Buffalos grazing on the Darß peninsula to a total of 84 in 2012. The male offspring are completely sold via the direct marketing structure of the enterprise, whereas the female offspring are raised for breeding and to increase the herd stock. According to data from the slaughterhouse and the prices from professional sales achieved there, a profit margin of 1,179 EUR per cow is achievable (Table 6.5). However, from this amount should be drawn the proportional costs of marketing (farm shop, mobile shop, sales personnel).

Table 6.5: Contribution margin of low-intensity Water Buffalo farming with and without attributed fixed costs for storage and slaughter, plus different prices and levels of fertility management.

Variants		Water Buffalo (summer grazing)			
Price level		Optimistic (13.20 € kg^{-1})		Pessimistic (5.97 € kg^{-1})	
Calving interval (CI)		430 days	730 days	430 days	730 days
Revenues from marketing*					
Bull 650 kg (364 kg carcass weight)	€ Cow^{-1} a^{-1}	1,970	1,129	891	511
Heifer (2,500 € per animal)	€ Cow^{-1} a^{-1}	1,025	588	1,025	588
Direct costs					
Rearing Water BSuffalo cow (feed, labour, litter, cattle tax, machine costs, veterinary, membership fees) [1, 2, 3]	€ Cow^{-1} a^{-1}	-1,000	-1,000	-1,000	-1,000
Fattening bull (285th to 900th day) [1,2]	€ Cow^{-1} a^{-1}	-493	-282	-449	-258
Raising heifers (285th to 720th day) [1,2]	€ Cow^{-1} a^{-1}	-323	-185	-323	-185
Contribution margin including labour costs (CM)	**€ Cow^{-1} a^{-1}**	**1,179**	**250**	**144**	**-344**
Shelter[3]	€ Cow^{-1} a^{-1}	-15	-15	-15	-15
Refrigeration, freezer room and slaughtering, butchering [3]	€ Cow^{-1} a^{-1}	-178	-178	-178	-178
CM with attributable fixed costs subtracted	**€ Cow^{-1} a^{-1}**	**986**	**57**	**-49**	**-537**

* annual revenue based on proportionally produced calves (total price bull sales ~ € 4,800, with CI 730*0.235, with CI420*0.41).
[1] Spindler (2008).
[2] own data collection at Gut Darß.
[3] KTBL (2004).

6.1.4 Outlook

Sabine Wichmann

In the previous sections, specific costs and opportunities for income generation from paludiculture were discussed. Several ways of production were compared with regard to profitability, advantages and disadvantages. Additional revenues, such as agricultural subsidies and payments within conservation schemes, were not considered.

The calculations are partially based on assumptions because of lack of large-scale, long-term practice. Further development, testing and improvement of the entire production chain – from harvesting to the final consumer, are needed to obtain more reliable data. Various running research projects on the use of wetland biomass contribute to this objective (Wichtmann & Couwenberg 2013). Driving forces for innovation are: the increasing demand for biomass as a renewable raw material and a fuel, as well as the search for useful utilisation of biomass from nature conservation areas to cofinance the necessary management measures.

Besides the production methods presented above, there are further options for biomass use from paludiculture, for which, though, the energy balance and climatic effect have to be further assessed. A promising plant species is Cattail (*Typha* spp.), because of its superior qualities as a material for insulation and for lightweight construction (Chapter 3.1.3, Chapter 10.8). Similarly promising is the cultivation of Peatmoss (*Sphagnum* spp.) biomass, which offers a sustainable alternative to fossil peat in professional horticulture (see Chapter 3.1.6).

Especially those production methods in paludiculture that result in high quality end products – such as Water Buffalo farming or harvesting reed for thatch, are economically successful. Other practices are, under the current economic conditions, not profitable in farming terms when solely based on sales revenues. Admittedly, the costs of many agricultural activities on drained peatland, including suckler cow farming or mere mulching (as practised in regions with excess grassland), can also only be covered with subsidies. In contrast to drainage based agriculture, paludiculture provides, apart from market products, also some other ecosystem services, which are required by the community but have currently no 'market' (Chapter 2.4 and 5). The challenge is to establish market-like instruments in order to regard climate protection, water quality protection, and nature conservation as services provided by agriculture. Remuneration of ecosystem services associated with wet peatland utilisation may take place via agri-environmental or agri-climate programmes (Chapter 7.2), voluntary carbon markets (Chapter 7.3.3) or higher willingness to pay for certified products (Chapter 6.2). Establishing paludiculture beyond a niche market – as an alternative large-scale land use practice, requires the remuneration of the full range of goods and services produced.

6.2 Certification of biomass from paludiculture

Tobias Dahms & Achim Schäfer

Given the increased ecological awareness of consumers (Rückert-John et al. 2013), producers and suppliers are challenged to market environmental friendliness as a characteristic of their products. The use of labels indicating environmental friendliness to improve sales has experienced a growing interest in recent years. Labels claim that products are climate neutral, environmentally friendly, sustainable, regional or promote nature conservation. In Germany, the Ecolabel Index (Bic Room Inc. 2013) mentions about 100 different labels claiming the product to be environmentally friendly, even though only the larger labels were included.

German consumers use these labels, information by way of guidance, but there is not much trust in these claims (Rückert-John et al. 2013). To tackle this issue, committed suppliers use labels that are verified and certified by an independent third party. Such certificates guarantee a quantified environmental improvement or compliance with specific production or product standards (e.g. organic farming). The term 'certification' derives from Latin *certus* (certain, definite) and *facere* (to do, to make).

In the following sections we will discuss certification of production methods and products as a means to enforce environmental policy, both within the enterprise and as a marketing strategy. In Germany, the best known certificates in this field are the Euro-Leaf (the European Union organic label), the German environmental certificate Blauer Engel ("Blue Angel") and the FSC (Forest Stewardship Council) certificate for wood products.

6.2.1 Certification: benefits, costs and standards

Lack of information often prevents consumers to use environmental criteria to guide their choice of products. Consequently, suppliers have no incentives to improve the environmental performance of both their products and production methods (Karl & Orwat 1999). Certification significantly facilitates obtaining information about the environmental impact of a product. Certificates also simplify complex relationships to one specific attribute. From a consumer point of view, the guiding criteria to choose certified products are ethics, quality and health (Rückert-John et al. 2013). By buying certified products, consumers can increase the demand for environmentally friendly products and, in this way, influence methods of production (Boström & Klintman 2008).

For producers, certification is a means to differentiate – by environmental performance, their product from that of competitors, with the aims of gaining higher market shares or higher prices, to create a market niche or to improve the image of the producer (LeBlanc 2003, Rex & Baumann 2007, Boström & Klintman 2008).

Certification may also take place to satisfy an external demand from the buyer, the wholesaler, or legislation. Examples are: the MSC label for sustainable fishery of the Marine Stewardship Council (Boström & Klintman 2008); the International Sustainability and Carbon Certification (ICSS); and the European Union Renewable Energy Directive (EU RED, Chapter 6.2.3, EU 2009).

In environmental policy, certification is used as a market-based and consumer-oriented control tool. By certifying products, the otherwise not-visible environmentally relevant characteristics of a product play a role in market competition. If certified products have a competitive advantage, companies are incentivised to reduce or eliminate the negative environmental effects of their products. This encourages the internalisation of otherwise external costs, which arise when a product has negative impacts on the environment (LeBlanc 2003, Boström & Klintman 2008).

In order to create incentives for producers to adopt voluntary certification, certification needs to generate a net benefit at all the levels it affects. Thus, certification does not only concern the producer but also all the other participants down the production chain (Simula 1996), as a certified product must be traceable. Increased direct costs are incurred by the producer, because of the costs of verification and use of the label. In addition, indirect costs derived from adjusting the business's operational model arise, with possible yield losses resulting from this adjustment. Further costs arise from planning, control and reporting (Simula 1996). In the subsequent stages of the production chain, extra costs arise from tracking the product and keeping it separate from non-certified produce. Additionally the producer has to pay the certificating issuer's fees. The certification includes establishing the required standard(s), marketing the certificate, and monitoring and controlling the production chain. As a result, all producers involved in the production chain are, at first, burdened with additional costs, whereas the retailer receives the benefits of additional revenues and increased demand (Rotherham 2005, Boström & Klintman 2008).

Certification schemes must fulfil specific demands. The ISO standard 14020 (ISO 2001) lists a number of criteria that have to be met:
- Claims need to be exact, verifiable, correct and not misleading;
- Labelling content must be based on scientific methods;

Colour pictures 57–86

Colour picture 58: Harvester for chopped biomass (Sporova, Belarus; Wendelin Wichtmann, Oktober 2011).

Colour picture 60: First generation wetlandtrack: One step harvesting (Netherlands; Van Stipdonk, 2004).

Colour picture 57: Reed harvester (Sumo-Quaxi) with mounted baler (Lake Neusiedl, Austria; Jürgen Krail, 2012).

Colour picture 59: Harvester for reed bundles for thatch (Rozwarowo, Poland; Sabine Wichmann, 2013).

Colour pictures 57–86

Colour picture 61: First step: Mowing and deposition of biomass in swaths by a tracked vehicle equipped with a disc mower (Murchiner Wiesen, Mecklenburg-West Pomerania, Germany; Sabine Wichmann, 2011).

Colour picture 62: Second step: Uptake, chopping and removal of the biomass with chopper and trailer (Murchiner Wiesen, Mecklenburg-West Pomerania, Germany; Wendelin Wichtmann, 2011).

Colour picture 63: Second generation wetlandtrack: Two step harvesting by mowing and uptake of biomass with a self loading trailer (Netherlands; Van Stipdonk, 2010).

Colour picture 64: Tracked vehicle with a mower at front and a self loading trailer for mowing, deposition and uptake of biomass in one step (Relzower Wiesen, Mecklenburg-West Pomerania, Germany; Christian Saul, 2013).

Colour pictures 57–86

Colour picture 65: First step: Tracked vehicle with disc mower (Stiftung Naturschutz Schleswig-Holstein; Germany; K. Wiese, 2009).

Colour picture 66: Second step: Tracked vehicle with pulled baler equipped with tandem axle and twin tyres (Stiftung Naturschutz Schleswig-Holstein; K. Wiese, 2009).

Colour picture 67: Third step: Transport of big bales by a front mounted fork (Stiftung Naturschutz Schleswig-Holstein, Germany; K. Wiese, 2010).

Colour picture 68: Third step: Transport of bales by a trailer based on tracks (Biebrza, Poland; Ireneusz Mirowski, 2011).

Colour picture 70: Harvest of reed for thatch and separate transport vehicle for big bales (Insel Fehmarn, Schleswig-Holstein, Germany; Enno Franck, 2012).

Colour picture 72: Trailer with tracks and crane for loading bales (Poland; L. Mucha, 2011).

Colour picture 69: Loading chopped biomass with an excavator on a landing craft for river transport (Peene, Mecklenburg-West Pomerania, Germany; Christian Schröder, 2011).

Colour picture 71: Unloading big bales (reed for thatch) by a tipping bunker (Hanze Wetlands, Netherlands; Sabine Wichmann, March 2013).

Colour picture 73: Parallel harvesting with a tracked forage harvester and tracked transport vehicle with a mounted bunker that can be tipped (Netherlands; De Vries Cornjum, 2010).

Colour picture 74: Reloading of biomass from a tracked trailer to a standard wheeled trailer for road transport (Netherlands; De Vries Cornjum, 2012).

Colour picture 76: The Softrak can be adapted for various applications, in this case to harvest bog myrtle (*Myrica gale*) for essential oils (Great Britain; Loglogic, 2004).

Colour picture 75: 'Elbotel' harvester mounted with baler and crane for loading the bales, used for wetland management (Association de la Grande Cariçaie, Switzerland; C. Le Nédic, 2012).

Colour picture 78: Damaged sward on field track (Murchiner Wiesen, Mecklenburg-West Pomerania, Germany; Christian Schröder, September 2011).

Colour picture 80: Driving lanes from previous years may still be visible in the next vegetation period (Biebrza, Poland; Lars Lachmann, Juli 2011).

Colour picture 77: Damaged sward caused by frequent manoeuvering nearby the biomass unloading site (Murchiner Wiesen, Mecklenburg-West Pomerania, Germany; Christian Schröder, September 2011).

Colour picture 79: Destroyed access road to the harvesting site (Murchiner Wiesen, Mecklenburg-West Pomerania, Germany; Christian Schröder, September 2011).

Colour picture 82: Mobile road plates to protect sensitve areas (Libnower Mühlbach, Mecklenburg-West Pomerania, Germany; Christian Schröder, August 2011).

Colour picture 84: Stabilisation of access point to a mobile pontoon bridge with metal ramp and log wood (Brandenburg, Germany; Bernd Lukask, 2013).

Colour picture 81: Crosswise arranged fascines from bushes can stabilize frequently used tracks (Biebrza, Polen; Lars Lachmann, September 2009).

Colour picture 83: Mobile pontoon bridges can be used to cross ditches and small streams (Brandenburg, Germany; Bernd Lukask, 2013).

Colour picture 85: ´Traditional´ harvest by hand for field research in the Peene Valley (Redoute, Germany; Christian Schröder, August 2011).

Colour picture 86: Transport of small high density bales by small tractor and sleigh (Schadefähre Island; Neidhardt Krauss, August 2011).

6.2 Certification of biomass from paludiculture

- Principles and operational sequences must be transparent;
- All parts of the life cycle have to be taken into account;
- Administrative costs should be kept low;
- Preparation of the standard should be participatory;
- The consumer must be provided with relevant traceable background information;
- No obstacles for trade should result (demand of the World Trade Organisation WTO).

These requirements aim, on the one hand, to establish trust and credibility on the side of the consumer, and on the other hand, to ensure acceptance from the producers of certified products. Additional key components of environmental certification are: the articulation of moral and ethical criteria (Zedler et al. 2009), and the independence of issuer and auditor.

FCS (in forestry) and MSC (in fisheries) belong to the best known established certification systems in their respective fields. Their introduction was in both cases facilitated by increasing public awareness regarding the environmental impact of production, the growing pressure of non-governmental organisations (NGOs), and the consequent development of the certificates by the NGOs, in cooperation with wholesalers and providers of wood and fish products (Boström & Klintman 2008).

6.2.2 Certification of paludiculture

Certification of paludiculture methods and products is possible by creating a specific certificate or by using existing certificates. Both variants have advantages and disadvantages. A certificate focused on new land use may adopt particular criteria regarding the use of land. On the other hand, the efforts for administration and monitoring are correspondingly high, and only feasible if large areas are involved. For niche products without established market, it is truly difficult to achieve a competitive product differentiation based on environmental performance.

Within existing markets (i.e. for biomass) it is impossible to identify whether a product originates from a peatland or not. Compared to drained peatland, production on rewetted areas indeed provides substantial environmental benefits. However, certification of paludiculture not only would bring no benefits on these markets compared to simple labelling (see below), but also would entail additional costs. Therefore, there are almost no incentives for producers to establish voluntary certification schemes, and more specifically, for paludiculture.

Essential advantages of established certificates are: their well-known profile, the consumers' trust already gained, and the already existing infrastructure. Costs will be lower because of their wider application. Certificates that could be used are those that deal with production (e.g. FSC), the product (e.g. 'natureplus') or the origin (e.g. regional labels like 'MV tut gut'). However, most of the relevant certificates lack criteria regarding land use on organic soils. Therefore, an additional labelling of products, which points out the production in rewetted areas, may be advisable.

6.2.3 Paludiculture and established certificates

In the following section, we introduce examples of established certificates that are suitable for the marketing of paludiculture products. Paludiculture belongs to the domains of forestry (e.g. alder cultivation) and agriculture. In Germany, in forestry, there are two certificates suited for paludiculture, namely the international FSC certificate and the German Naturland Wald-Zertifikat. Their suitability resides in prohibiting both drainage and maintenance of drainage infrastructure. In contrast, in other countries (such as Finland) drainage-based land use is eligible for FSC certification, as the FSC criteria are controlled at national level.

For the production of food (e.g. Water Buffalo meat, spices) on wet peatland, the organic certificates granted by the European Union or by organic farming associations, such as Bioland or Biopark, are suitable. These certificates lack requirements for land use on organic soils, but do regulate elements of land use – e.g. the use of pesticides and fertiliser (EU 2008, Bioland e.V. 2011, Naturland 2012). The aforementioned certificates are, however, unsuitable to certify climate-friendly land use. For instance, there are many certificated organic farms that operate on drained peatlands.

Further examples of product-related certificates are the International Sustainability and Carbon Certification (ISCC) and the ISCC-Plus System (ISCC 2013). The ISCC was established to implement sustainability requirements for biofuels of the EU RED (EU 2009), and was extended to include food and material use of biomass (ISSC plus). Regarding organic soils, the EU RED and ISCC's only requirement is that the soils should not be drained deeper than the levels they had on January 1st 2008.

Examples of product-related certificates include the Blauer Engel (RAL 2013a) and the EU Ecolabel (RAL 2013b), which apply to various product groups, including wood chips, insulation materials, and horticultural growing media. Construction materials could also be certified with natureplus (natureplus e.V. 2013). The criteria of this certificate focus on health aspects but also consider, to some extent, the production of the raw materials. For example, natureplus certifies insulation materials and construction boards. The Cradle-

Table 6.6: Comparison of selected certificates/labels.

	Land use				Environmental label			Other			
	Forest Stewardship Council (FSC)	Naturland (forestry)	Organic farming (EU)	ISCC	Blauer Engel	EU Ecolabel	natureplus	Grünes Gas Label	Cradle2Cradle	Regionalsiegel
Target	Sustainable forestry	Sustainable forestry	Organic farming	Sustainable biomass	Environmentally friendly products	Environmentally friendly products	Sustainable building products	Environmentally friendly biogas	Closed material cycles	Regional products
Object of certification	Timber production (forestry)	Timber production (forestry)	Product and processing	Biomass production	Product and processing	Product and processing	Product (and processing)	Biogas	Product and processing	Product (and manufacturing)
Products	Timber	Timber	Food	Energy	Miscellaneous, e.g. wood products, construction products, wood chips	Miscellaneous, e.g. wood products, soil improvers, growing media	Construction materials		Material use e.g. insulation material, biodegradable plastic moulded parts, and so on	Miscellaneous, food, energy, construction materials
Used in	Global, regionally differentiated	Germany	Global	Global	Germany	European Union	Global	Germany	Global	Regional
Social criteria	Yes	No	No	Yes	Partially	Partially	Yes	No	Yes	
Environmental criteria	Yes	Yes	Yes	Yes	Partially	Partially	Partially	Yes	Yes	Partially
Paludiculture relevant criteria	No drainage and ditch maintenance	No drainage and ditch maintenance	-	No biomass from peatlands that were drained after 2008	-	No peat*	-	-	-	-
In paludiculture applicable to	Alder, willow	Alder, willow	Water Buffalo, Sweet Grass (*Hierochloe odorata*) etc.	Plant biomass	Alder, willow, reed, cattail	Alder, willow, peatmoss	Reed, cattail, alder, willow	Reed, sedges, Reed Canary Grass	Reed, cattail, alder, willow	Reed, alder, Water Buffalo
Source	www.fsc-deutschland.de	www.naturland.de	www.bio-siegel.de	www.iscc-system.org	www.blauer-engel.de	www.eu-ecolabel.de	www.natureplus.org	http://www.gruenerstrom-label.de/gruenes-gas-label/	www.c2ccertified.org	

2Cradle certificate for high-tech construction materials based on biomass, such as bio-plastics and moulds, as well as oriented strand (OSB) and medium-density fibre (MDF) boards (C2C 2011), aims at recycling the product or its raw materials at the end of its life cycle. Also, the 'Grünes Gas Label' ('green gas label') ("Grüner Strom Label e.V." 2013) for environmentally-friendly produced biogas, could be used for products from paludiculture.

Additionally, an origin-related labelling could be used for marketing paludiculture products. Regional labels are assigned to products from a specific geographic region. Examples are 'MV tut gut' (for products originating from the German federal state Mecklenburg-West Pomerania) or those of biosphere reserves (Kullmann 2012). Consumer acceptance for regional labels is particularly pronounced in the field of food products (Wirz & Klingmann 2012). Regional labels are also used for marketing regionally produced energy (SooNahe 2011). An overview of existing certificates is given in Table 6.6.

6.2.4 Shortcomings of established certificates

The claim of sustainable land use is incompatible with land use of drained peatlands. Nevertheless, established certificates do not concern enough (or not at all) the environmental impact of drained peatlands (Table 6.6). In Germany, exceptions are the forestry certificates Naturland Wald and FSC. In contrast, in countries with a large proportion of peatlands such as Finland, land use of drained peatlands is permitted – also under the FSC certificate (Finnish FSC Association 2010, FSC Working Group Germany 2012). Although EU regulation 834/2007 and the organic farming associations include the protection of soil, water quality and climate as their objectives (Bioland e.V. 2011, EU 2008, Naturland 2012), they refrain from establishing adequate criteria for peatlands.

The ISCC, which uses the European directives for sustainable biofuels (EU 2009), also permits the production of biomass on drained peatlands if these were drained before the 1st of January 2008 (see Joosten 2009, 2012). If the emissions resulting from these drained peatlands were adequately accounted for, the ISCC criteria could not be fulfilled at all. This is because greenhouse gas emissions from biomass produced in this way surpass those of equivalent fossil fuels many-fold (Couwenberg 2007, BioGrace 2013).

This contradiction is aggravated by the fact that the Renewables Directive (2009/28/EC) erroneously considers peatlands as a category similar to forests (Joosten 2012). If forests are turned into agricultural areas, the carbon loss from forest clearance is a once-off event (which also can be reverted, in the long term, by reforestation). However, on drained peatlands, greenhouse gas emissions occur perpetually until the total peat layer is lost. This also implies that, after rewetting, the greenhouse gas emissions are permanently reduced compared to the continuation of drainage based land use. Established certificates do not take adequately into account this typical peatland dynamics; this is a substantial failure. In order to justify a claim of site-adapted, sustainable land use, the established certificates need to be complemented by criteria for organic soils.

6.2.5 Conclusions and outlook

A separate certification system for paludiculture would be very costly without yielding corresponding benefits. Therefore, it is more sensible to pursue a simple labelling of products from rewetted peatlands, as well as to adapt established certification systems that include criteria for organic soils.

As a tool of environmental business management (Förtsch & Meinholz 2011), certificates have little effect on the use of organic soils because the proportion of 'green' consumers is rather small. Even successful certification systems (e.g. for organic products) get a market share of a few percentage points only (BÖLW 2012). Consequently, the requirements of soil and water quality protection, as well as nature conservation, must primarily be tackled by legislation.

Additionally, the particularities of organic soils need to be effectively considered in the established certificates, in order to correct the current deceptive claim of environmental integrity. An important step in this direction would be the inclusion of environmental impacts of land use in the life cycle assessments of the certificates, for which science must provide the data.

6.3 The creation of regional value

Till Holsten

Rural areas around the globe usually show disparity regarding their economic performance (OECD 2006). Germany is not an exception. Whereas some rural areas are prosperous, others are economically underdeveloped and subject to socio-economic stagnation and decline (Friedel & Spindler 2009, BMELV 2012). One reason for this unequal development is the declining importance of the agrarian economy. Thus, alternative opportunities of value creation, such as the generation of renewable energy (Chapter 6.4), are becoming an important economic driver to counteract economic decline in rural areas (George et al. 2009, Kosfeld & Gückelhorn 2012). The term 'Regional value creation' denotes the economic value that a community or a region achieves (Hirschl et al. 2011a), and represents the macroeconomic performance of enterprises within a region, where resources are generated, transformed and refined into products. Each stage of processing creates added value (Hirschl et al. 2011b), as do all entrepreneurial activities within the region. This contributes to regional value creation.

6.3.1 Added value creation by paludiculture

In the long term, drainage of peatlands causes a loss of agricultural area and thus, a loss of productive land. Paludiculture particularly contributes to create regional value by ensuring the availability of productive peatlands – also on the long term. The best way to create regional value from paludiculture is to process the biomass for material use within the region. Material processing normally requires multiple working steps, and the final product can be sold for higher market prices (FNR 2006). Such process increases value creation and creates more employment per tonne of biomass. To some extent, the biomass can be used in a multiple sequential way (BMELV 2009a). The range of material paludiculture products is diverse (Chapter 3.4), whereas the examples illustrate that products from paludiculture biomass are marketable (see Box 3.3).

Apart from regional and national product sales via wholesalers and retailers, there is the option of direct marketing – i.e. selling Water Buffalo meat in farm shops and to restaurants; selling insulation material to house builders (Box 3.7); or selling reed directly to local thatchers.

There are several methods to use paludiculture biomass for energy generation (Chapter 3.5). The rising demand for biomass fuels (Ammermann & Mengel 2011) fosters the sales of paludiculture biomass. The proportion of renewable energy produced from biomass in Germany was 67.4% in 2011. Biomass reached a share of 8.4% of total German energy consumption, and the trend is rising. Most renewable energy (43.7%) is used for the generation of heat, for which biomass is particularly well suited. Only 12.3% of the biomass is used for electricity generation (BMU 2012). Thus, to increase the utilisation of biomass for energy, further biomass resources have to be made available. Paludiculture provides opportunities to diversify the spectrum of energy biomass, and may in this way contribute to the decentralisation of the energy sector. The increasing tendency to re-establish shared ownership in the energy sector (i.e. buy-back of communal electricity grids, energy trade, regional energy generation, Hirschl et al. 2011a) provides additional potential for value creation from paludiculture – e.g. integrated in energy autarchic bioenergy villages or 'renewable energy regions'.

Independently of the production chain, positive economic effects from paludiculture can be expected. Revenues, wages of employees and taxes will contribute directly to value creation (Hirschl et al. 2011a). Furthermore, opportunities for indirect added value creation exist by linking paludiculture to other branches, such as ecotourism or environmental education.

6.3.2 Development of regional added value chains

Key players driving the implementation of paludiculture are crucial to enforce its contribution to sustainable regional development (Chapter 8.2). These key players may come from nature conservation, agriculture or politics, and play an important role as initiators. Cooperation amongst regional businesses not only facilitates the creation of regional products but also contributes to building a regional identity, which involves the identification with regional products (Henckel et al. 2010). An example of this dynamics is the distinctive large-scale use of reed for thatch in northern Germany. The formation of networks and links amongst partners may increase the acceptance towards implementation of paludiculture, while strengthening stability and cooperation (Böcher 2009, Chapter 8.4). Therefore, all stakeholders of paludiculture are important, although their respective interests may differ. Profitability is the key motivation for foresters, farmers and all further participants of an added value chain (crafts, energy producers, butchers, restaurants and so on). Most important issues for conservationists are biodiversity and the natural functions of peatlands, whereas for tourism, the attractiveness of the landscape prevails.

6.3.3 Perspectives for regional rural development

In order to maintain the rural area as a place for living and also for working, politics aim at strengthening economically disadvantaged regions and supporting the

region specific potentials (BMELV 2009b, BBSR 2012). The main tool of the Common Agricultural Policy (CAP) to promote economic development and support rural areas in the European Union (EU) is the 'European Agricultural Fund for Rural Development' (EAFRD). The EAFRD comprises the following six funding priorities for rural development (EU 2013):
1. Knowledge transfer and innovation in agriculture, forestry, and rural areas;
2. Enhancing farm viability and competitiveness of all types of agriculture in all regions, as well as promoting innovative farm technologies and the sustainable management of forests;
3. Promoting food chain organisation, including processing and marketing of agricultural products, animal welfare and risk management in agriculture;
4. Restoring, preserving and enhancing ecosystems related to agriculture and forestry;
5. Resource efficiency and supporting the shift towards a low carbon and climate resilient economy in the agriculture, food and forestry sectors;
6. Social inclusion, poverty reduction, and economic development in rural areas (EU 2013).

Most EAFRD funding schemes focus on the agricultural and silvicultural sector (Grajewski et al. 2011). This matter specifically relates to the focal points 2 to 5. Consequently, companies that practise paludiculture are eligible for these schemes if they are recognised as agricultural or forestry enterprises. In this context, it has to be taken into account that rewetted peatlands are often regarded as unsuitable for agriculture and forestry, and are therefore not eligible for EU direct payments (Kowatsch et al. 2008; Chapter 7.2). Nonetheless, funding opportunities for non-agricultural enterprises have been strengthened for the funding period 2014–2020 (Grajewski et al. 2011).

For instance, funding opportunities may exist for small enterprises that process paludiculture products, in terms of establishment, product diversification, and production facility expansion (Art. 19 EAFRD Council Regulation). Projects eligible in the area of tourist-oriented environment could, for example, include the establishment of an information centre (e.g. on sustainable peatland use) or a signposted trail with guided tours on demand (Art. 20 section 1 e, Art. 35 section 2 b EAFRD Council Regulation). In general, the design of EAFRD eligible projects and the grantable portion differ among the European Union regions, – as it happens in Germany among the federal states (Fährmann et al. 2008). This variation is a result of trying to meet region-specific needs and capabilities.

For instance, the predominantly rural federal state of Mecklenburg-West Pomerania features a characteristic northern German architecture, which is based on the traditional utilisation of reed for thatching. This architectonic feature can be seen in preserved historic village centres, manors, farm houses and fisherman cottages. Paludiculture projects would be eligible for EAFRD funding to support the preservation and revival of the cultural heritage as a regional characteristic, and also, as a unique touristic selling point, which can be used in the framework of integrative regional development projects.

Since 2014, a common strategic framework (CSF), which combines the funding strategies of the EU member states (EU 2013), facilitates paludiculture projects and enterprises – e.g. by supporting business start-ups.

6.3.4 Conclusions

Paludiculture is a sustainable type of land use on rewetted peatlands, which contributes to the diversification of forestry and agriculture. Paludiculture links economic, environmental and social interests. Its potential for regional economic development arises from the restricted transportability of paludiculture biomass. The latter is, however, also its main constraint.

Cooperation between stakeholders from various economic and public sectors (agriculture and forestry, nature conservation, tourism, energy sector) may benefit the greater good that paludiculture can be, even though it might not be the ideal solution for every individual stakeholder (e.g. utilisation of rewetted peatlands is not always preferred by conservationists, see Chapter 8.2). Regional development through paludiculture therefore requires a willingness to cooperate (Chapter 8.4). Land users and biomass processing industries will need financial incentives that stimulate them to switch to paludiculture. Especially in structurally weak regions, like Northeast Germany with its abundance of drained fen peatlands, paludiculture may provide alternative sources of income. Networks of local stakeholders plus existing industrial, technical and social infrastructures should be involved in implementing new ways of value creation, as well as in establishing synergies to contribute to a sustainable and cohesive regional economic development.

6.4 Welfare aspects of land use on peatland

Achim Schäfer

The economic effects of various land use options for peatland can be quantified by cost-benefit analysis (Freeman 2007, Mishan & Quah 2007). Cost-benefit analysis constitutes a useful decision-support tool to answer the question whether and how future land use on peatlands should take place. Under the current conditions (Chapter 6.4.1), abandonment of drained peatlands is unlikely (Schroeder 2012). Accordingly, possible alternatives with respect to peatland use include:
- Continuation of current drainage based agricultural land use;
- Rewetting without land use;
- Rewetting and paludiculture.

In order to quantify the effect of peatland rewetting on public welfare, all costs and benefits of these land use alternatives have to be considered comprehensively – i.e. including the external effects, as well as the marketable and unmarketable goods and services (Turner et al. 2000).

6.4.1 The economic assessment of ecosystem services

In cost-benefit analysis, costs are approached as 'opportunity costs' – i.e. as the depletion of goods and services that no longer can be provided because of the delivery of a good or service targeted alternatively. Cost-benefit analysis aims at recording all the effects of the alternatives, independently of who eventually bears the costs.

The consumption of common production factors such as labour, soil and capital is calculated by way of prices oriented to marginal costs (Hampicke 2009). Such prices are an important indicator of scarcity. Market pricing steers the scarce production factors into those areas where they are most urgently needed or provide the largest benefits.

Figure 6.6 demonstrates the relation between ecosystem functions and welfare relevant ecosystem services. Ecosystem functions encompass all ecosystem performance that can be objectively recorded and quantified. For instance, the climate effect of peatlands can be quantified by assessing the greenhouse gases fluxes in tons of CO_2 equivalents per year using the GEST approach (Chapter 5.1, Box 5.4).

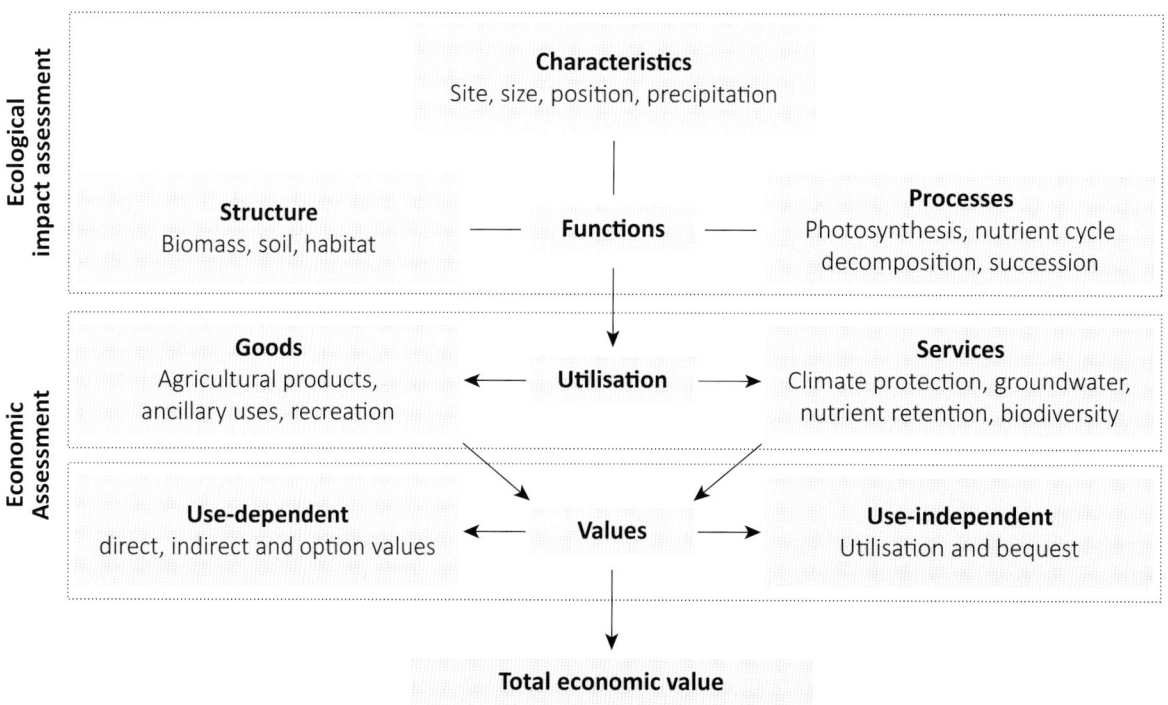

Figure 6.6: Economic evaluation of economic goods and services (based on Turner et al. 2000).

6.4 Welfare aspects of land use on peatland

Only by the provision of goods and services that satisfy human needs, ecosystem functions become ecosystem services (e.g. climate change mitigation), which can be evaluated economically. An economic valuation of ecosystem goods and services is only possible if these are scarce in the sense that they are not available anytime and everywhere in the desired quality and quantity. Ecosystem services often originate from the interaction between production factors (labour, soil, capital) and ecosystem functions to create useful goods (e.g. agricultural products) or services (e.g. GHG emission reduction by rewetting of peatlands) (Figure 6.6).

Benefits and costs of ecosystem services can be expressed in monetary terms via prices and preferences (Turner et al. 2008). The value of agricultural products can be quantified with market price based 'value added' methodologies (Ewers & Schulz 1982).

The added value is the operating farm income, that is, the amount of money that is available to pay all factor income (e.g. wages, interest, profits and rents) used in a company for the production of a good or service. In the national accounts, the operating income of a company corresponds to its contribution to the national product (Chapter 6.4.3). The operating income is a more appropriate indicator of the long-term result of an enterprise, because – in contrast to the cost accounting approaches (e.g. marginal costing, Chapter 6.1), it also includes the medium and long-term fixed costs. The operating income can also be compared with indicators of national accounting (cf. BMELV 2011).

Because market prices of agricultural products are distorted by subsidies and the exclusion of external effects, these distortions need to be corrected when welfare effects are to be adequately displayed.

The monetisation of ecosystem services (e.g. water quality protection) is often based on the costs of alternative methods to prevent damage (Schäfer 2012). The public appreciation for non-marketable goods and services (i.e. biodiversity) can be assessed by socio-economic valuation approaches – e.g. by analysing people's willingness to pay (Endres & Holm-Müller 1997, Turner et al. 2008, Hanley & Barbier 2009). These approaches try to establish how people value an ecosystem service through assessing their preferences for improvements in the provision of that ecosystem service. The preference of the population for a specific measure to improve the ecosystem service is assessed by means of interviews and statistical analysis. This allows the economic evaluation of even small ('marginal') changes in utility.

Socio-economic methods can also be used to quantify non-use values – e.g. the willingness to pay for the existence of entities or processes for their own sake ('existence values'). The combination of all these value components form the total economic value (Pearce 1993, Turner et al. 2000, Figure 6.6).

A comprehensive cost-benefit analysis must consider two elements: a) the effects on the production of marketable products; b) the welfare effects of changes of other ecosystem services and biodiversity, as induced by the change in land use. The latter include non-marketable goods and ecosystem services that

Table 6.7: Decision making framework of cost-benefit analysis.

Alternative	Ecosystem services	Method	Indicator	Evaluation
Continuation of current land use	Production	Method of value creation, Net value creation	€ ha^{-1} a^{-1}	Comparison of damage costs and net value creation
	Climate	Negative external effects, Damage costs	€ t^{-1} CO$_2$ e a^{-1}	Costs of failed political regulation
Rewetting without land use	Climate	Positive external effects, Avoided damage costs and mitigation costs	€ t^{-1} CO$_2$ e a^{-1}	Comparison of abatement costs
	Biodiversity	Analysis of willingness to pay	€ ha^{-1} a^{-1}	Willingness to pay plus additional ecological and social benefits
Rewetting and paludiculture	Production and climate	Positive external effects, Avoided damage costs and mitigation costs	€ t^{-1} CO$_2$ e a^{-1}	Comparison of abatement costs
	Biodiversity	Analysis of willingness to pay	€ ha^{-1} a^{-1}	Willingness to pay plus additional ecological and social benefits

were or have become rare (such as the politically determined limited sink capacity of the atmosphere for greenhouse gases) as far as they are affected by the production of agricultural products.

After analysing the initial situation (Chapter 6.4.2), an economic valuation of the alternatives listed above is conducted using methods and indicators mentioned in Table 6.7.

6.4.2 Analysing current use

The economic valuation of alternatives does not merely consist of comparing costs and benefits. The benefits derived from economic goods and ecosystem services depend to a large degree on the extent and distribution of rights of disposition and use – i.e. rights that refer to purchase, use, encumbrance and transfer of goods. These include the existence of property rights for peatland use, and the analysis of how regulatory and economic instruments affect the sustainability of both peatland use and the associated ecosystem services.

In economics, 'ownership' is understood as a set of property rights that regulate the relationships between people regarding scarce resources. Property rights are usually divided into usus, usus fructus and abusus, which go back to Roman law (Furubotn & Pejovich 1974). The distribution and determination of property rights with respect to peatland use has developed over time. In former times, peatlands were drained with great efforts to be used for agricultural production (Succow 2001b, Chapter 2.1). At that time, people were not aware of the resulting GHG emissions or of the association between GHG emissions and climate change. As politics now wants to limit the GHG sink capacity of the atmosphere (the 'two degree target'), an ecosystem service that formerly could be used free of charge has now become a scarce ecosystem service.

The example shows that property rights as well as scarcity may change over time. Currently, it seems that the right of use inherent to ownership permits the use of the soil (*usus*), the collection of yields resulting from land use (*usus fructus*), and the destructive use of the peat body (*abusus*). In this context, the historically developed property rights can be seen as a root cause of the currently unsustainable use of peatlands (Schäfer 2004).

Drainage of peatlands has resulted in an all-embracing allocation of property rights, which institutionally ensured the production of marketable agricultural products.

From an economical point of view, the existence of private land ownership is justified by the assumptions that private land use provides the best incentive for its economic use, and that self-interest will ensure the sustainability of the production base (Held & Nutzinger 1998). The unsustainable use of peatlands, however, convincingly demonstrates that the supposed self-interest that increases efficiency does not exist: the GHG emissions that result from drainage-based agricultural peatland use generate costs that have to be borne by society. Three centuries after the introduction of the sustainability concept in forestry, and several decades after the 1992 United Nations Earth Summit in Rio de Janeiro demanded a critical evaluation of the institutional framework regarding sustainable land use, the negative impacts of drainage based agriculture on peatland (Chapter 2.3) must no longer be accepted as tolerable side effects whose costs have to be borne by society. From an economical point of view, it seems more sensible to reduce these negative external effects, as far as possible, by rewetting. This would yield important ecosystem services for climate change mitigation, which are welfare relevant because they avoid the accrual of costs caused by damage related to climate change.

An important contribution for climate change mitigation is to preserve the existing carbon stock of organic soils. In Germany, the basic principles of good agricultural practice (German: Gute fachliche Praxis), namely the rights of use of soil, are regulated by the German Federal Soil Protection Act (German: Bundesbodenschutzgesetz). This Act prescribes the 'sustainable protection of soil fertility and productivity as a natural resource' (article 17, paragraph 2) but proves to be 'a toothless tiger'. The available regulatory tools are unable to ensure sufficient protection of soil functions or the preservation of the carbon stock in organic soils. The reference to downstream agro-political instruments (i.e. Frenz 2004) reveals the helplessness, and illustrates the urgent regulatory need for action (Chapter 7.3).

6.4.3 Value added and costs of conventional agricultural land use

When considering the continuation of current land use, two questions arise, namely which benefits may result from current agricultural production, and what its negative environmental impacts are.

When peatlands are rewetted and land use is abandoned, as has happened in Germany over the last 20 years, society foregoes the economic value of agricultural products. Simultaneously, however, the costs avoided by the change in land use, which would otherwise accrue from environmental damage, can be accounted.

In the following section, the benefits of agricultural production are quantified by calculating the value added based on market prices (Ewers & Schulz 1982). In order to assess the economic benefits of production, the business economic results of a representative selection of agricultural enterprises from the Ger-

6.4 Welfare aspects of land use on peatland

Table 6.8: Average value creation of dairy and cattle farming in Mecklenburg-West Pomerania 2006–2012 (after LFA M-V, lfd.).

	Dairy farms	Cattle farms
	EUR ha^{-1} a^{-1}	
Value creation	658	341
allowances and subsidies	376	406
Net value creation	282	-65

man national test enterprise network (BMLV 2011) are used. Peat soils in Mecklenburg-West Pomerania are predominantly used by mixed farms for dairy cow and grazing stock farming. The intensively operating dairy farms require deeply drained land (soil moisture class 2+/-, for more information on soil moisture classes see Box 5.3). Beef producing and grazing stock enterprises (with predominantly calves, weaners and suckler cows) operate on sites where water conditions are easy to regulate (soil moisture class 3+/-). Low-intensity farming (permanent pasture with low cattle stocking) is practised by agricultural enterprises on hydrologically challenging peatland sites (soil moisture classes 4+ to 3+) in order to maintain eligibility for EU direct payments (Müller & Heilmann 2011).

In Mecklenburg-West Pomerania, high-intensity dairy farming enterprises generated over the years 2006 to 2012 an average annual business income of 658 EUR per hectare of agriculturally used area (AUA). When conducting an economic cost-benefit analysis, the received allowances and subsidies of 376 EUR per ha of AUA need to be subtracted because these unilateral transfer payments, to a large extent, do not involve a depletion of scarce production factors – i.e. are paid without return of services. Without transfer payments, dairy farms generated an added value of 282 EUR per ha of AUA. Pasture based, beef producing farms generated an average business income of 341 EUR per ha of AUA, but received slightly higher allowances and subsidies than dairy farming (406 EUR per ha of AUA, Table 6.8), resulting – without state allocated compensatory payments – in a negative value creation (LFA MV, continuous).

In contrast to business economic direct cost calculation – which in case of low intensity pasture based beef farming definitely may result in positive profit margins, these figures demonstrate that profitability of low intensity farming on peatland is only achieved with state-allocated payments (predominantly direct payments and agri-environmental programmes). This decisive dependency on temporary transfer payments discourages the agricultural enterprises from making the necessary continuous investments, and forces them to live on their own capital. Unlike transfer payments that are granted without service in return, payments within agri-environmental programmes promote biodiversity conservation or remunerate services that are requested by society (Chapter 7). A distinction between pure transfer payments and economically relevant expenditures for products or efforts is not possible with the aggregate data of the national test enterprise network.

In economic cost-benefit analysis, also the external costs of agricultural production need to be taken into consideration. These costs caused on the farm level have to be borne by third parties or society. Conventional agricultural land use on peatlands accrues external costs, whereas elsewhere in the economy, considerable financial means are required to reach climate protection targets and to avoid significant welfare losses for future generations, which result from climate change related damage. An economic evaluation compares the added value of conventional peatland agriculture with the costs of climate-related damage. Damage costs reflect the current value of future climate-related damage that results from nowadays emission of greenhouse gases (Umweltbundesamt 2007). These costs are calculated as marginal costs, which result from the emission of an additional ton CO_2 equivalent.

The methodological convention used by the German Federal Environment Agency (German: Umwelt-

Table 6.9: Damage costs of agricultural production methods on drained fen peatland sites (after Schäfer 2012).

Use category[1]	GHG emissions [2]	Damage costs [3]	Value creation [4]
	t CO_2 e ha^{-1} a^{-1}	EUR ha^{-1} a^{-1}	EUR ha^{-1} a^{-1}
Dairy farms	24.0	1,920	282
Cattle farms	8.5–15.0	680–1,200	-65

1) Use categories after Müller & Heilmann 2011.
2) GHG-emissions after Couwenberg et al. 2008.
3) Following Umweltbundesamt 2012.
4) Calculation after LFA Mecklenburg-Vorpommern, lfd. (Table 6.8).

bundesamt) to economically evaluate environmental damage requires to include external costs in economic cost-benefit analysis (Umweltbundesamt 2007). For the damage costs, a default value of 80 EUR per ton CO_2 equivalent is proposed (Umweltbundesamt 2012). Using this value, the minimum climate-related damage costs of conventional agriculture on peatland amount to 680 to 1,920 EUR per hectare and year (Table 6.9, Schäfer 2009). The costs may be significantly higher because the applied evaluation models only fragmentarily reflect singular extreme events with major consequences as well as the costs of climate change adaptation (Tol 2005, Stern 2007). Furthermore, it needs to be considered that conventional peatland agriculture causes further external costs (e.g. by water pollution) that so far have not been monetised.

In accordance with the 'polluter pays' principle, the environmental costs that result from economic activities are to be attributed to the polluter. Whereas in other economic sectors suitable instruments (e.g. ecotax, trade of certificates) lead to an internalisation of external costs, the environmentally damaging agricultural use of peatlands is even encouraged by the government through transfer payments. This does not refer to the often invoked contrast between economy and ecology nor to market failure (Fritsch et al. 2007), but it refers clearly to a policy that turns the 'polluter pays' principle upside down.

The relevant problem for economic policy is that the existing system of incentives enables land users to generate private income that is not based on an adequate service in return. As the provision of ecological services is so far remunerated only to a limited extent, no incentives exist for the land user to change over to more environmentally-friendly land use (Chapter 8.3).

The question how we should use peatlands in future can already partially be answered: in a more sustainable way than until now and also economically more efficient. In Chapter 7.3 we address how these goals could be achieved.

6.4.4 Costs and financial requirements of peatland conservation and paludiculture

For the empirical part of a cost-benefit analysis, the costs and benefits associated with a change in land use need to be collected or calculated.

At this point, attention must be drawn to the difference between an economic cost-benefit analysis and a fiscal approach. Financial requirements are an important parameter both in public budget planning and political decision making to shape structural and agri-environmental measures on the basis of real costs. On this basis, public financial expenditures are calculated for implementing peatland conservation measures under the prevailing institutional, political and economic conditions.

The previous calculations were based on a solid database (e.g. BMELV 2011) and on numerous studies from climate impact research (e.g. Tol 2005, Stern 2007). Conversely, the assessment of investments required for rewetting and implementing paludiculture is based on anticipatory (ex ante) model calculations with a high degree of uncertainty. However, such uncertainty will be insignificant compared to other public investments (e.g. several major building works in a number of German cities such as Berlin, Hamburg, Stuttgart and elsewhere).

As a result of limited information, the financial requirements for rewetting and subsequent cultivation

Table 6.10: Cost elements of rewetting without land use and with paludiculture. VOB = German Construction Tendering and Contract Regulations.

Costs	Rewetting		Quantification
	without land use	with paludiculture	
Coordination and communication	✓	✓	Difficult
Research and investigations	✓	✓	VOB
Planning of measures	✓	✓	VOB
Land consolidation		?	Partially costs of procedure
Land acquisition	✓	?	Possible loss of agricultural land after rewetting
Planning procedure/planning permission	✓	?	Difficult
Conservation assessment	✓	?	Difficult
Constructional implementation	✓	✓	Construction and dismantling costs
Infrastructure development		✓	Construction costs for regular paludiculture
Efficiency control, monitoring	✓	✓	Generally possible

can only be partially assessed. Some cost elements listed in Table 6.10 are not directly effecting the expenses, or are difficult to quantify – i.e. preparatory coordination, project- associated communication and the statutory tasks of authorities (e.g. nature conservation administration).

Financial requirements for paludiculture include, next to hydrological measures for rewetting, also additional costs for: infrastructure development (e.g. for constructing tracks, turning facilities and crossings, Chapter 9.4); maintenance (e.g. of dikes and pumping stations); and the development of the necessary machinery.

Peat soil conservation and paludiculture will generally be implemented in areas that are currently in agricultural use and, after rewetting, will involve low intensity land use under wet site conditions. Part of the rewetted areas may be too deeply inundated to allow any use, including paludiculture. Besides, possible limitations to land use in neighbouring areas must be taken into account.

To calculate the financial requirements for rewetting (section 6.4.1) one can resort to approximate figures from nature conservation projects. In Mecklenburg-West Pomerania, land procurement was largely (circa 60%) paid by public funds from both the agriculture budget of the EU (75%) and the federal state itself (25%). Land procurement costs include, besides the purchase of the land, the early termination of lease. In recent years, the purchase value of agricultural land in Germany has significantly increased, from on average 3,695 EUR between 1995 and 2004 to 5,961 EUR per ha of AUA in the period 2005-2010. In 2012, the purchase value was already 12,675 EUR per hectare (Statistisches Amt MV 2013).

A number of factors, namely the substantial increase in land prices, the changed level of co-financing in the current EU funding period, and the constraints of public budgets have prompted the question of how land procurement for rewetting could be financed in the future and justified towards the tax payer. This issue brings forth another question: Should land owners be rewarded for decades of predatory exploitation of natural capital, and for devastating productive land for private profits? Questions like these and others of intergenerational relevance cannot be answered within the framework of a cost-benefit analysis, but relate to political issues of distribution that have to be addressed in socio-political and public debate.

Particularly relevant for cost-benefit analysis are the planning and construction costs. In 31 rewetting projects implemented in Mecklenburg-West Pomerania before 2003, the planning and construction costs strongly depended on project size (Table 6.11). With a sector-specific price increase of on average 2% per annum (Statistisches Jahrbuch, BMELV 2009c), the rewetting costs are currently estimated at between 1,243 and 1,815 EUR per hectare. This price corresponds very well with the costs of a project in the Blindower Wiesen located in the Ucker river valley near Prenzlau (federal state of Brandenburg, Germany). After rewetting an area of 130 ha, an additional 192 ha of reedbeds bordering the initial area will be taken in use. The costs of track building, hydrologic and civil engineering, installing a pontoon bridge, and further costs (including initial construction and compensation payments) amount to in total 376,911 EUR or 1,171–2,897 EUR per ha (Chapter 9.3, Hasch et al. 2012). The planning and construction costs for rewetting polder Kieve within the MoorFutures® programme were 2,100 EUR per ha.

Table 6.11: Planning and construction costs of rewetting measures.

German federal state	Project	Target	Area size [ha]	Costs [EUR ha^{-1}]
Mecklenburg-West Pomerania	14 Projects < 100 ha [1]	Peatland conservation	837	1,815
Mecklenburg-West Pomerania	17 Projects > 100 ha [1]	Peatland conservation	6,498	1,243
Mecklenburg-West Pomerania	Polder Kieve [2]	Climate change mitigation	65	2,100
Brandenburg	Blindower Wiesen [3]	Paludiculture	130	2,897
Mecklenburg-West Pomerania	47 Projects [4]	Peatland conservation	15,456	2,824
Brandenburg	6 Projects >20 ha [5]	Peatland conservation	658	3,198
	All Projects		**23,644**	**2,363**

1) Based on data of the Coordination Centre for Peatland Conservation for rewetting projects before 2003.
2) Based on data of the Landgesellschaft Mecklenburg-Vorpommern.
3) Hasch et al. 2012.
4) After Landgesellschaft MV 2011 for rewetting projects between 2009–2011.
5) Based on data of Biota, Institute for Ecological Research and Planning LLC, Bützow.

Table 6.12: Comparison of CO_2-abatement costs.

Action	Source	Mitigation costs [EUR t⁻¹ CO_2]
Rewetting without agricultural land use	Own calculations	10–15
Rewetting with reed cultivation	Schäfer 1999	0
Rewetting with alder cultivation	Schäfer & Joosten 2005	0–2
Hydro- and wind power generation	Geiger et al. 2004	22–70
Wood chips (from short rotation coppice)	Isermeyer et al. 2008	-11
Co-firing of straw (in coal-fired heating plant)	Isermeyer et al. 2008	45
Maize based biogas (with and without thermal use)	Isermeyer et al. 2008	267–378
Maize based biogas (fed into the grid and combined heat and power plant)	Isermeyer et al. 2008	316

Between 2009 and 2011, 47 peatland rewetting projects – covering in total 15,456 ha, were implemented by the 'Landgesellschaft Mecklenburg-Vorpommern', in the framework of the 'Directive for the promotion of the sustainable development of water bodies and wetland habitats in Mecklenburg-West Pomerania (FöRiGeF)' (Landgesellschaft MV 2011). With an investment volume of 54.1 million EUR, the costs were on average 2,824 EUR per ha. Costs of land procurement cannot be extracted from these aggregated numbers. Envisaged costs of six rewetting projects in the German federal state of Brandenburg that have not been completed yet (Table 6.11) are based on estimates or on average construction prices from the years 2012 and 2013.

It is not yet possible to determine exactly to what extent large-scale paludicultures can be economically self-supporting. Various model calculations from pilot projects (based on direct costing or full cost-accounting) showed that paludiculture can be profitable without continuous funding (Schäfer 1999, Schäfer 2004, Schäfer & Joosten 2005, Schätzl et al. 2006, Wichmann & Wichtmann 2009, Wichtmann & Wichtmann 2011). Further business economic calculations for viable production methods were conducted in Chapter 6.1, using model calculations. Future research must clarify which paludiculture systems are profitable, and under what conditions paludiculture is economically competitive. It also needs to be determined whether additional investments in rural development measures for regular paludiculture are profitable for individual farmers. Subject to further research, the budget presented in Table 6.11 seems to be an appropriate basis for calculating the financial requirements of rewetting projects and paludiculture.

6.4.5 Welfare effects and societal appreciation of rewetting and paludiculture

Rewetting drained peatlands used as arable land and grassland restores important ecosystem services for the protection of climate, water quality and biodiversity (Chapter 2.4). Avoidance of GHG emissions compared to conventional agricultural land use reflects an ecosystem service with welfare effects. The emission avoidance contributes to the target 'to hold the increase in global average temperature below 2°C above pre-industrial levels', agreed in 2010 by the 192 parties to the United Nations Framework Convention on Climate Change.

The abatement costs are an important measure for assessing the economic efficiency of climate change mitigation action. Abatement costs are opportunity costs of climate measures, which allow comparison with measures of other economic sectors (e.g. energy sector), and selection of cost-efficient measures of climate change mitigation. In order to calculate the abatement costs, the one-off costs of rewetting need to be transformed into an annuity. The annual costs are 221 (152) EUR per ha, assuming a generously rounded upper cost limit for rewetting of 3,000 EUR per ha (see Tab 6.11), an interest rate of 4% annually, and an amortisation of the investments after 20 (40) years. If rewetting is implemented for climate change mitigation only, and GHG reduction is 15 t CO_2 equivalent per ha and year, the avoidance costs amount to 10.13-14.73 EUR per t CO_2 equivalent. Comparison with measures of climate change mitigation in other economic sectors (Geiger et al. 2004), or in bioenergy production (Isermeyer et al. 2008) demonstrates that rewetting of peatlands is

6.4 Welfare aspects of land use on peatland

Table 6.13: Costs and benefits of agriculturally used and unused peatlands.

	Costs [EUR ha^{-1} a^{-1}]	Benefits [EUR ha^{-1} a^{-1}]	Benefit-to-cost ratio
Conventional peatland use	595 to 1,680 [1]	-66 to 282 [2]	-0.11 to 0.17
Rewetting			
without land use	152 to 221 [3]	3,835 [4]	17.35 to 25.23
avoided damage costs		5,035	22.78 to 33.13

[1] External costs for climate-related damage.
[2] Value creation without subsidies.
[3] Planning and building costs without land acquisition, 3,000 EUR per ha, Annuity t = 40 years, interest rate 4% p.a.
[4] After Meyerhoff et al. 2012.

a very cost-effective way of climate change mitigation (Table 6.12).

Among the benefits for society are the avoided damage costs resulting from the reduction of GHG emissions. Rewetting results in benefits of 1,200 EUR per ha and year, assuming an emission avoidance of 15 t CO_2 equivalents per ha and year and damage costs of 80 EUR per t CO_2 equivalents.

An additional benefit is the societal appreciation, which in the case of Mecklenburg-West-Pomerania was calculated on the basis of a representative survey about the willingness to pay for peatland conservation and paludiculture (Meyerhoff et al. 2012). In a cost-benefit analysis for the implementation of the German National Biodiversity Strategy, the welfare effects of concrete actions and land claims were quantified for six types of land use (peatlands, floodplains, xeric sites, woodland, arable land and grassland) (Wüstemann et al. 2013). The package for peatland conservation included:
- The promotion of natural development of intact peatlands (70,000 ha);
- The restoration and management of wet heaths by grazing (50,000 ha);
- The raising of water levels in peatlands used by forestry and agriculture (173,000 ha).

Furthermore, the cost-benefit analysis involved the stopping of all arable use of peatlands and the establishment of site-adapted land use (150,000 ha), even though these actions are not part of the national biodiversity strategy. Additional land claims were explicitly included in the recommended programme of action, with the objective to achieve more ambitious targets in peatland conservation (Spielmans et al. 2012).

In the survey, the willingness to pay for the individual programmes for the six types of land use and for the overall programme was determined. The annual willingness to pay for the entire programme was between 2.6 and 13.9 billion EUR. For the implementation of the peatland conservation measures, an average willingness to pay was determined of 149.76 EUR per household and year, which results in a national willingness to pay of 1.61 billion EUR per year. In comparison, the financial requirements to implement the peatland conservation programme outlined in the survey were calculated to be 88 million EUR per year (Meyerhoff et al. 2012). If the willingness to pay for the overall programme is interpreted as an upper budget limit and the willingness to pay for the six individual programmes as proportional preferences. Thus the willingness to pay for the peatland conservation programme is 3,835 EUR per ha and year.

The benefit-to-cost relationship indicates how high the benefits of peatland conservation are compared to its costs. If the quotient is higher than one, implementation of the measure is beneficial from an economic point of view. Conventional drainage based peatland use has a benefit-to-cost relationship lower than one (Table 6.13). If the above mentioned peatland conservation programme would be implemented according to the revealed preferences of the consulted households, the benefit-to-cost relationship between 17.35 and 25.23 for each Euro invested (Table 6.13).

The benefit-to-cost relationship improves if the avoided damage costs of climate change related damage are taken into consideration; then, it amounts to between 22.78 and 33.13 per EUR invested.

6.4.6 Conclusions

The property rights regarding land use on peatland permit a destructive use and do not meet the requirements of sustainable use of natural sources, as postulated in article 17 of the German Federal Soil Protection Act (German Bundesbodenschutzgesetz, Chapter 6.4.2). A possible approach to correct this loophole could be a substantiation of the concept of 'best practice' in this Act (Chapter 7.3).

The climate-related damage costs resulting from conventional agricultural land use far exceed its value creation (Chapter 6.4.3). Nevertheless, this type of land use of peatlands is encouraged by State allocated compensatory payments. The main problem of economic policy is that the elementary 'polluter pays' principle is interpreted the other way: land users are rewarded for environmentally damaging activities and have no incentive whatsoever to change this land use. From an economic point of view, the resulting welfare losses have to be interpreted as costs of political (in-)activity, which might be justified with distributional but not with welfare-economic arguments.

The calculated abatement costs (Chapter 6.4.5) show that rewetting and paludiculture are cost-effective climate protection measures, which fulfil very effectively the welfare-economic postulate of efficiency. The willingness to pay certainly proves that there is public appreciation of peatland conservation and paludiculture (Chapter 6.4.5). The very good benefit-to-cost-relationship shows convincingly that every Euro devoted to peatland conservation is very well invested, and can be welfare-economically legitimised with the preferences expressed by society.

7 Legal and political aspects of paludiculture

Land use and land use changes never occur in a vacuum; rather, they depend on the prevailing boundary conditions. In this chapter, we analyse the often complex subsidies, legislative (Chapter 7.1), and agricultural regulatory frameworks (Chapter 7.2) that may have an effect on (the introduction of) paludiculture in the European Union as well as in Germany, and more specifically, in the federal state of Mecklenburg-West Pomerania.

The legal and political grey area that surrounds paludiculture makes it essential to work closely with the responsible authorities and agencies when implementing paludiculture in practice. This may be the case if paludiculture management goes beyond humid and wet meadows, and new cultures like reed or cattail plantations are set up. In general, it cannot a priori be assumed that paludiculture can be implemented. Therefore, it is essential to try and modify the boundary conditions to the extent that they do not frustrate the implementation of sustainable peatland management.

This Chapter outlines the complications and conflicts related to paludiculture, as well as possible solutions within existing law, using the cultivation and utilisation of reed in paludiculture as an example. The goal is to put paludiculture on equal footing with other forms of agricultural land use. Moreover, there is a fundamental need to provide farm enterprises with sufficient long-term planning security to switch to sustainable production. Examples from other land use innovations (short rotation coppice, cultivation of biofuel crops) have shown that such conditions can be met by adjusting existing rules and regulations, provided that the political willingness exists (Chapter 7.3).

7.1 The legal framework

Detlef Czybulka & Laura Kölsch

Paludiculture ranges from crop establishment and raising the water level via regular cultivation of rewetted peatlands up to the utilisation of the biomass produced. The various practices span many legal areas, which mainly pertain to water, land, nature conservation and agri-environment law. These judicial areas comprise legal constraints and loopholes, due to the fact that the climate-friendly use of organic soil through paludiculture has hardly been considered in national or European law. On the other hand, some vegetation types (e.g. reed beds) are protected under nature conservation law as habitats and ecosystems. Consequently, paludiculture may also touch upon the legal aspects of biodiversity conservation.

As it was said above, this Chapter deals with the legal framework surrounding paludiculture as an agricultural practice exemplified by the cultivation (including planting) and winter harvest of reed on fen peatlands in Germany (cf. Chapter. 7.2). The Chapter explains the legal aspects of rewetting and agricultural land management, recognizes the related legal obstacles, and presents suggestions for minimizing these constraints. Furthermore, the transferability of these suggestions to other forms of paludiculture is clarified. Problems in relation to the use of biofuels, especially the strict air quality requirements when burning grasses, are not considered.

7.1.1 Water law in relation to peatland rewetting

Whenever a course of action affects a 'water body', in Germany water law applies. Water law covers water bodies that range from drainage ditches to the groundwater (§ 2 Section 1 German water management law, WHG). Federal states have the authority to exclude small water bodies from the stipulations of the WHG (§ 2 Section 2 WHG), i.e. 'drainage ditches that only serve one parcel of land' (§ 1 Section 4 Brandenburgisches Wassergesetz – BbgWG, similarly § 1 Section 2 Wassergesetz des Landes Mecklenburg-Vorpommern – LWaG M-V).

Water bodies (including groundwater) have the status of an entity under public law. The water authorities have discretionary power over management practices, which means that someone wanting to use a water body has no entitlement to obtain a license or permit. However, such permission is in general required because the raising of the water level – which is implicit when rewetting, requires permission (cf. § 9 Section 1 No. 2 and Section 2 No. 1 WHG).

If a proposed paludiculture project demands the creation, elimination, or alteration of a water body or its shorelines ('Gewässerausbau', § 67 Section 2 WHG), water authorities cross check through the necessary planning approval procedures (§ 68 Section 3 WHG; Chapter 9.5) the compatibility of the water level rise, with all legal rights concerned. However, the permissibility of land cultivation (crop establishment, mowing) goes unchecked. Due to the complications of planning approval procedures under water law in case of private interests, the project proponent should try to influence the management objectives of the water authorities at an early stage of official water resource planning (e.g. programme of measures, river basin management plans) and to call for active participation of the farmers involved (cf. § 85 WHG). Therefore, early contact with the appropriate water authorities in the project area (§§ 106 f. LWaG M-V, 124 ff. BbgWG) is strongly encouraged. If a paludiculture project does not fully meet the conceptual plan laid out by the water authorities, there is little chance of the project to be approved.

Insofar the rewetting project does not require alteration of a water body, a license under water law is anyhow required for raising the water level (§ 10 Section 1 Alt.1 WHG). The permission to raise the water level of a water body will be refused if harmful changes in the water body are to be expected, which cannot be avoided nor compensated by ancillary provisions. As a rule, paludiculture projects without harmful impact on water bodies should be permitted. A revocation of such permits can, however, occur 'at any time' for objective reasons of water management. If an alteration of a water body or a rise in water level are not necessary, the water bodies should still be maintained 'in a state that with respect to the discharge or retention of water corresponds to the needs of water management' (§ 39 WHG). For water bodies of secondary rank, maintenance is the responsibility of the regional water and soil boards.

If the project area is located in a water protection area, the specific legal requirements can sometimes favour the implementation of paludiculture, especially when fertilisers and pesticides are not used.

7.1.2 Nature conservation law in relation to peatland rewetting

Rewetting measures also need to be reviewed under nature conservation law. The intervention and compensation regulation (§§ 13 ff. Federal Nature Conservation Act – BNatSchG) is the instrument by which conservation law aims to avoid, minimize and compensate for possible negative impacts from interventions in nature and landscapes. The goal of this regulation is to maintain a total ecological balance (Wagner & Czybulka 2012). Any change in the groundwater that is connected to the active soil levels can be considered an 'intervention', depending on whether the performance and functionality of the ecosystem (including landscape scenery) are significantly affected (§ 14 Section 1 BNatSchG). This can only be conclusively determined on a case-by-case basis through relevant land assessment and evaluation studies. The Federal Compensation Regulation, which was to be put in place according to § 15 Section 7 BNatSchG, was not yet adopted in May 2015. When applicable, an intervention approval must also be obtained in the framework of water law permission procedures, or also separately (according to § 40 Naturschutzausführungsgesetz Mecklenburg-Vorpommern, – NatSchAGM-V).

Land use according to proper agricultural practices is, as a basic principle, exempt from the intervention regulation (§ 14 Section 2 BNatSchG). This exemption, however, only applies to direct primary agricultural production (Fischer-Hüftle & Czybulka 2010). Measures that are indirectly related to the harvesting of agricultural products, such as water management, are not considered part of agricultural land use (Deutscher Bundestag 1996).

Conversely, raising the water level can also increase the functionality of nature and thus be used to compensate for impacts of other activities (road construction and so on). In this respect, the costs of hydraulic engineering could be borne by the party responsible for such impacts.

If land for prospective paludiculture projects happens to lie in conservation areas (cf. § 20 BNatSchG) – such as nature reserves or landscape protection areas, the permanent rise in water level associated with paludiculture should be evaluated against the goals of that conservation area. Areas under conservation law are generally designated through statutory order.

In areas protected under the EU Birds or Habitat Directives (which are included in the European Natura 2000 network), the maintenance and (if necessary) restoration of a favourable conservation status for wild birds, their natural habitats and all FFH-habitat types resp. species of wild fauna and flora (Annex II) must be ensured (Art. 3 Section 1 Birds Directive, Art. 3 Section 1 Habitat Directive). The necessary conservation and development measures must be laid out by EU member states. In Germany these measures are implemented through reserve regulations, which govern the conservation goals, the 'do's and don'ts' of their enforcement, as well as the exceptions. Furthermore, management plans are set up for larger areas following Art. 6 Section 1 Habitat Directive and § 32 Section 5 BNatSchG. All actions that may significantly affect the conservation objectives or the purpose of protection of a Natura 2000 site or its constituent parts are inadmissible (Art. 6 Section 2 Habitat Directive, § 33 Section 1 BNatSchG). Additionally, a Habitat Directive compatibility test must be performed when it cannot be guaranteed with scientific certainty that no disruptions or disturbances will occur.

However, conservation areas are subject to a so-called absolute ban on change (§ 23 Section 2 BNatSchG) 'in accordance with more concrete regulations', which are usually the protected area ordinances. All actions that may lead to degradation of the protected area or any of its components are to some extent either prohibited or allowed (§ 23 Section 2 BNatSchG). In landscape protection areas, in contrast, actions are only prohibited if they indeed harm the character of the area or its specific conservation goals – as laid out in the respective protected area ordinance (§ 26 Section 2 BNatSchG).

Many protected area regulations include peatland rewetting as one of their objectives. The admissibility of a water level rise in these areas must be assessed in that context. On the other hand, rewetting breeding and feeding grounds for birds (e. g. meadow birds) protected according to the respective protected area ordinance, could adversely affect these populations and would thus not be allowed.

7.1.3 Agricultural management and nature conservation law

Agricultural land use is not seen as an 'intervention' in terms of the intervention and compensation regulation (§ 14 Section 2 Sentence 1 BNatSchG), insofar as the goals of nature and landscape conservation are taken into account ('agricultural clause'). This is the case (in accordance with § 14 Section 2 Sentence 2 BNatSchG) if 'good practice' is abided by. The term 'agriculture' usually refers to normal cultivation of arable land, as well as conventional management of meadow and pasture for grazing purposes (Fischer-Hüftle & Czybulka 2010). Still, it could be argued that paludiculture should also be regarded as a form of agriculture, because it involves the direct and commercial extraction of agricultural products (see above and Chapter 7.2). In a former decision, the Federal Administrative Court has denied that traditional reed mowing is an agricultural activity. It argumented that the mowing of reeds lacks the regularity of traditional agriculture, because throughout the year does not take place crop establishment, or active management nor soil cultivation (BVerwG 1997). Though, paludiculture should be seen from a different perspective because, in fact, it entails a planned cultivation of land through targeted planting, management, and mowing. If the implementation of paludiculture is consistent with the principles of 'good practice', it poses no real negative impact. Specific 'good practice' requirements for this type of farming do not exist in § 5 Section 2 BNatSchG/§ 17 Section 2 Federal Soil Conservation Act (BBodSchG). Through winter mowing over sufficiently frozen soils plus the implementation of peat-protecting harvesting techniques, any serious soil compaction can be avoided. Consequently, an official approval or notification is usually not required, and the legal consequences of the intervention and compensation regulation do not come into play.

However, it is important to note that this privilege does not apply to legal habitat protection (see below), nor to territory protection, because these are not tied in with the interventions outlined under § 14 Section 1 BNatSchG, but rather with actions that are described as change, harm, destruction or impacting activity (Fischer-Hüftle & Czybulka 2010).

In connection to the intervention and compensation regulation, § 14 Section 3 BNatSchG should be kept in mind. In case of contractual agreements relating to agri-environment or early compensatory measures, this section of the law allows for reinstituting conventional farming without management restrictions within ten years. This rule can be both an incentive and a safeguard for farmers to participate in such programmes (for the similar provisions of § 30 Section 5 BNatSchG; see below).

However, the use of reeds on rewetted fen soils can conflict with the legal protection of 'reed bed' and 'peatland' habitats (§ 30 Section 2 No. 2 BNatSchG). Since reed paludiculture includes the establishment of areas dominated by reed, and the legal protection extends to all relevant habitat types regardless of their origin (Kratsch & Czybulka 2010) including plantings, paludiculture areas that include reed beds and marshes can fall under legal protection if they meet the characteristics specified for these habitat types under federal and state law.

Under closer consideration, the harvesting of reed is mostly not in conflict with the federally (§ 30 Section 2 Nr. 2 BNatSchG) regulated protection of reed beds because this regulation only concerns water reed stands (reed on the banks of standing and flowing waters) (Deutscher Bundestag 2001). As far as paludiculture reed beds – similar to land reeds, are established above the mean water level of water bodies or in waterlogged but not regularly inundated areas, they will – according to federal law – constitute no protected stands.

A more differentiated legal situation appears in federal state legislation because some states (e.g. Schleswig-Holstein, Brandenburg, and Mecklenburg-West Pomerania) have put more habitats under protection using the statutory authority of § 30 Section 2 Sentence 2 BNatSchG. According to § 32 Section 1 No. 2 of the Nature Conservation Act, in conjunction with point 2.3 of the Habitat Protection Regulation Brandenburg, 'land reeds as dominant stands of reed dwelling species on peat or peaty land that is characterised by groundwater levels, which are permanently or temporarily close to the surface' are placed under habitat protection. The legal situation in the federal state of Schleswig-Holstein is factually the same because there 'areas on moist or wet soils with plant communities that are widely characterized by reed plants' are legally protected habitats (§ 21 Section 1 Naturschutzgesetz in connection with No. 2c of the Biotope regulation of Schleswig-Holstein). Reed stands on rewetted sites in both Brandenburg and Schleswig-Holstein can thus, after some time, develop to legally protected sites. In Mecklenburg-West Pomerania, this is not necessarily the case. In accordance with § 20 Section 1 No. 1 in connection with Appendix 2, No. 1.4 Naturschutzausführungsgesetz Mecklenburg-Vorpommern (NatSchAG M-V), 'land reeds ... in waterlogged areas (reed stands on fallow arable and meadow land)' are protected. However, planted reed beds did not spontaneously establish by ecological succession on fallow land. Therefore, it can be argued that reed beds established by planting do not fall under the definition of reeds that are protected under state law.

This does, however, not rule out that planted reed beds in rewetted fen peatlands in Mecklenburg-West Pomerania (and potentially in other federal states as well) are subject to legal habitat protection as 'peatlands'. Under federal law 'natural and semi-natural ecosystems on peat soils – including some degradation and regeneration stages, that depend on mineral soil water' including among others 'intact (...) non-forested fens, e.g. (...) reed beds on peat soils' are protected as peatlands (Deutscher Bundestag 2001).

The definition of peatlands under state law (Appendix 2, No. 1.1 of NatSchAG M-V) does not differ much from the definition under federal law. The latter also includes reeds on peat soil that may function entirely (natural peatlands) or partly (semi-natural, degraded, or regenerating peatlands) as habitat for a community of wild species (definition of biotope, § 7 Section 2 No. 4 BNatSchG). Since paludicultures may indeed comply with these conditions (depending on their stages of development), in these cases they must be considered to be legally protected peatlands.

§ 30 Section 1 BNatSchG puts as a mandatory general principle 'certain parts of nature and the landscape that are of particular significance as habitats' under immediate legal protection. It also forbids 'actions that could lead to the destruction of or significant damage to protected habitats' (§ 30 Section 2 Sentence 1 BNatSchG). This wording implies that the mere threat of an adverse effect is sufficient to provide protection, since no actual negative impact is required (Endres 2011). Exemptions from this ban can be granted under the condition that the habitat impairment is compensated (§ 30 Section 3 BNatSchG). 'Compensation' in this case means replacement by a similar habitat within close spatial proximity. In contrast to the intervention regulation, in case of habitat protection other substitutions or compensatory payments are not allowed.

If it is assumed that planted reed beds meet the requirements of legally protected habitats, the question remains whether winter mowing adversely impacts the habitat sufficiently to be prohibited by law. In this context, an impact would already be significant if it could endanger the habitat function for one currently or potentially occurring wild species (Kerkmann 2010). Here it is sufficient to concentrate on the effects on the reeds as components of peatlands because the significant deterioration of a habitat component already implies the deterioration of the entire habitat. Planted reed beds can be valuable – from a nature conservation point of view, in all stages of development, especially for species that are restricted to reeds. While high and dense reed beds serve as breeding grounds for birds (e.g. Marsh Harrier and Little Crake) and as overwintering grounds for some species of duck, butterflies, and insects (e.g. Goldeneye, Flame Wainscot, Gall Midges), slightly buckled reed stands provide potential breeding grounds for Great Reed Warbler and Bittern (Blab 1994, Krägenow & Wiesehöfer 1999). Sparsely growing reed beds that are cut in winter can serve as breeding ground for Aquatic Warbler the following year (Tanneberger et al. 2009, Chapter 5.2, Chapter 10.2).

The effects of winter mowing can thus vary widely. If breeding grounds are to be created for Aquatic Warbler and mowing follows nature conservation instructions, no special permit or compensation would be required. If nature conservation, however, aims at the creation or maintenance of diverse reed bed structures, winter mowing can have an immediate impact on overwintering species, as well as on the breeding grounds for the following year. Winter mowing is then (similarly to the traditional mowing of reed beds) only allowed if an exemption is granted, and then only under the condition that the harm done can be compensated for. Compensation – i.e. the restoration of impaired ecosystem functions, must replace the degraded functions of the original habitat in close proximity to the original habitat (§ 15 Section 2 Sentence 2 BNatSchG). If an area serves as a habitat for species that rely on mowing, there is no need for compensation as long as mowing takes the nature conservation instructions into account. If the goal is to create or maintain reed beds with a buckled layer of old stalks, the effects of mowing must be compensated for. In case of planted reed, compensation could be achieved by section-wise or low-frequency mowing of the reed beds. When establishing new reed beds, it should be planned from the outset to refrain from mowing at least parts of the planted area.

The establishment of reed beds could also be regarded as an 'early compensatory measure', which could be stored as a credit in a so-called 'eco-account'. Such credits could then be used to compensate for the mowing and for other impacts (§ 16 BNatSchG). Because of the still unclear appraisal and handling of such situations, it is advisable to agree in advance with the responsible nature conservation authorities on the proper procedure for granting exemption. In any case, proper documentation of the initial condition of the land is required (§ 16 Section 1 No. 5 BNatSchG).

In case of section-wise harvesting, the conclusion of contractual agreements between landowners and the relevant nature conservation authority is recommended ('contractual conservation', § 30 Section 5 BNatSchG). § 30 Section 5 BNatSchG allows the impairment of protected habitats that have developed during the contract period within ten years after the end of the contract, if resumption of former agricultural activities is preferred.

This rule reflects the idea of 'temporary nature conservation'. It aims to prevent farmers from refraining from voluntary land use restrictions because they are afraid of having to conform to new nature protection requirements if they resume the original use (Deutscher Bundestag 1998a). In this way, it can be assured that the resumption of original use and the consequent habitat elimination would not be prohibited or could only be realized with an exemption and the associated obligatory compensation.

The winter mowing of planted reed beds must also satisfy the requirements of both general and specific species protection law. General species protection also encompasses the protection of the habitats of all wildlife (§ 39 Section 1 No. 3 BNatSchG). In accordance with § 39 Section 5 S. 1 Nr. 3 BNatSchG it is 'forbidden to cut back reed stands between March 1 and September 30; at times other than this, reed stands can only be cut back in sections'. For the protection of reed stand dwelling wildlife, the cutting of reed stands is forbidden during the growing season. Since many species are dependent on standing stalks from the previous year, it must furthermore be secured that adequate amounts of stalks remain (Deutscher Bundestag 2009).

§ 44 Section 1 BNatSchG regulates species protection and controls access bans with respect to specially and strictly protected species, respectively. Birds that need reeds for breeding are – like all European bird species, specially protected (§ 7 Section 2 No. 13 b BNatSchG), not, however the afore mentioned butterfly and other insect species. Winter mowing can affect the breeding sites of these species (§ 44 Section 1 No. 3 BNatSchG), when the planted reed beds have become regularly used breeding grounds in a verifiable way. Regularly used breeding grounds are regarded as permanent reproduction areas if the nests lie so close together that no unprotected areas can be designated between them. (Niedersächsisches Oberverwaltungsgericht 2011, Gellermann 2012). Then, such areas are year-round protected, if it can be assumed that the birds use these areas regularly, even if they are temporarily absent. If, however, regular reed mowing has created suitable breeding grounds for Aquatic Warbler, it is conceivable that commercial use of this area could continue without violating the access ban for strictly protected wildlife.

'Agricultural land use' does not violate the access-, possession-, or trade ban for specially protected species, as long as this use meets the principles of 'good practice', and the conservation status of the local wildlife population is not impaired by the management of the land (§ 44 Section 4 BNatSchG). This legal variant does not focus on individual animals, but rather on the conservation status of the 'local population' of a species, which may not be compromised by agricultural use.

Whether winter mowing harms the local population of a bird species in the following year can only be determined on a site-specific basis. First of all, a local population of a species protected in Europe needs to be present in the area for the regulation to have any bearing. In the evaluation, the 'local' population is not determined by the borders of land ownership, but rather according to the boundaries of the respective habitat (Gellermann 2007). In specific cases, local populations may be unaffected if uncut reed beds in the vicinity are available as alternative habitats. This is usually the case when harvesting is performed on a section by section basis (for possible further requirements, see §44 Section4 Sentence 3 BNatSchG).

In summary, it can be concluded that planted reed beds on fen peatland may develop into habitats protected by nature conservation law. Winter mowing of such areas is only allowed with a special permit and, when appropriate, with compensatory measures. For this reason, a prior agreement should be reached with the responsible nature conservation authorities. Mowing between March 1 and September 30 is usually forbidden. Outside of this time frame, reed beds can only be cut section-wise. In some cases, reed mowing can violate the restrictions of specific species protection, even if mowing is considered agricultural land use. However, this situation can be avoided through clear management instructions from the responsible conservation authority – for example, by abstaining from mowing part of the area and, if necessary, by appropriate awareness-raising activities and contractual arrangements (cf. §44 Section 4 Sentence 3 BNatSchG).

The following suggestions can be given: The identified barriers to paludiculture can be eliminated through a change in the legal basis for the protection of habitats and the general protection of species. The new regulations should dictate that reed beds on agricultural land are not reed beds in terms of nature conservation law, and as such do not represent protected habitats and do not require section-wise harvest. In this way, land owners could be successfully motivated to switch to paludiculture. If such amendment is not feasible, additional plantings and deliberately refraining from mowing part of the reed area could be used for (early) compensation. Section-wise mowing creates and maintains diversely structured reed stands, which are generally more valuable for nature conservation than the initial area. Documentation regarding the initial condition of the land in question is required. This type of sustainable agriculture connects conservation with production. The unmown areas could – as replacement breeding sites, prevent the deterioration of the conservation status of local populations that depend on unused reed beds, and thus allow paludiculture mowing to be exempted from restricted access to breeding sites. Using this alternative, it would be unnecessary to change the law about prescribed section-wise mowing according to §39 BNatSchG. Also, no agreement would be required in the sense of §30 Section 5 BNatSchG to resume the original agricultural use in the event of termination of paludicultural use.

The transferability of these results with respect to habitat and species conservation law to other paludiculture crops (e.g. reed canary grass, cattail, or sedges) remains to be investigated in detail. For instance, it has to be verified whether cattail stands in individual federal states are legally classified as protected reeds, whether their habitat function is significantly affected by paludiculture practices, and whether they would fall under the ban on summer mowing or be subjected to the mandatory section-wise winter mowing of reedbeds following §39 BNatSchG. 'Tall sedge reeds', or 'plant communities dominated by tall sedges on groundwater-influenced sites' (Deutscher Bundestag 2009) are in any case legally protected habitats since the amendment of BNatSchG in 2009 (§30 Section 2 No. 2 BNatSchG). Also, small sedge communities may be subject to legal habitat protection in some federal states (incl. Mecklenburg-West Pomerania, cf. §20 Section 1 No. 1 in connection with App. 2 No. 1.4 NatSchAG M-V).

The planting and winter mowing of reed beds in protected areas has additionally and primarily to be evaluated against the legal regulations of the respective protected area. In some Natura 2000 areas, reed beds can indeed be considered a priority habitat under the Habitat Directive – e.g. brackish reedbeds along estuaries (Habitat type 1130); but in general, reeds on fen soils do not classify as such. In habitats of Annex II species (such as beaver and otter, which could be affected by the mowing of reeds near water bodies) water levels of reed beds appropriate for mowing will generally be too low to accommodate for these protected species. If the preservation of reed-dependent birds and their habitat elements (e.g. nesting or feeding grounds) is a goal of conservation policy of an area, winter mowing may imply a considerable impairment of these areas. Here, again, a potential hazard suffices.

In nature reserves, biosphere reserves and national parks, actions that might otherwise constitute good agricultural practice may be restricted. The conversion of permanent grassland is generally forbidden, and this can stand in the way of paludiculture implementation. An exemption to this rule will – in accordance with federal state regulations for the maintenance of permanent grassland (see below), not be granted. In protected landscapes, agricultural use is normally not prohibited. Whether barriers to reed production and management exist for reasons of area conservation, can only be concretely determined on the basis of the respective reserve regulations.

7.1.4 Agri-environmental law

The establishment of paludicultures on permanent grassland may be in conflict with the general ban on ploughing grassland. This prohibition should certainly be considered if paludicultures are treated as permanent crop under the Common Agricultural Policy (cf. Chapter 7.2), and their establishment indeed requires ploughing up the sward.

Since 2015, there is a total ban on conversion of permanent grassland in Natura 2000 sites (environmentally sensitive permanent grasslands, Art. 45 Reg. (EU) 1307/2013). Outside of these areas a conversion of permanent grassland must be approved. As a rule, in case of approval, the establishment of new permanent grassland in another location is obligatory.

Besides individual approval procedures, also the regional maintenance of permanent grasslands is regulated. If the portion of permanent grassland declines more than 5 % compared with a reference in 2015 (Art. 45 Reg. (EU) 1307/2013), no further approvals for conversion can be given in the region concerned.

Permanent grassland on mineral soil contributes to carbon sequestration in the soil, to landscape water retention and to the preservation of biodiversity (Klinck 2012). Paludicultures may have similar or even larger positive effects on climate, biodiversity and water quality (at least in case of section-wise mowing and refrainment from using fertilisers; see Chapter 5). However, the regulations for the preservation of permanent grasslands until now do not consider such equivalent forms of utilisation (with the exception of afforestation). Therefore, EU regulations should be amended to include an exemption from the ban of ploughing of permanent grasslands, if such equivalents are implemented. Alternatively, an exemption could be made from the obligation to establish permanent grassland as a compensation measure.

The information above makes clear that, although some legal obstacles to the implementation of paludiculture do still exist, these obstacles are not insurmountable.

7.2 Agricultural policy

Laura Kölsch, Simone Witzel, Detlef Czybulka & Theodor Fock

The introduction of paludicultures into the landscape should preferentially be focused on peatlands that are currently being used for conventional drainage-based agriculture. Paludiculture represents a new form of agriculture and land management, which should benefit from the same legal promotion as conventional agricultural practices. Therefore, the current and future agricultural policies and legal frameworks have a substantial influence on the introduction of paludiculture.

7.2.1 Current conditions

The large scale establishment of paludiculture requires both the continuation of direct payments under the First Pillar of the CAP, as well as the granting of subsidies for rural development (i.e. agri-environment and climate schemes, investment aid) under the Second Pillar. In Germany, direct payments are until now granted as regionalized rates – i.e. the annual payment depends on the region and on the area of land that is managed by the applicant. Each parcel of land ('field-block') is classified as either arable land or permanent grassland. In 2013, the payments did not differentiate between these types of land use, and the payment received was – in national average, 340 EUR per hectare per year. Despite the same level of direct payments, the conversion of permanent grassland to arable land remains economically attractive, since the profitability of the latter type of land use is often much higher. The ban on conversion of permanent pasture aims to minimize the conversion of grassland into arable land. Arable land can, however, easily be used as grassland. If grassland utilisation takes place without interruption for at least five years, the area is reclassified as permanent grassland. This is usually avoided by ploughing and reseeding at least every five years. From the perspective of the agricultural enterprise, a loss of value takes place when the land is reclassified from arable land to permanent pasture. If arable land on peat soil – after implementation of paludiculture, would be reclassified to permanent grassland, the imputed loss of value would be relevant.

In December 2013, after many years of negotiations, the new CAP directions for the subvention period 2014–2020 have been decided, which in principle will apply from 2015 onward. Meanwhile, the concrete formulation of the CAP in the EU Member States has been decided for the years 2015–2020. Direct payments remain the central element of income policy in the CAP, although the level of payment will sink in

Germany to 265 EUR per hectare per year (not including extra amounts for small farms and young farmers). Therefore, direct payments will continue to play an essential role when deciding on land use alternatives (e.g. the rewetting of peatland). To date, areas with very high water levels lose their status as 'agricultural land', by which the entitlement to claim payments is lost. If reeds spread on waterlogged land, this entitlement may be lost in case of regular winter harvesting, if reed is used as an indicator of abandonment or is not recognized as an agricultural crop. This risk should be avoided if the introduction of paludiculture is to be more widely accepted.

Under the Second Pillar of the CAP, a package of provisions is promoted through the European Agricultural Fund for Rural Development (EAFRD-Council Regulation (EU) 1305/2013), of which especially agri-environment subsidies could be pivotal for promoting paludiculture. In Germany, agri-environment subsidies are issued by the federal states and co-financed by the EAFRD. Many agri-environment measures are also co-financed by the federal government in the framework of the Joint Task for the Improvement of Agrarian Structures and Coastal Protection (German abbreviation: GAK). When shaping directives for schemes of subsidies, the GAK framework (which is coordinated between federal states and the German government) must be considered. This means that the factual creative scope on federal state level is limited. In principle, the subsidized amount for agri-environment measures is calculated on the basis of yield reduction or additional expenses, in comparison to conventional agricultural land use. To this end, profitability has to be calculated on a national level, or for another appropriate region for conventional agriculture that works according to the prevailing legal conditions ('good practice', Chapter 7.1). The difference is then used as a basis for the subsidy. In case of organic farming, the difficult economic situation associated with the conversion of land use can be addressed by higher subsidies in the first years. A transfer of this funding model to paludiculture would make sense, since – similarly to organic farming, the first years of crop establishment are associated with higher costs and reduced yields. An additional incentive may not be granted, according to EU subsidy legislation. However, a subsidy that simply compensates for lower yields and/or higher costs is not really an economic incentive to promote a change to more environmentally friendly farming. Still, the use of average profitability calculated at the state or regional level may create an incentive, if local income levels are below the regional or state average. In Mecklenburg-West Pomerania, for example, ecologically friendly grassland management is until now subsidized with – in average, 200 EUR per hectare per year. Under the Second Pillar of the CAP, individual enterprises and corporate initiatives can also receive a subsidy for business investments of up to 30 % (Chapter 7.2.2).

7.2.2 Grant funding perspectives for paludiculture land

Under the First Pillar of the CAP, funding can be granted for agricultural land that is used for agricultural activities (Art. 34, Section 2 a, in connection with Art. 4 Regulation (EU) 1307/2013 on direct payments to farmers). The direct payment regulation defines the eligibility requirements as follows: agricultural land is to be used as arable land, permanent grassland or permanent cropland. Permanent crops are 'non-rotational crops other than permanent grassland that occupy the land for five years or longer and yield repeated harvests, including nurseries and short rotation coppice'. Permanent grassland is defined as 'land used to grow grasses or other herbaceous forage naturally (self-seeded) or through cultivation (sown) and that has not been included in crop rotation for five years or longer'. 'Grasses or other herbaceous forage' means all herbaceous plants traditionally found in natural pastures. One of several definitions of agricultural activity describes it as 'the production, rearing or growing of agricultural products'.

'Agricultural products' are those products listed in Annex I to the Treaty on the Functioning of the European Union (TFEU). Annex I TFEU contains selected chapters of the list of goods of the European tariff nomenclature (cf. the list in Annex to Commission Regulation (EU) 1006/2011). Therefore, to receive funding, a product must be assigned to a European tariff nomenclature chapter. This chapter, in turn, has to be listed in Annex I TFEU.

It is conceivable to classify Reed, Reed Canary Grass, sedges, and cattails as permanent crops, because they occupy the same land for five years or longer and yield repeated harvests. Whether these paludiculture crops could be incorporated into a Chapter of the tariff nomenclature (which is listed in annex I of the TFEU), remains up to now unclear. Reed is explicitly listed in Chapter 14 ('Vegetable materials of a kind used primarily for plaiting') of the tariff nomenclature, but is not listed in Annex I TFEU. Consequently, the German Government/States Working Group for direct payments (which is an administrative council of the Agricultural Ministries of the federal states and the government for nationwide coordination of direct payment practice), meanwhile classified reed as a non-agricultural product, which is not eligible for direct payments. This and the classification of other – not explicitly listed, paludiculture crops such as cattail need to remain subject of a coordination process with the agricultural administration. The still unclear classi-

7.2 Agricultural policy

fication basis of *Miscanthus*, which is eligible for direct payments, could be helpful in this matter.

If the land had already the status of permanent grassland in its former drained condition, it is feasible that (after rewetting) also the established Reed and Reed Canary Grass can be classified as such. Reed and Reed Canary Grass are 'grasses or other herbaceous forage' in the sense of the definition of permanent grassland. For the purposes of that definition, they belong (like other forage grasses) to the plant family Poaceae (formerly Gramineae). The EU Commission considers all herbaceous plants as 'herbaceous forage' (Silva Rodriguez, personal communication 2011). Reed and Reed Canary Grass are anyhow considered to be herbaceous (non-woody, Strasburger & Sitte 1998), conventional plants of moist grasslands. Moreover, nutritional values of Reed and Reed Canary Grass are documented in agricultural literature (Klapp & von Boberfeld 1995).

Unlike arable or permanent crops, no classification of individual plant species in the tariff nomenclature is required in case of permanent grassland, because the definition of grassland itself (see above) determines which vegetation is eligible for direct payments. This makes grass and other herbaceous forage directly agricultural products.

In addition to agricultural land, also those rewetted lands are eligible for the receipt of direct payments, which were already entitled in 2008 but are no longer suitable for agricultural use (i.e. as permanent cropland, grassland or arable land) due to the application of the Water Framework Directive (WFD, (EC) 60/2000) (Art. 32 Section 2 b i Regulation (EU) 1307/2013).This option would come into consideration if the classification of paludiculture as permanent grassland and cropland fails. However, because the categorisation under the WFD is constrained both conceptually and in time terms, it is to be preferred to have the granting of aid under the general regulations of agricultural use as permanent crops or permanent grassland. The recognition of paludiculture as being 'production of agricultural products on agricultural land' is also a prerequisite for obtaining area-based subsidies under the Second Pillar of the CAP. Farmers can obtain additional grant money for agri-environment measures – i.e. voluntarily entered contractual obligations on agricultural land that go beyond the cross-compliance commitments and other requirements of agricultural law (Art. 28 EAFRD Council Regulation). Under the presumption that paludiculture is recognized as permanent grassland, it is, for instance, possible in Mecklenburg-West Pomerania to have paludiculture included in the 'programme funding of conservation-oriented grassland management' (e.g. wet grassland). The agri-environmental achievement that would go beyond legal stipulations could imply the total refraining of using fertilisers, which would surpass the 'good practice' outlined by the federal Fertiliser Ordinance (FO) (§ 3 Section 5 FO). In other federal states, similar programmes with identical or similar conditions exist for low-intensity grassland utilisation (Thomas et al. 2009).

Because of the legal protection of reed beds, a section-wise reed harvest in winter may be appropriate (Chapter 7.1). The lower yields that result from this practice could be settled financially through agri-environment programmes (e.g. as part of a separate programme for paludiculture as permanent cropland). In such cases, the inadmissibility of double funding must be observed.

Another important consideration for the successful implementation of paludiculture is whether capital investments (e.g. in adapted harvesting or processing equipment, storage buildings, downstream processing plants as for pellet or briquette production, and so on) will be subsidized. The federal states award farms investment subsidies for the funding areas 'agricultural investment' and 'diversification', following the principles laid out by the EAFRD Council Regulation and the Joint Task for the Improvement of Agrarian Structures and Coastal Protection (GAK). With regard to agricultural investment, funds are provided for investment in durable goods that support the production, processing or direct marketing of products listed in Annex I of the TFEU (Art. 17 EAFRD Council Regulation, GAK -Investment in agricultural holdings- AFP). With respect to diversification, investments can also be subsidized for non-agricultural purposes, as long as they are not exclusively limited to the production of products referred to in Annex I to the TFEU (Art. 19 EAFRD Council Regulation).

7.2.3 Prospects and conclusion

Decisions concerning the exact content of the CAP for 2014–2020 have been taken at the EU level (April 2015) – particularly new provisions for the implementation of 'greening' and of measures in the Second Pillar of the CAP. The approval of the programmes for rural development on the level of the federal states is in progress, so conditions for the coming subsidies are more or less clear. As direct payments remain a central element of income policy at the EU level, and the largest proportion of funds will be used for this purpose, it is, for the introduction of paludiculture, of utmost importance that payment entitlements of an area are maintained (e.g. when cultivating reed).

Regarding paludiculture, climate measures, which supplement agri-environment measures, are also interesting because they subsidise voluntarily entered commitments to a climate-friendly management (Art. 28 EAFRD Council Regulation). The provision of such programmes is mandatory for EU member states. The subsidy amount is calculated analogously to other agri-environment measures in order to compensate for lower yields and higher costs.

Securing added value generation and employment opportunities in the rural area as part of the CAP design, needs to be aligned with the objectives of sustainable peatland use. As EU funding for subsidies under the Second Pillar of the CAP in Germany has declined, active competition for available funds among different subsidy aims may be expected. EU subsidies and the reallocation of funds from the First to the Second Pillar of the CAP will work out differentially for the various federal states, which are responsible for designing the specific subsidy policies. In some states (e.g. Schleswig-Holstein and Lower Saxony) available funds are clearly increasing, while in others (e.g., Mecklenburg-West Pomerania and Brandenburg) they will decrease. The responsible state authorities have proposed subsidies for the economically viable utilisation of peatlands with high water levels, especially for low-intensity grazing and paludiculture (Jensen et al. 2012). However, these proposals have not (yet) been incorporated in the development programmes of the federal states. Still, even under the given conditions, an attractive set of subsidies could be created to foster the introduction and continuance of paludicultures.

In conclusion, it should be noted that the eligibility of paludicultures for subsidies should be maintained both in the form of direct payments under the First Pillar of the CAP, and as agri-environment or climate protection subsidies under the Second Pillar. Additionally, subsidies for investments, such as for innovative technology or water management are possible. Further possibilities have been created with the European Innovation Partnership (EIP), a newly established agricultural policy and subsidy framework consistent with the existing system. paludiculture could be supported within this programme by pilot and demonstration projects. Paludiculture can now be put on equal footing with other forms of agriculture, and by doing this, utilise the same subsidy policy framework. To introduce paludiculture on the largest possible scale, further technological and economic development is necessary to increase the competitiveness of paludiculture against conventional agriculture. This could be achieved by encouraging research and development, as well as by implementing pilot projects for the hydrological reconstruction of agricultural areas on peatland.

7.3 Control mechanisms and incentives for paludiculture

Achim Schäfer

The previous two chapters have shown that farms face several land use restrictions if they aimed to rewet land and to introduce paludiculture. Furthermore, it has been learned that agricultural policies will, in general, remain the same. This means that currently the incentives to change for farmers are low. The costs associated with the continuation of conventional agricultural practices on peatlands contribute to welfare losses (Chapter 6.4), and will have to be borne by taxpayers, society, and future generations. In the awareness that peatland rewetting and the introduction of paludiculture are hindered by a variety of politically motivated subsidies and complex, inappropriate legal rules and regulations, a question emerges: What regulatory and taxation reforms are needed to facilitate a sustainable utilisation of peatlands?

7.3.1 Economic policy

The problem with current economic policy is that the elemental 'polluter pays principle' has been turned upside down, and the recipients of current subsidies have no incentive at all to introduce more sustainable management practices (Chapter 6.4, Schäfer 2004, Schäfer 2009).

From an economic policy standpoint, the question arises as to how the existing incentive systems must be changed in order for paludiculture to be an attractive option for farmers. The most important regulations for peatlands are the ones that pertain to property rights (Demsetz 1967), to income policy under the First Pillar of the CAP, and to the integration of land use in a consistent climate change mitigation regime (e.g. carbon markets).

Retrospectively, it should be noted that the reform proposals for the CAP over the past few years have been of little impact and importance (e.g. Bissels & Oppermann 2011, Grajewski et al. 2011, Hebauer et al. 2011, Forstner et al. 2012, Lakner et al. 2012, Tangemann & Cramon-Taubadel 2013, SRU 2009, WissenschaftlicherBeirat 2010). It is hard to understand why this has been called a 'reform', when in fact everything stays in the same way as it was in the past. The pillar concept of the CAP and the associated redistribution of taxpayers' money without specific performance standards, increasingly lose societal acceptance. The Scientific Advisory Board for Agricultural Policy at the Federal German Ministry for Food, Agriculture, and Consumer Protection refers to the direct payments system as an 'outdated instrument', and advocates that the money would be put to better use if it were dedicated to environmental and climate protection services (Isermeyer & Weingarten 2012). Also, the European Court of Auditors (ECA 2011a) has observed that direct payments do not contribute to the majority of objectives of EU agricultural policy; indeed, such payments contradict these objectives in many ways, and the costs are disproportionally high compared to the minor benefits they achieve. The court also points to the lack of effectiveness of many agri-environment policies. Another point of critique is the political willingness to promote nationwide greening measures (as new environmental standards for granting direct payments from 2015 on), which are rarely effective, rather that implementing specific and effective measures at a regional level (ECA 2011b). A more targeted use of public funds would require a gradual reduction of direct payments and an increasing funding of differentiated and effective action (Forstner et al. 2012).

The standard to ensure a 'balance between irrigation, drainage and water table replenishment' – originally submitted by the European Commission, was not retained by the Council of the European Union, 'although it addressed one of the main environmental problems of most Member States' (ECA 2009). This certainly applies to peatland areas where drainage and the consequent 'vicious cycle of peatland use' (Chapter 2.3.1) makes agricultural land use increasingly difficult.

Paludiculture strives to create a more sustainable use of peatlands. Incentives for its implementation are, however, not yet available to the extent necessary, despite the fact that the possibilities to mitigate negative external effects (Chapter 6.4) are being outdated (Pigou 1920, Coase 1960). These instruments include clearly defined liability rules, the consequent application of the polluter pays principle, and the introduction of incentive based tools (OECD 1997).

What is needed is a firm establishment of the concept of 'good practice' – for example in the Federal Soil Conservation Act, which has to be clearly defined for the conservation of organic soils. In addition, there is an urgent need for regulating liability for land use on peatlands that is harmful for the climate. Currently, the general public have to purchase the rights of landowners to emit greenhouse gases, that is, the taxpayer must pay so that activities that are harmful to the environment and society are *not* undertaken.

A consequent application of the polluter pays principle requires the immediate abolition of environmentally harmful subsidies, and the taxation of environmentally harmful activities. Additionally, beneficial ecosystem services should be rewarded – e.g. through a bonus-malus system with positive and negative incentives, which could act as a mechanism to ensure that the desired behaviour is achieved.

Rational economic policy requires a consistent and complete set of goals, to achieve a high success rate through the targeted use of regulatory and economic instruments (Fritsch et al. 2010). The principles of sound financial management demand that policy goals should be SMART: Specific, measurable, achievable, relevant and time-bound. Additionally, performance indicators need to be determined to verify the achievement of the objectives against a reference situation (Art. 27 Section 3 EU Financial Regulation).

Subsidies for conventional agriculture on peatlands represent the exact opposite: they are perverse subsidies that reward environmentally damaging behaviour. In Germany alone, the direct payments for conventional agricultural use on peatlands amount to over 300 million EUR per year. With a political shadow price of 80 EUR per tonne CO_2 (Maibach et al. 2007, UBA 2012b), the yearly release of 45 megatonnes of CO_2 (UBA 2012a) would translate to economic losses of at least 3.6 billion EUR per year. With respect to the legitimization of this government's irrational way of subsidising, the agricultural economist Folkhard Isermeyer asks the socio-politically important question: 'Can we continue to prosper when policy decisions are made that cost a lot of money and that – according to scientific knowledge – do not contribute to the attainability of declared political goals?' (Isermeyer 2012).

7.3.2 Regulatory policy

According to §1 of the Federal Soil Conservation Act (BBodSchG), 'harmful soil changes are to be avoided (...) and precautions shall be taken against adverse impacts to the soil. Impacts on the soil that affect its natural functions and its function as an archive of natural and cultural history should be avoided to the greatest extent possible'. Harmful soil changes include 'the impairment of pertinent soil functions that may result in dangers, significant disadvantages or significant annoyances to individuals or the general public' (§2 Section 3 BBodSchG).

The requirements for 'good agricultural practice', as described in §17 BBodSchG, remain unclear. In the real world, the above mentioned requirements seemingly do not have to be implemented. According to §17 Section 2 Line 4 BBodSchG, 'soil erosion should, where possible, be avoided by means of site-adapted use.' Independent from how site-adapted use is concretely defined, the question arises how a harmful change in soil conditions 'where possible' can be avoided. With the formulated 'significant disadvantages or significant annoyances to individuals or the general public', §2 Section 3 BBodSchG provides a broad tolerance limit that leaves the door open for opportunistic interpretations. Therefore, good practice regulations must be interpreted to mean that soil functions don't have to be conserved for the full 100 %, but that the level of conservation is dependent on what an individual financially can be expected to bear.

In a technical position paper, the Federal and State Working Group on Soil Conservation rightly asked the Standing Committee for Legal Affairs to examine the climate relevant aspects of soil protection legislation. The committee came to the conclusion that the Soil Conservation Act 'in its current form [can] not by itself guarantee sufficient protection of the soil functions nor of the carbon sink function of soils, as for climate protection no specific legal instruments for soil conservation are available' (Martin et al. 2011). The committee proposed to include the climate protection function of the soil, and 'measures that contribute to climate change mitigation and adaptation, in particular for safeguarding the humus balance and the carbon and water retention capacity of the soil' explicitly in the Federal Soil Conservation Act.

Furthermore, the proposal was made to compliment in the Federal Soil Conservation Ordinance that 'the use of hydromorphic soils may not be changed and that exceptions may only be made if their function as a carbon store is not affected'. With these changes, agricultural land use standards could be normalized and enforced by ordinance (Martin et al. 2011). These changes in soil conservation legislation would not only contribute to urgently needed legal certainty, but would also be an important milestone for the introduction of alternative environmentally friendly peatland utilisation strategies.

7.3.3 Climate policy

In the field of climate policy, the comprehensive integration of land use and land use change is only envisaged for a post 2020 climate regime (European Parliament 2013). In their comments on the CAP reform, Isermeyer & Weingarten (2012) stress the importance of peatland rewetting for climate change mitigation by stating that 'in order to make significant strides toward climate protection goals, it is of utmost importance to give priority to the rewetting of drained carbon rich soils and to stop arable use of these areas.' The current possibilities for including peatlands in a comprehensive climate protection policy are, however, still quite limited. In December 2012, the Conference of Parties (COP 18) of the United Nations Framework Convention on Climate Change (UNFCCC) set new guidelines for the reporting and accounting of greenhouse gas emissions in the land use sector (Land Use, Land Use Change and Forestry, LULUCF). This included the introduction of a new activity ('Wetland Drainage and Rewetting') in the Kyoto Protocol, which enables the industrial states to account for peatland rewetting as a positive factor in their greenhouse gas balance. In May 2013, the European Parliament decided that EU Member States must mandatorily account for greenhouse gas emissions from cropland and grazing land within the EU by January 1, 2021 (European Parliament 2013). Such a decision specifically names 'paludiculture practices' as a measure for improving the agricultural management of organic soils (European Parliament 2013).

Peatland rewetting must already be accounted under the Kyoto Protocol in case of rewetting of drained forests (under forest management) and afforestation of rewetted peatlands (under afforestation) (Barthelmes et al. 2015).

Against the background of inadequate options available at the national level, a voluntary carbon market has evolved in recent years, which since spring 2011 also covers the conservation and rewetting of peatlands. Pioneer is the Verified Carbon Standard (VCS), which has created its own module for peatland projects (Peatland Rewetting and Conservation, PRC) (VCS 2011).

At the same time, a new regional carbon credit system called MoorFutures® was developed in Mecklenburg-West Pomerania and brought to market (Permien & Ziebarth 2011, Schäfer et al. 2012, Joosten et al. 2015). These MoorFutures® can act as a financial instrument for advancing the rewetting of peatlands, but have not been designed for financing unprofitable paludicultures.

7.3.4 Funding opportunities for paludiculture

As a part of the CAP reform process of recent years, proposals for the conservation of peatland have been made from different angles. One of these proposals of interest is a policy position paper from peatland-rich federal states in Germany, which focuses on an intensified effort for rewetting drained peatlands (Jensen et al. 2012).The Scientific Advisory Board for Agricultural Policy at the federal Ministry of Food, Agriculture and Consumer Protection (BMELV) (WBA 2010) recommends the federal government to reallocate appropriations recommitted by the reduction of direct payments to targeted measures for combatting climate change or loss of biodiversity, including the rewetting of peatlands.

Chapter 7.2 discussed the possibilities of funding within the current agricultural policy framework, and how policies need to change to make paludiculture eligible for subsidies. Important are the maintenance of direct payments for paludiculture under the First Pillar, and the promotion of paludiculture under the Second Pillar of the CAP. Additionally, the introduction of paludiculture requires conceptual changes in the structure of the subsidies, including:

1. A change in the way how subsidies for agri-environment measures are calculated: reduced yields

7.3 Control mechanisms and incentives for paludiculture

and additional costs compared to conventional agriculture are no appropriate reference point, as it would implicitly and incorrectly accredit drainage-based management of peatlands as 'good practice'.
2. The promotion of typical peatland biodiversity on paludiculture areas through agri-environment measures, which can be achieved by restricted use (e.g. late and section-wise mowing, the creation of landscape elements and so on).

For that purpose, an extension of agri-environment schemes is necessary. The duration of the schemes should extend beyond the respective funding period to secure (similarly to primary afforestation grants) the long-term concerns of the farms – e.g. be hedged in accordance with the reforestation premium.

In summary, it can be concluded that the proposals for peatland conservation submitted under the CAP reform processes contribute only in a limited way to the sustainable agricultural use of peatlands. The retention of the agricultural subsidies under the First and Second Pillars of the CAP is hardly suitable for stimulating the widespread implementation of paludiculture (Chapters 7.1 and 7.2). There is an urgent and fundamental need for reform. Moreover, what is needed is a consistent application of the polluter pays principle in relation to environmental degradation through irresponsible use of peatlands. Above all, environmentally harmful subsidies must be stopped; a start can be made by the greening and the formulation of cross-compliance regulations for peatlands. Regardless of how the reforms to direct payments will play out, a concretization of 'good practice' for organic soils is needed. To this end, appropriate changes in the Federal Soil Conservation Act are required, which therewith would automatically become effective in the subsidy regulation of cross-compliance.

8 Social aspects of paludiculture implementation

In the past, people had ambivalent feelings towards mires (Chapter 8.1). Mires were avoided, seen as inhospitable and threatening places. Equally, mires were also landscapes with a mystical, hidden, and even beautiful side. By drainage and cultivation, mires lost their characteristic appearance. Many people today are not aware that the drained grassland next to the road is in fact a peatland. Some people still remember the efforts and hardships required to turn mires into usable land. Yet, some others feel bad about the loss of characteristic peatland fauna and flora as a result of intensified land use. When peatlands are rewetted for paludiculture, many people feel this change again as a dramatic loss of familiar landscape and as a capitulation to nature. Thus, conflicts of interest are inevitable.

Planning and decision making have to take the views and interests of residents and land users into account to avoid conflicts of interest (Chapter 8.2). Stakeholders must be given the chance to voice their opinion, and their concerns should be taken seriously. All information about the needs and consequences of land use change should be freely available and taken into consideration.

The implementation of paludiculture involves many risks and difficulties at the operational level, which have to be identified and addressed (Chapter 8.3). Answers to problems should be found by cooperation between research, politics and administration. Experiences from other innovative land use approaches can show what to take into account to implement paludiculture successfully. Besides political frameworks such as the EU Common Agricultural Policy and the German Renewable Energy Act, consulting services and advanced training of land users play a major role in implementing and expanding paludiculture (Chapter 8.4). Pilot projects must demonstrate its practical suitability, and further develop the concept. Such pilot projects could be initiated by local networks with the aim to produce and consume the biomass at a regional level.

8.1 The relationship between humans and mires over time

Steffi Deickert & Jenny Piegsa

Generally mires were – and partly still are – seen as incontestable and mythical, but also as places that had to be conquered. In more recent times, they have been deeply drained and changed into cultivated land. In Western and Central Europe, only a tiny minority of pristine mire has survived as last refuges of 'wilderness'. Mires provoke feelings of both unease and fascination. Throughout the centuries, the perception of mires has been very diverse. Furthermore it has been changing immensely, parallel to the progressing knowledge about these ecosystems. The following Chapter gives an insight about changing attitudes towards conservation and utilisation of mires from the 17th century onwards.

8.1.1 From curse to blessing: peatlands under cultivation

'By God's punishing hand they have been commanded, as a curse for the people that lived in this land in the old times and as a warning for us, their descendants'. This is how the German-Dutch reverend Johan Picardt of Coevorden explained the occurrence of mires in 1660. With this statement, he reflected exactly the contemporary perception of people with regard to mires. Farmers regarded mires as a hindrance that frustrated tillage, animal husbandry and travel, besides isolating the people living in their surroundings (Krüger 1916). The lowlands of northern Germany were particularly affected by this 'curse' that had to be lifted. It was king Frederick the Great (1712–1786) who decisively initiated large-scale agricultural development and colonisation of what was considered as 'wasteland' (Chapter 2.1).

Such standpoints prevail until today, but the perceptions of people regarding mires have become more complex. How did people relate to mires in the past and how do they currently do? How did the perceptions change over time? The attitudes and feelings towards mires are reflected in contemporary art and literature in a wide range of ways (Colour pictures 87 and 88). In poetry, words like 'shudder' and 'unease' alternate with 'captivation' and 'love' for the homeland, as in Hugo Kaeker's poems 'Heath of Pomerania', ''Moorspuk'' (meaning mire spook) and 'Will o' the wisp' (Kaeker 1919). In his poem 'On the mire', the Pomeranian poet Hermann Kasten (1906) expresses uncanny and uneasy feelings, and inspires the creation of ghastly stories and legends.

In the story 'Vagabond' by the Norwegian author Knut Hamsun (Hamsun 1928) the mire in his native landscape ranges from the 'murky pool of hell' to a 'blessing'. The first image refers to the mire in its original form; the second image describes the mire after reclamation. The message is clear: only a drained peatland is a good peatland. Similar is the way the Pomeranian artist Paul Holz (born 1883, near Pasewalk) documented the reclamation of Pomeranian mires at the turn of the 20th century through a series of ink drawings (Figure 8.1, Figure 8.2) (Förster 1998). Unlike the 'Vagabond', the novel 'The Jeromin Children' (second volume, 1947) by Ernst Wiechert, associates positive emotions with mires. In this novel, the mire is strongly linked to the love for the homeland and offers refuge in times of war: 'A mire can provide stable ground in unsteady times' (Wiechert 1947).

The views that painters, illustrators, poets or writers had about mires might have been quite different from those of a settler, whose task consisted in turning that dreadful space into a cultural landscape. The question arises as to why people did not avoid mires completely if the sites made them feel anxious and fearful. One of the main reasons for accessing mires was population growth. New areas for settlement and food production were needed, especially during the population expansion after the 30 Years' War in the 17th century, as well as after World Wars I and II, when large numbers of refugees arrived, and soldiers returned to what nowadays is Germany (Hammer & Leng 2008).

Besides, people knew how to use the gloomy ambience and inaccessibility of mires to their advantage. Mires offered protection from unwanted attackers (protection against external threats) and were used to

Figure 8.1: 'Moorlandschaft' ('Peatland landscape') by Paul Holz, 1936/37.

8.1 The relationship between humans and mires over time

isolate and deter socially problematic groups, such as prisoners (protection against internal threats) (Hammer & Leng 2008). The colonisation of mires not only happened out of necessity but was also done intentionally. The protective proximity of mires was used to establish settlements like Demmin, which is situated at the border between West Pomerania and Mecklenburg (Succow & Jeschke 1990).

Mires play a role in the perception of, and the identification with, the native landscape, as can be inferred from Pomeranian names of places. The word 'Luch', for instance, means swamp or mire (Holsten 1940). Holsten (1940) explains in his article 'How the Pomeranian does view his native land?' that the words 'spuk' and 'grugel' (meaning spook and horror), together with the colour black, were used to denote sinister places and also mires. In contrast, mires are also often associated with the colour white: 'However, in our bogs, the cotton grass (*Eriophorum*) and birch grow; one with the white fibres of its fruit and the other with the white bark of its trunks, shrouding the mire with a white veil' Holsten (1940). Holsten's descriptions show a duality regarding mires: whereas mires generate negative associations as stories, novels and poems denote, they also are perceived as bright and friendly places.

The 'Urbarmachungsedikt' (Reclamation Edict) of the Prussian king Frederick the Great in 1765 played a crucial role in the increased colonisation and agricultural reclamation of the mires in Northern Germany. The edict commanded the state-enforced colonisation and cultivation of mires (Hammer & Leng 2008). Frederick the Great invested huge amounts of money in colonising the land, because he held the view that the power of a state depended on the size of its population (Knobelsdorff-Brenkenhoff 1988; Pantenius & Schönert 1999). Consequently, mires and swamps became a target of this 'land relief', as these 'useless wildernesses' did not conform to the contemporaneous orderliness and the general obsession to improve, as king Frederick stated in 1775: 'He who improves his lands, reclaims wilderness and drains swamps, conquers these from barbarism, and provides livelihood to colonists. These then work, as they are now able to marry, in good spirits on the procreation of the human race, and increase the numbers of industrious citizens' (Preuß 1834).

Figure 8.2: 'Moorlandschaft mit Kühen' ('Peatland with cows') by Paul Holz, 1936/37.

Many families found a new home in the peatland colonies. However, the vision of Frederick the Great of a densely populated, prospering state had also a downside: The recruitment of colonists was often based on false promises (Gilsenbach 1971) because initially, the colonist families had a life full of deprivations. Nonetheless, after several generations, the hostile mires were transformed into 'blooming settlements with neat houses, wavy grain fields, and green pastures crowded with healthy livestock' (Conwentz 1916).

The view that natural mires are worthless, and that transforming them to productive land is a major agricultural achievement, persisted well into the 20th century (Hammer & Leng 2008). Integrating peatlands into the agricultural sector was always a political priority. In Germany, the pressure exerted on peatlands as a 'land reserve' was reinforced by the efforts to achieve self-sufficiency in food production during both the Third Reich and the German Democratic Republic (GDR) periods.

In the GDR, peatlands were substantially modified by 'Komplexmelioration' (Chapter 2.1) in the pursuit of self-sufficiency in food production (Könker 2007). This concept departed from the assumption that it is possible to identify all environmental limitations to production and to improve (latin *meliorare*) the site conditions with a variety of measures (Könker 2007). This belief resulted in an extreme disruption of mires during the 1960s. Projects of gigantic dimensions – like the melioration of the Friedländer Große Wiese mire complex and the Oderbruch, reflect an attitude towards nature that had its basis in the 'malleability of everything' ideology of socialism (Behrens & Rösler 1999). Projects were declared 'youth projects', in which students and young workers were summoned to participate as 'volunteers' and were attracted by the provided entertainment. Many former participants still recall today the collective achievements of former times (Behrens & Rösler 1999). The reclamation of the Friedländer Große Wiese even provided the basis for a novel and a film entitled 'Egon and the eighth wonder of the world' (Wohlgemuth 1962, Steinke 1964).

The main driver behind the intensification of peatland use was the state (Könker 2007, Mohr 2007). It went so far as to introduce new study courses on Komplexmelioration. However – especially in the case of peatlands, the 'improvement' proved soon to be counterproductive. Critical voices were initially ignored but became strong enough to challenge and eventually stop socialist progress euphoria. Such criticism was part of the general rise in environmental consciousness in the late 1980s, and proved to be right: in the long term, the intensified cultivation of peatlands was not progress. Instead of improving productivity sustainably, the peatland soils were irreversibly damaged (Succow 2001a, Chapter 2.3). Supported by a general change of approach towards nature conservation and green politics, the attitude towards peatlands took a new turn.

8.1.2 Mires under protection: from natural heritage to the saviour of climate

The alarming rate of change that occurred in intensively used peatland areas lead to the first considerations of mire conservation at the end of the 19th and increasingly at the beginning of the 20th century. Compared to the 200-year history of nature conservation in Germany, mire conservation started somewhat late. Furthermore, focus and justification of mire conservation changed over time. Initially, mires were protected as natural monuments to preserve the memory of the original German landscape. Today, their importance for the global climate is a major argument for their conservation.

The history of mire conservation can be divided into three stages (cf. Fansa 1999). The early stage (1900–1948) was started by C.A. Weber with the essay 'On the conservation of mires and heaths of Northern Germany in their pristine state, as well as on the restoration of pristine forests' (Weber 1901). Shortly before Weber's publication, the Prussian Ministry of Agriculture had invited the expert opinion of scientists and scientific associations on the question 'in which ways animal and plant species, as well as vegetation stands with an interesting physical appearance, whose existence is threatened by the growing expansion of cities, the destructive impact of industry, the continuously consolidating net of railways, the canalisation of rivers, and the agri- and silvicultural meliorations, can be locally preserved for future generations?' (Weber 1901). Weber stated: 'Regretfully, it cannot be denied that researchers in the northwest German lowlands – one of regions with the densest concentration of mires in the world, who engage in the numerous questions that especially bogs are posing, now already struggle in vain for their solution. In a few years, this will not be possible anymore in Germany, considering the speed with which one aims to sacrifice the last traces of Nature on these interesting formations, to utility.' He proposed to put, amongst others, the Zehlaubruch mire in Eastern Prussia (today in Kaliningrad Oblast, Russia) under protection, a status that was indeed granted in 1910. In 1904, Hugo Conwentz expressed the wish 'that here or there an individual mire remains secured against every human impact and protected for study purposes'. In 1907, the Plagefenn, a terrestrialisation mire near Chorin, received protection as the first nature reserve in northern Germany. The move was initiated by forest superintendent Max Kienitz, as well as by Hugo Conwentz, who, since 1906, was the head of the newly founded Administration for the Maintenance of Natural Heritage (Engel 2007).

In reaction on the growing claims on mires during World War I, especially as labour camps for prisoners of war, Conwentz convened in 1915 a conference specifically dedicated to the conservation of mires (Schoenichen 1926). By that time, many functions of mires had already been recognised – for instance, their role in regulating regional hydrology (Jaekel 1922), as deposits of archaeological artefacts, as objects of scientific study, as habitats for specialised plant and animal species and as carriers of aesthetic and ethical values (Beltz 1916a, Conwentz 1916, Geinitz 1916). However, the willingness to conserve mires was restricted to small areas, which should be conserved for future generations and should preserve the original natural scenery (Conwentz 1916, Beltz 1916b). Only Weber (1901) had dared to propose 'to withdraw a several square kilometres large landscape from cul-tivation'.

In the National Socialist era, the exploitation of mires continued in spite of the enactment of the Reich Conservation Act ('Reichsnaturschutzgesetz') and the increasing attention for species protection (Frohn 2012). The rationale involved an attempt to achieve self-sufficiency in food production during this time, which focused on highest agricultural yields. Prisoners of concentration camps and convicts were forced to drain peatlands 'explicitly without the use of machines, thus enduring enormous physical and psychical hardships' (Berg 1999). Songs like the 'Börgermoorlied' (the song of the peat bog soldiers) and the factual antifascist report 'The peat bog soldiers' are impressive testimonies of those events (Langhoff 1935).

The above mentioned early stage of peatland conservation was followed by the so-called latent stage (1949–1989). In the 1950s, the predominant view was that all technical means should be used to cultivate peatlands. Conservationists were regarded to stifle the development and to be working against progress (Frohn 2012). Apart from this attitude that hardly tolerated any objections, the hardship that came immediately after the war put further pressure on the use of mires. Western Germany received millions of refugees after World War II from former German territories, and from the Soviet Occupied Zone. Those people needed food supply (Frohn 2012). In the Federal Republic of Germany, giant steam-powered ploughs (developed by the company Ottomeyer) were used to cultivate raised bogs in the Emsland region (Emslandplan). Those giant ploughs turned bogs into agricultural land irreversibly (Berg 1999). In the GDR, specific legislation on 'scheduled design of socialistic land improvement' (Landeskulturgesetz) was enacted in 1970, with the aim to fundamentally and permanently improve the productivity of agriculture and forestry soils (Christoph et al. 1973). However, the utilisation of peatlands based on this legislation did not prove to be sustainable. The melioration of peatlands resulted quickly in soil degradation instead of increased productivity (Chapter 2.2).

In 1971, the Ramsar Convention was signed in Iran, which was of international importance for wetland conservation (Succow & Jeschke 1990). Gradually, the views on nature conservation and environmental protection became more positive, and mires subsequently benefitted from that process. From the late 1980s onwards, landscape ecology was taken into account and integrated into measures of melioration (Könker 2007).

The implementation of successful mire conservation only started in the third stage of emancipation (after 1990), when the significant importance of peatlands as carbon stocks was increasingly recognised (Chapter 10.1). Since 1990, the topics of global climate change and biodiversity are frequently present in the media and environmental policy, and peatlands receive increased attention for their ecosystem services. This in turn led to the development of mire conservation programmes at federal state level in Germany, which included large scale rewetting schemes (Länderfachbehörden 2012, Ullrich & Riecken 2012, Box 2.2). This development benefitted from the decreasing demand for agricultural land, because both productivity and the costs to work the land have increased. Additionally, restoration projects were funded by the European Union. However, the efforts to restore and foster natural development were only implemented on a small part of the overall degraded peatland area. More than 90 % of the entire peatland area in the lowlands of northern Germany is still used and continue to be subject of peat degradation.

8.1.3 Reconciling conservation and use: Paludiculture as an opportunity

Mires can only be conserved and utilised if the water level stays close to the surface. Rewetting often meets critique and opposition, because the rise of the groundwater table may affect residential areas and can also result in the dieback of trees (Chapter 8.2). For that reason, people stick to drainage-based agricultural land use, which is neither sustainable nor viable in the long term. In contrast, for the implementation of paludiculture, groundwater levels are raised up to the ground surface. This preserves the peat and regenerates new habitats for mire-specific animals and plants. Harvesting biomass from paludiculture enables to combine peatland conservation with their economic utilisation. However, paludiculture may also leads to substantial changes of the landscape shape, and it is not clear how people will perceive this new form of economic activity with, for instance, large-scale reed fields. The question arises whether paludiculture will change people's perception of mires, or not. Existing conflicts of interest should therefore be identified by involving all stakeholders in the conception and implementation of paludiculture.

8.2 The integration of stakeholders and the public

Box 8.1: Stakeholders of paludiculture

Silke Kleinhückelkotten & Horst-Peter Neitzke

To identify the individuals and institutions relevant to the implementation of paludiculture, it is recommended to carry out a stakeholder analysis for the respective region. This analysis should include a comprehensive group of stakeholders that range from the local municipality to the European Union. The analysis can be conducted using internet research, and qualitative methods (e. g. personal communication and telephone interviews with al-

Table 8.1: Stakeholders and their potential role to implement paludiculture in the region West-Pomerania (o / + / ++/ +++: none / medium / high / very high (potential) importance; (+), (++): indirect importance).

Stakeholder	Type of stakeholder	Experts	Decision makers	Multipliers	Beneficiaries	Affected individuals	Pioneers	Opponents
Federal State of Mecklenburg-West Pomerania (MWP)	Member of MWP State Parliament	o	++	+	o	o	++	++
	Ministry of Agriculture, the Environment and Consumer Protection	++	+++	++	o	o	+++	+++
	State Office of the Environment, Nature Conservation and Geology	+++	++	++	o	o	+++	+++
	State Office of Agriculture and the Environment West-Pomerania	++	++	+	o	o	++	++
	State Forestry Commission	+	+	+	(+)	(+)	+	++
	Landgesellschaft MV mbH	+	o	++	o	o	++	o
	Foundation for the Environment and Nature Conservation MWP	++	o	++	o	o	++	+
	Farmers Association MWP	+	o	++	(+)	(+)	++	++
	Environmental and nature conservation associations	+	o	++	o	o	+	+
Administrative district (Landkreis)	Members of the District Council	o	+	+	(+)	(+)	++	++
	District Administrator	o	++	+	(+)	(+)	++	++
	Environmental Office	++	+	+	o	o	++	++
	Regional Planning Association of West-Pomerania	o	+	+	(+)	(+)	+	+
	Farmers Association of West-Pomerania e. V.	+	o	++	(+)	(+)	++	++
Municipality	Council members	o	o	o	(+)	(++)	++	++
	Chief Administration Officer	o	++	++	o	(+)	+	+
	Mayor	o	++	++	(+)	(++)	++	++
	Municipal Administration	o	+	+	(+)	(+)	+	+
Areas for potential implementation	Water and Soil associations	+++	o	+++	+	+	++	++
	Nature park administrations	+++	+	+++	(+)	(+)	++	++
	Regional land associations	++	+	++	(+)	(+)	+	++
	Environmental and nature conservation associations	++	o	+	+	++	+	++
	Enterprises							
	Paludiculture implementing agricultural businesses	+	o	o	+	o	++	o
	Agricultural businesses adjacent to paludiculture plots	+	o	o	o	+++	o	++
	Landscape conservation enterprises	++	o	+	+	+	+	o
	Biomass processing plants	+	o	+	+++	o	++	o
	Population							
	Local residents in the vicinity of paludiculture plots	o	o	o	o	+++	o	++
	Local citizens' groups and associations	+	o	+	o	+	+	+++

> ready identified stakeholders). Depending on their (potential) function, several stakeholder categories have to be considered, namely (Table 8.1):
> - Specialists with the required knowledge and expertise of conceptual development and implementation of paludiculture;
> - Decision makers setting the political and legal framework for paludiculture, or with responsibility for its implementation;
> - Multipliers that may spread the concept of paludiculture;
> - Beneficiaries that may directly profit from paludiculture implementation;
> - Individuals that could be negatively affected by the implementation and do not benefit directly from it;
> - Pioneers paving the way and actively supporting the idea of paludiculture;
> - Opponents with the intent and the potential to hamper the implementation.
>
> The results of a stakeholder analysis in the German federal state of Mecklenburg-West Pomerania are shown in Table 8.1 (Behrendt et al. 2012). The table does not include the national and EU levels. Every stakeholder listed in the first column is allocated to a respective category (column 2–8). This allocation depends on the influence that each particular stakeholder can exert on the direct implementation of paludiculture. Some stakeholders could, depending on their position, act both as pioneers or as opponents for the implementation of paludiculture.

resulting from successfully implemented paludiculture projects. However, it can be very difficult – even impossible, to convince people who are emotionally attached to landscape appearance, by using ecological or economic arguments. This especially applies to people who grew up in the native countryside or whose professional activities are related to supplying the population with food from drainage based peatland agriculture. Similar difficulties arise with stakeholders (including local politicians) who feel negatively affected, sometimes only for the reason that they were not sufficiently informed and not adequately involved in the planning process. These deeply rooted reservations pose a substantial challenge for the implementation of paludiculture projects.

8.2.2 Communication and participation in land management

Prior to the implementation of paludiculture projects a communication process should be initiated involving the various stakeholders (Box 8.1). The involvement of socially and professionally relevant stakeholders as well as of the local population is a precondition that is necessary but not sufficient to gain acceptance. To approach the various stakeholders, the format of events and the form of communication needs to be specifically adjusted to each target group (about the importance of stakeholder participation in planning processes and target group oriented communication, see for example Bischoff et al. 2005, Kleinhückelkotten & Neitzke 2005, Kleinhückelkotten & Wegner 2008, Sinning 2005).

Symposia can be used to inform professional stakeholders from policy making, nature protection as well as from the agriculture and energy sector about the idea, aims and methods of paludiculture. Symposia allow stakeholders to discuss opportunities and risks posed by paludiculture to the development of rural areas. Additionally, symposia can be used to identify bottlenecks and opponents that frustrate the turn to paludiculture. Furthermore, such events may serve to develop ideas to address specific problems (e.g. high drainage costs). In symposia on the prospects of paludiculture in Mecklenburg-West Pomerania, for example, possibilities to improve the framework of agricultural policies, as well as the introduction of carbon credits to support the rewetting and management of wet peatlands have been discussed.

Talks with specific stakeholders can be used to spread and facilitate the exchange of knowledge about paludiculture. Another way to transfer knowledge is by organising workshops for local stakeholders, where specific concerns can be addressed. A more detailed discussion is needed when it is necessary to identify possible conflicts of interest regarding paludiculture (Chapter 8.2.1). In West-Pomerania, such talks and workshops were also used to identify areas that could be suitable for paludiculture projects. The target groups for this type of events in Germany include (Table 8.1):

8.2 The integration of stakeholders and the public

Box 8.2: Conflicts of interest related to land use change

Silke Kleinhückelkotten & Horst-Peter Neitzke

Figure 8.3 illustrates the range of reactions from various stakeholders on the suggestion to switch land use on a particular peatland from pasture to paludiculture. In the particular case, parts of the field are already wet and its use as pasture is only possible by continuously pumping the water out. Other areas of this field are a nesting habitat for meadow-breeding birds. As an alternative to current land use, it was suggested to turn the field into a reed stand for the production of fuel or construction materials, under the precondition that groundwater level would be raised by restricting – or even stopping, drainage. The reactions of the stakeholders (four people working either professionally or as volunteers in nature conservation, two farmers, several residents, and an owner of local holiday apartments) are presented in an approximate way.

The example shows that members of the same stakeholder group, with the same attitudes towards nature and landscape, have different ideas of how that area should be used in the future. The two farmers took attitudes that are very common. One farmer had an economic and rational attitude to the potential project whereas the other had a strong attachment to his identity as a producer of food. The feelings of the local residents are strongly influenced by discussions about past fen rewetting projects in the region. In this particular case, several arguments were made against the switch to paludiculture, and referred to potential conflicts. These reasons were also quoted in several other talks and events about the implementation of paludiculture. The objections not always represented the personal opinion of the participants but rather their personal assessment on how affected people might see the subject matter (Table 8.2). Parts of the arguments were technically wrong, which shows that ignorance plays a role in creating conflicts.

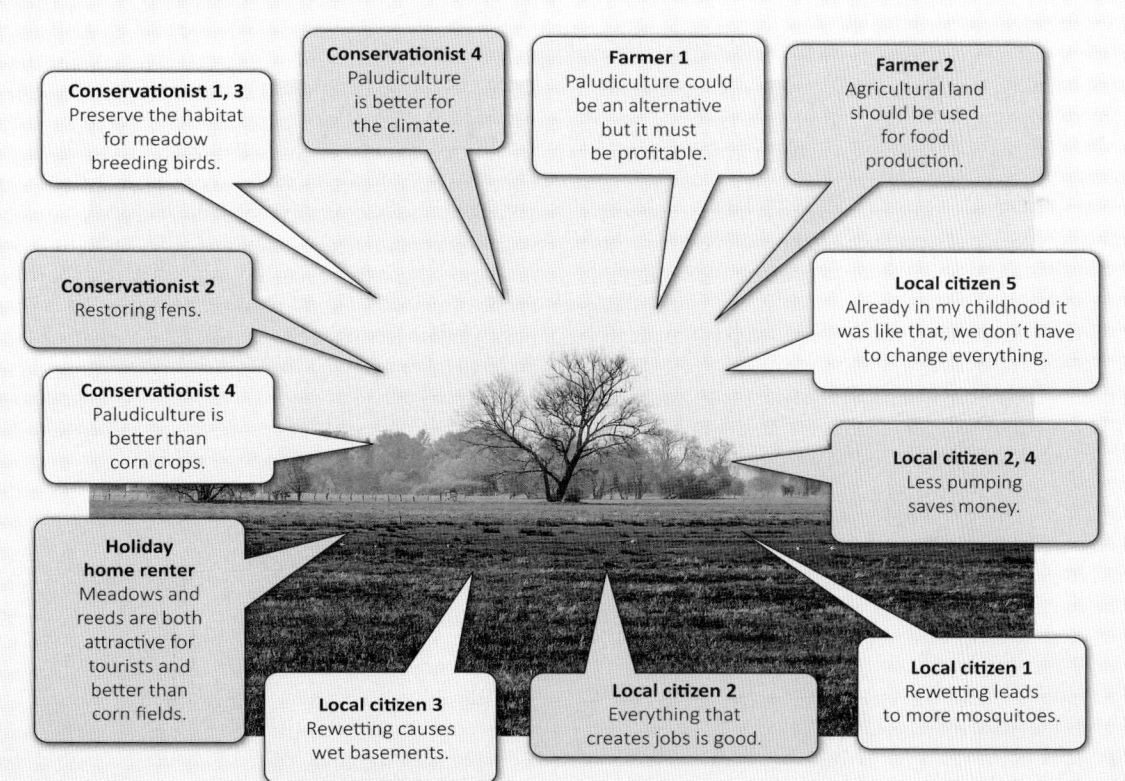

Figure 8.3: Reactions of various stakeholders to the suggestion to rewet a drained peatland and cultivate reed in paludiculture.

8.2 The integration of stakeholders and the public

Silke Kleinhückelkotten & Horst-Peter Neitzke

Paludiculture is a form of land use hitherto unknown to the public. In particular, the associated rewetting potentially involves conflicts of interest. The implementation of paludiculture therefore requires:
- Information provision to relevant stakeholders about the aims and methods as well as the advantages and possible risks of paludiculture including a comparison to conventional drainage-based land use;
- Participation of stakeholders, who with their knowledge and relevant skills may contribute to analysing the potential of paludiculture, as well as to develop strategies for its implementation;
- Stimulation of initiators of paludiculture projects;
- Integration of stakeholders with professional and financial capacity for developing and implementing paludiculture projects;
- The establishment of regulatory, financial and political frameworks, particularly with respect to agricultural subsidies and nature conservation legislation;
- Public acceptance of this new form of land use, and the landscape changes involved.

In order to achieve the above mentioned objectives, many different target groups have to be approached and involved in ways specifically adjusted to their interests. The first two steps in that direction are the identification of important stakeholders (Box 8.1) and the assessment of conflicts of interest that can arise with the implementation of paludiculture. Taken the above mentioned steps as a basis, strategies can then be developed for communication and participation.

8.2.1 Competition for land and conflicts of interest

As previously stated, land use is often associated with conflicts of interest. Competition occurs if several stakeholders have different interests regarding the use of a specific plot of land. If a particular stakeholder asserts his ideas against the interests of other stakeholders, conflicts usually arise. A conflict describes a situation where the actions or intentions of a stakeholder violate or threaten the interests of another stakeholder.

A change in land use towards paludiculture is a decision that is not only made at business level. Consequently, potential conflicts of interest have to be identified and considered early in the planning process. It should also be taken into account that the decision to adopt paludiculture in a particular area is a long-term one – and in some cases, irreversible. In this regard, a change to paludiculture differs substantially from conventional changes in land use, such as crop rotation or a change in cultivation methods. It is therefore important to involve people who are able to assess the views of the local population and other local stakeholders on change in both land use and landscape appearance. As these attitudes can be very heterogeneous, it does not suffice to have only a talk with local elected representatives like the mayor. If large-scale paludiculture projects are planned, it is sensible to carry out polls among the population and consult individuals in key positions (Box 8.1). It is important to involve all socio-demographic groups in such polls, and consider all ways in which stakeholders could be affected. An idea of the general inclination of people towards land use change can be obtained from the local media. Box 8.2 (Kleinhückelkotten & Neitzke 2013) illustrates competing interests and conflict potential for a site in Vorpommern (North-eastern Germany), where paludiculture could be the dominant scenario of land use.

Large conflicts may arise when higher groundwater levels are re-established. In particular, the word 'rewetting' provokes negative feelings in many places. Notably, local residents of the areas to be rewetted are likely to offer considerable resistance, because of fearing that rewetting will cause material damage and affect the quality of life. Strong opposition may also be expected where woodlands are affected, as forests are generally associated with high economic, ecological and aesthetical values (Kleinhückelkotten et al. 2009). Problems for the implementation of paludiculture also arise out of mistakes made in earlier rewetting and restoration schemes, which has caused strong prejudices towards such projects. Shortcomings have especially occurred regarding communication between project proponents/executors and the local population, when people felt insufficiently informed and excluded from decision making processes.

Scepticism towards paludiculture also emerges from the idea that this form of land use is not economically viable, that it is implemented at the expense of conventional agriculture, and leads to changes in the familiar appearance of the landscape (Table 8.1). In many places grasslands have already been converted to arable production of bioenergy crops, thus the remaining meadows and pastures on peat soils are increasingly seen as valuable habitats that need to be preserved.

There are some reservations about paludiculture because of lack of knowledge. Such problems can be solved by providing information to the local population and to stakeholders – e.g. on greenhouse gas emissions and ground subsidence in both conventionally used peatlands and sites used under wet conditions. A significant argument could be the economic benefits

Table 8.2: Ecological, social and economic arguments against paludiculture and rewetting of agricultural land.

		Ecological	Social	Economic
Land use		The habitat of rare species such as meadow breeders is destroyed.	Agricultural areas that would be needed to mitigate hunger and rising food prices are abandoned.	
			Agricultural areas that have been reclaimed for agriculture with great effort are simply abandoned.	
			A historically grown landscape is destroyed.	The economics of paludiculture are questionable.
			The familiar appearance of the landscape is destroyed.	
Rewetting		Rewetting leads to anaerobic decay of plant matter which, in turn, leads to the emission of methane, a potent greenhouse gas.	In case open water bodies originate, nuisance by mosquitoes will increase.	The rise of the groundwater table causes moisture-related damage to adjacent houses.
			Climate change may facilitate the spread of mosquitoes that carry pathogens of infectious diseases such as malaria.	
		Many rewetting or restoration projects were not successful in ecological terms.	In the past, rewetting projects were implemented against the wishes of the parties concerned.	
		The rising water table causes death of trees in forests adjacent to the rewetted plots.		

- Representatives of the federal, regional and municipal authorities;
- Associations responsible for water and soil management;
- Nature and environmental conservation associations and local nature conservationists;
- Farmers associations and landscape management organisations;
- Nature conservation and forestry administration.

By the method of activating interview, stakeholders are introduced to topics that might be interesting for them. This method was applied in West-Pomerania to inform important stakeholders about the topic of paludiculture, to familiarise them with this new form of land use, as well as to explore the prospective of support.

The experiences resulting from special events, stakeholder talks and interviews in West-Pomerania show that the acceptance for paludiculture projects by the local population, their municipal representatives plus local action groups is of great importance. The involvement of local people in the discussions about paludiculture poses a challenge. The objective should be to involve not only individuals that are already active, but to consider all view points of the population concerned. One way of doing so is to use a citizens' panel (German: Bürgerforum) (Box 8.3).

Box 8.3: Citizens' panel about paludiculture

Steffi Deickert & Christian Schröder

A citizens' jury is a specific and ambitious form of participatory action research. The experts, whose opinion and assessments usually dominate decision making processes, remain in the background but share their knowledge with the lay members of the citizens' panel (Ley & Weitz 2004). Up to 25 individuals gather information, facts, and discuss opinions and positions. Throughout the workshop, participants take part in an opinion-forming process. The result is a final assessment of the topic by the citizens' panel.

Here we discuss the experiences of a citizens' panel that was held in the coastal municipality of Usedom-Süd, in West-Pomerania. Two thousand citizens were randomly selected and invited to participate in the process. The invitation included the plea to participate and to submit a binding registration. Thirty individuals signed on to participate in the citizens' panel. The participation figure was 1.5%, – well within the expected 1–7% of invitees. Later, five of the individuals that initially had registered were unable to participate for personal and health reasons, lowering the final number of participants to 25 (of which one quitted after the first meeting). Over three weekends, the group discussed the future land use of the Thurbruch peatland. This 1,600 ha large peatland is currently being used for low intensity farming (grazing and fodder production). This form of land use is drainage-based and therefore, the peat soil has already subsided partly by more than 2 metres over the past 250 years. As the area is currently situated only about 0.5 m above sea level, measures to prevent it from being flooded by the sea are urgently required. However, the implementation of paludiculture is not yet planned in this particular example. The participation of citizens in the early stage of a planning process is exemplary for future measures in other fen areas, as the implementation of paludiculture is not only technically challenging, but should also take into account social aspects as well as the ideas and knowledge of local citizens.

During the first and second meetings of the citizens' panel, the participants were informed about the various subject areas. At the third meeting, the participants made the assessment 'The future of the Thurbruch – a life with the peatland' (Authors' team 2013). This assessment includes recommendations for measures that may serve as a basis for decision making to politicians and the public in future planning. The assessment of the citizens' panel was presented in a press conference and handed over to politicians, local government, scientists, and the media.

Some quotes from the citizens' panel's assessment include:
- 'Our assessment aims to inform all citizens of the region on the status quo of the Thurbruch peatland, and should animate to conceive and employ new ways of open-minded cooperation.'
- 'We need a future-orientated and long-term sustainable approach to land use of the Thurbruch peatland.'
- 'The principal objective should be that neither property nor people will be damaged nor harmed.'
- 'We aim for a diverse landscape in the Thurbruch peatland, where it is possible to integrate farming, paludiculture, and nature conservation.'
- 'The aim is to equally develop farming, tourism and nature conservation.'
- 'There is a consensus that paludiculture can contribute to climate regulation, and that renewable raw materials can be used to produce renewable energy and new products. Paludiculture can provide alternative income, as well as employment'.
- 'Paludiculture can contribute, at some extent, to the conservation and use of peatlands, and in this way, complement conventional agriculture.'
- 'By balancing the water management, the conflict of interest between farming, mire conservation, tourism, and flood prevention can largely be solved. It is important to prevent further drastic impact on the peatland, which implies that farming methods have to ensure that peat is preserved. This could be implemented by maintaining the current form of land use or by implementing paludiculture.'
- 'These ideas can pave the road for a new way of thinking with respect to other peatland areas.'

The citizens' panel's assessment has shown that local people have in-depth knowledge of the region and deploy creativity to solve problems, whilst regarding the interests of various stakeholders. The citizens wish for more self-responsibility and transparency about decisions, and are able to reach consensus even in case of complicated controversies.

8.3 Acceptance and implementation at the producer level

Michael Rühs, Achim Schäfer & Christian Schröder

Conventional peatland use entails undesired environmental effects and management problems. Nonetheless, a large part of the peatland areas – particularly in Northeast Germany, continues to be used for agriculture, although in the last few years large-scale extensification of farming did occur. Many farms are dependent on agricultural subsidies and restrict their activities to the minimally required maintenance. Low-intensity cattle farming (suckler cows), a widespread form of land use in drained peatlands, is unprofitable and can only be maintained by substantial subsidies (Müller & Heilmann 2011, Chapter 6.2). The lack of success of these funding programmes in reaching their objectives has often been criticized. As often more than 30% and in some cases over 50% of the revenue of extensive farming is dependent on subsidies (Rühs et al. 2005, Weber 2015), the future of this form of land use is determined largely by external decisions about the future of agricultural subsidies. The consequence is that many farms with this form of low-intensity land use have an uncertain future, and are held for disposal at least in the medium term.

The continuance of land use on peatlands requires the introduction of innovative and environmentally friendly methods of production. An important precondition is to use all opportunities provided by the transition to paludiculture, and remove all obstacles to achieve this aim. Methods of land use under wet site conditions have been so far tested only during pilot projects, or alternatively, remain as relicts of traditional land use. More practical experience regarding large scale paludiculture is still required. In addition, the implementation of paludiculture requires an adjustment of operational structures and processes. This adaptation is also necessary for the downstream-processing industry that manufactures marketable end products. At operational level, technical equipment and production processes have to be adapted, and it has to be assessed in what ways paludiculture may influence and improve business development. In this matter, one can draw on innovation processes in other land use sectors.

8.3.1 Innovations in land management

The path from the origin of an idea or new technology – via pilot projects, to its social acceptance and wide-scale establishment, usually entails a lengthy process involving different stakeholders. The process of development of social acceptance of new ideas or technologies and its posterior dissemination is called 'diffusion'. Similarly, the term 'adoption' refers to the implementation of new ideas by individuals, and its integration at the business level. The diffusion of innovations in the agricultural sector was analysed with respect to the introduction of new technologies (e.g. automatic milking systems), renewable raw materials, as well as bioenergy by Roos et al. (1999) and McCormick & Kåberger (2007).

Opinion leaders of different social strata are important drivers of diffusion because they are trusted and regarded as professionally competent. Equally, change agents are individuals that play an important role in convincing opinion leaders to support and spread an innovation, either out of personal belief or out of economic interest. If the process of diffusion is driven by a broad movement and network of many different social strata, it is called a bottom-up process. If the process is rather driven by a small group of professional individuals who are responsible to set the framework, then it is a top-down process. Diffusion of innovations can carry characteristics of both processes.

An early adopter or first mover is a person or business that embraces new technology before most people or businesses do the same. Early adopters take risks and advance into areas that are outside societal mainstream. This stance requires a high degree of independence. If early adopters successfully introduce a new idea or technology, the dynamic of diffusion may gain momentum in its surroundings. As the process of diffusion unfolds, the new technology becomes more acceptable and the risks more calculable for both society and the individual entrepreneur, and others will more willingly adopt the new technology.

The acceptance of innovative technological ideas does not usually follow a straight path. At business and operational levels, the need for many adaptations has to be considered before new ideas or technologies are implemented. Diffusion and adoption at business level can be driven by a change in consumer demand (bottom-up process) or benefit from a supportive environment. The implementation of new ideas/technology at business level can also be politically promoted by changing frameworks, or by offering incentives (e.g. subsidies or guaranteed prices). This would be a top-down process. The introduction of organic farming (bottom-up) as well as the spread of biogas plants (top-down) can serve as examples of how to promote paludiculture.

In Germany, over several decades, consumers have come to accept organic farming as a trustworthy alternative to conventional agriculture. Its market share has increased considerably and organic products are now commonly found in most supermarkets and retail shops. The search for alternatives to high intensity farming (which has become increasingly environmen-

tally destructive, and also affects the quality of the food products negatively) has prompted the development of organic farming. For instance, in western Germany, organic farming was part of the cutting-edge environmental movement, which was supported by large parts of the population throughout the 1970s and 1980s. The diffusion of organic farming occurred over many years as a typical bottom-up process in many parts of Germany and other countries. Several factors contributed to its long-term success:
- A high awareness among consumers;
- The creation of new brands and markets for a premium price;
- Networking between local and regional stakeholders;
- Exertion of influence on political levels to fund the transition to organic farming.

Politicians and consumers were eventually convinced that organic farming would decrease environmental impact, because it implies abandoning the use of environmentally harmful technology and substances. They were also prepared to pay for the extra financial effort and to initiate relevant funding programmes.

Yet, the boom of biogas production was the result of specific financial incentives that were introduced by the German Renewable Energy Act (top-down process). The technology to produce biogas had been already known for a long time, but had not spread widely. Originally, the objective of using biogas technology was to address the problem of what to do with the large quantities of liquid manure from intensive livestock farming, as well as other waste products. A useful option was on-farm energy production. The following elements lead to the success of biogas technology:
- Newly founded interest groups (e.g. the German Renewable Energy Association), which successfully represented the interests of the bioenergy sector;
- Exertion of influence on the development of the regulatory framework regarding financial attractiveness and long-term commitment;
- Provision of the necessary investment security for the biogas plants.

In the case of biogas technology, the implementation of incentive schemes resulted in financial support of this innovation and subsequently to its success. The guarantees to purchase the produced electricity at a premium price led to the adoption of biogas production within a short time. Furthermore, this approach created a new industry sector. However, the sludge based technology became less important and corn (which is most profitable as a substrate for biogas) was increasingly grown – also on peat soil. In areas where ploughing-up grassland is forbidden, it is lucrative for farms to intensify their grassland farming and use the grass cuttings as a substrate in biogas plants. In the case of biogas production, the top-down process that was initially strongly supported by interest groups went into widespread diffusion with some undesired side effects (see below).

The concepts of organic farming and of biogas plants were initially exemplarily successful, but now fail to expand. Organic farming is weighed down by a disastrous fall in market prices, whereas the price level of products by conventional agriculture simultaneously rises. In general, this situation results in declining rates of growth in the organic farming sector. In Mecklenburg-West Pomerania, the numbers of organic farms, as well as the organically farmed area, are in decline (Kuhnert et al. 2013). The sector of biogas production is increasingly subject to criticism. The long-term guaranteed purchases of electricity, regulated by the German Renewable Energy Act, face a decrease of social acceptance. The reason for this are the environmental impacts, including erosion, nitrogen pollution of groundwater, homogenisation of the appearance of the landscape, as well as leaks of methane from biogas plants and deposition of fermentation residues (WBA 2007, Hampicke 2013).

It remains to be seen how politics will counteract this development. To secure the continuation of organic farming, new funding programmes are discussed. The biogas industry introduced a number of measures, namely regulations regarding the handling of fermentation residue containers, adaptations of the German Renewable Energy Act, and research into new energy crops. If paludiculture is to be introduced at a large scale, negative developments should be avoided, and mechanisms for regulatory readjustments should be integrated.

The examples of organic farming and biogas production have shown that consumer demand, as well as financial incentives, can contribute to the successful adoption of an idea or specific technology. The advantage of the bottom-up process is that it is less prone to risks of dysfunction but requires the formation of a broad and decentralised network (Chapter 8.4.3). With regard to paludiculture, it is essential to introduce an innovative method of land use and at the same time to develop innovations in product development as well as in the handling and processing of raw materials. Supply and demand of biomass must be stimulated simultaneously because their development influences each other mutually. There are good reasons, both in business and political terms, to implement paludiculture. For a farmer, it is important that the land use method secures long-term profitability. For political decision makers, it is relevant that paludiculture contributes to socially desired objectives, such as climate protection.

8.3.2 Opportunities

Farmers do not evaluate paludiculture on the basis of results of scientific research but on the basis of subjective experiences and assessments. Polls amongst agricultural managers in Mecklenburg-West Pomerania show that many of them are aware of the low profitability of land use on drained peatlands. Moreover, this form of land use is highly dependent on agricultural subsidies and therefore, farmers are receptive towards new forms of land use. This is, in particular, the case in areas where land users have to carry the cost for continuous drainage themselves. When a farm is taken over by a new manager or the next generation farmers, both the disposition to question conventional land use (which implies continuing degradation of the peat layer) and the interest in alternatives will increase greatly. The handing over of the farm to the next generation usually carries a time of innovation and investment, which offers opportunities for a reorientation towards diversification regarding management and farming methods.

Especially energy-intensive production branches want to decrease their energy dependence, which is reinforced by increasing prices for energy and raw materials. Therefore, farmers often see chances and opportunities to use their own paludiculture biomass for direct thermal utilisation. This tendency is supported by recent developments in the renewable energy sector, which encourages farmers' acceptance of new business fields such as energy provision. With respect to using paludiculture biomass as a raw material for industrial use or manufacturing, the performed interviews convey a good deal of scepticism. Even though a potential for products is recognised, markets to sell these products are considered less developed, regionally limited or non-transparent. Also, compliance with product standard requirements during biomass harvest and processing is considered to be problematic, as well as the dependency on the further processing and production chain.

8.3.3 Obstacles and reservations

Many farmers are sceptical to the rewetting of peatlands, which is the precondition to paludiculture. Pointing at already implemented rewetting measures, farmers deplore the loss of economically valuable land. Moreover, there is doubt about the necessity of rewetting. The legal framework for the utilisation of reedbeds is ambiguous and could cause potential conflicts with nature conservation (Chapter 7.1). Furthermore, the irreversible rewetting of land is seen as a restriction of future land management options (Chapter 8.2).

Despite those reservations, farmers are mostly willing to consider wet forms of land use. In surveys, both innovative as well as traditional-minded farmers express their concerns about two issues: risks regarding the profitability of paludiculture and the uncertain outcome of investing in it. In connection with risk, there is uncertainty regarding structures of production, prices of raw materials, as well as existing markets and model businesses. Many farmers are sceptical about the feasibility of harvesting wetland biomass, both in technical and economic terms. Another concern is the low weight-bearing capacity of peat soils, albeit this is also often seen as a technical challenge. Further doubts exist regarding the long-term productivity of sites without additional fertilisation. Many agricultural businesses have little financial capital, and uncertain agricultural policy frameworks complicate investments in new business fields.

Next to economic concerns, farmers also have reservations towards paludiculture from an operational perspective – e.g. that long-established traditions have to be abandoned. Some farmers also favour livestock farming to produce food for human consumption, a farming method that provides more jobs than solely harvesting biomass.

8.3.4 Options to increase acceptance and implementation at the producer level

The obstacles and reservations towards paludiculture, outlined in the previous section, are to be found within the areas of legislation, economics and social affairs. The legal framework should provide certainty for the desired innovation. This implies that paludiculture and the established reedbeds must be given a clear position within agricultural subsidy policies (Chapter 7.2). This involves:

- The admission of Common Reed (*Phragmites australis*) and Cattail (*Typha latifolia* and *T. angustifolia*) as agricultural crops;
- Securing the status of paludiculture land as agricultural land;
- The flexible application of habitat conservation legislation in case of newly established reedbeds for paludiculture;
- The simplification of fuel certification and admission regulations.

The farmer himself has no influence on the legal framework that also affects economic aspects. In order to cover economic risks, it is essential to clarify the following points (Chapter 7.2 and 7.3):

- Equalisation of paludiculture with other agricultural land uses;
- Securing the receipt of direct payments;
- Development of agri-climatic schemes for paludiculture;
- Development of agri-environmental schemes for species and habitat protection within paludiculture plots;

- Extension of German Renewable Energy Act support to include thermal energy production from herbaceous crops;
- Specification and implementation of cross-compliance regulations concerning peatlands (for example, prescription of peat conserving minimum groundwater levels);
- Accessibility to investment funds;
- Establishment of demonstration plots and pilot enterprises.

The establishment of explicit legal and policy frameworks is a prerequisite for investments to switch to alternative forms of peatland use. Considering the observed societal reservations, it is necessary to provide correct information and training to farmers and the rural population, in order to widen the business opportunities for investors (Chapter 8.2, Chapter 8.4).

8.4 Transfer of knowledge

Henning Holst & Christian Schröder

Applied research is an important driver of innovation (Fritsch et al. 2008). The required transfer of knowledge increasingly develops into an independent research field with its own niche, particularly concerning integrated transdisciplinary research projects with respect to land use. Both science and practice benefit from close cooperation: the researchers gain from increased practice-orientation, the practitioners profit from access to the results of latest research. Cooperation can so result in competitive advantages. The experiences gained can also be used to share the innovations and promote their implementation to a large scale.

The following sections illustrate two ways of knowledge transfer relevant for the introduction of paludiculture. Large-scale transfer of new knowledge via multipliers and training courses enables reaching many land users, as well as obtaining broad feedback on viability and transferability of the innovation. Small cooperating networks, in contrast, enable the introduction of specific goal-oriented pilot projects to provide solutions to particular challenges.

8.4.1 Knowledge transfer about innovations in land use management

'An effort should be made to better convert knowledge into socioeconomic benefits. Therefore, public research organisations need to disseminate and to more effectively exploit publicly funded research results with a view to translating them into new products and services.' (EU Commission 2008).

This statement of the EU Commission brings new demands to science. Science must – in cooperative networks of a multitude of stakeholders, pursue an effective diffusion of its research findings by overcoming professional, social, cultural and psychological blockades. These challenges are difficult to meet by single individuals and could result in excessive demands on researchers. The scale and complexity of the task therefore require interdisciplinary cooperation (Figure 8.4).

A specific feature of knowledge transfer in the field of sustainable land management is the mutual dependence of practice and research. On the one hand, practice benefits from research, whereas on the other hand, science receives the assignment to solve actual problems from practice.

The transfer of knowledge is of particular importance for the implementation of paludiculture. Its objective is not to change crop rotation methods or to optimise the profitability of land use. Rather, the introduction of paludiculture as a form of land use is a change of paradigm (Kuhn 1996). Peatlands must now be used economically under more natural (i.e. wet) conditions. The site conditions are no longer adapted to the requirements of a very special agricultural practice; conversely, humans rise to the challenge

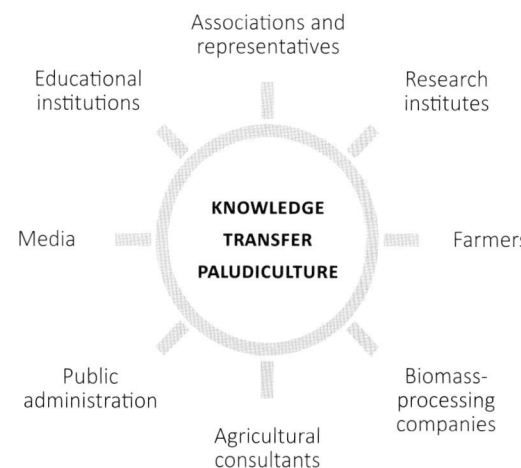

Figure 8.4: Structure of networks and target groups with regard to the diffusion of research findings into practice.

and adapt their land use to the site conditions. Such change does not only affect land use. As a new form of land use, paludiculture influences the interests of people more than most previous changes in land use (Chapter 8.2). Multiple interests have to be considered and, if possible, legitimised through democratic forms of participation (Box 8.3). Farmers, their representatives, as well as local residents and decision makers in politics and administration have to come to the fundamental understanding that this change in paradigm is necessary.

The implementation of sustainable land use in peatlands requires close cooperation between regional and supra-regional participants. Furthermore, a science-based approach is needed to link the questions of agriculture and water management to the challenges of nature and environmental conservation (World Bank 2006). A smooth flow of information is thus essential.

8.4.2 Transfer of knowledge through training of multipliers

The training of multipliers is aimed at interested stakeholders whose function is to convey knowledge. Trainings and seminars about paludiculture also target innovators which can set examples to followers. Multipliers of knowledge in the field of land use are training and consulting institutions, as well as consultants, associations, representatives, practical farmers, administrators, business enterprises and research facilities (Figure 8.4). The various multipliers differ with respect to their role and possibilities of knowledge transfer, namely:

Public and private educational institutions have their own active networks and can efficiently distribute new ideas like paludiculture. The questions, feedback, ideas, responses and proposals generated in such extended networks may strongly influence and improve the quality of applied science. As a further advantage, educational institutions are highly experienced in knowledge transfer. They benefit from cooperation with science by shaping a scientific market profile. Within collaborative research projects, special joint training programmes on paludiculture can be elaborated.

Agricultural consultants have close personal contacts with agricultural enterprises, and thus constitute an essential link between science and practice within the land use sector. As multipliers, consultants do not only have personal contact to potential users of new forms of land use, but are also familiar with the problems and obstacles they face. For instance, farmers who face increasing economic difficulties with drainage-based peatland use are reluctant to speak out these problems in public. Consultants and farmers have often developed long-term relationships which are based on mutual trust and are characterised by respect, honesty and confidentiality.

Associations and representatives play an important role in transferring knowledge via their networks as well as acting as promoters and mediators towards decision makers in politics and administration. As a voice of a larger group, they are legitimised to claim support for new forms of land use directly from the authorities. For example, farmers associations are very familiar with the problems and the state of affairs of agricultural businesses; consequently, they can discuss new forms of land use with land users while also representing their interest. They will provide useful advice and critical feedback for further development of paludiculture, and help to identify knowledge gaps and need for action. Nature conservation associations are legitimised by society to represent the interests of their members, act as a bridge between regional and supra-regional politics, and integrate citizens via information provision and public awareness raising.

The farmers themselves are those who must implement the innovations. Paludiculture implies an irreversible long-term investment decision for the farm enterprise that may also have a bearing on future generations. Therefore, farmers should approach this topic with caution and thoroughly consider a number of issues, such as business and labour economics, the availability of adequate machinery and the extent of grant funding. Farmers would only participate in training when it is useful, – i.e. when they can profit from it. Having practical farmers as multipliers is particularly important to promote paludiculture, as they can credibly transfer their knowledge and are able to inspire other land users to follow their example.

Public administration is responsible that regulations and legislation are followed, as for instance those linked to the award of subsidies. Further responsibilities of public authorities are the implementation of political frameworks set at EU, national and federal state levels besides informing and advising farmers about funding and applications. Within this field, public administrations have much room to manoeuvre, in particular at a regional level. Some administrators are in direct contact with agricultural businesses, which allows them to address the latter's concerns and needs in shaping political frameworks.

Research institutes develop theories and concepts. However, questions regarding practical applications often can only be answered insufficiently. The knowledge required for developing new forms of land use is very complex, therefore needs an interdisciplinary approach to address the challenges. This implies that knowledge gaps and research questions have to be discussed and processed by integrating other areas of expertise, which tackle those matters within their own fields of research and arrive at new conclusions. In this way, an interdisciplinary approach will contribute with important findings to the big picture. Another task of science is the dissemination of results, making them available to the stakeholder target groups by publication. In particular, application-oriented publications about paludiculture can reach all above mentioned stakeholder groups.

Biomass-processing companies implement innovations that result in new products, production methods and business models. These pioneers need a good intuition of new things to come plus the ability to implement these innovations, especially if, like in the case of paludiculture, no complete market for raw materials has developed yet. Pioneering companies are the first to profit from innovation in land use while also taking greater risks. However, if these pioneers are optimistic about the success chances of an innovation, this will accelerate the provision of investments and foster development of the new concept towards practical maturity. With respect to paludiculture, the demand for the raw material produced is essential to promote the required land use change, because without demand there is no reason to produce. The development of demand and production has to run parallel. If a market is created via an industrial company, this will stimulate the supply of raw materials and therewith the implementation of paludiculture.

Media are the most important players in large-scale diffusion. Involvement of editors of daily newspapers or the specialised periodical press will happen if press representatives consider the topic of paludiculture noteworthy enough. Long-term close personal cooperation is advantageous in this matter.

The greatest challenge to knowledge transfer of landscape-ecological and agricultural research findings can be expressed as: 'Education requires examples.' This implies that the transfer of knowledge relies on pilot and model projects. Specifically in the area of land use, farmers need to see practical examples in order to assess whether an innovation can provide an opportunity for their agricultural business.

8.4.3 The importance of networks and local cooperation for knowledge transfer

Henning Holst, Anke Nordt & Christian Schröder

A network is formed when people and/or organisations with the same aims and objectives are connected within clear structures. In land use, local networks are of major importance because they address actual challenges pragmatically by bringing together partners with different special knowledge.

Cooperation is essential in paludiculture in order to foster confidence to the suppliers, processing plants and end users of the biomass. As there is yet no real market for biomass from paludiculture, alternative ways have to be found to link supply and demand. Only then, biomass suppliers and consumers can – on economic grounds, decide for the implementation of paludiculture and consequently, regional value chains can develop.

The shortest form of value chain is using the biomass within the producing enterprise itself. For example, the heating demand from another production sector within the same enterprise (e.g. rearing piglets or aquaculture) can be satisfied by thermal use of home-grown paludiculture biomass. Furthermore, the self-sufficiency at business level can be extended to local and regional levels by integrating further participants. Bio-energy villages, for example, obtain their fuel (for electricity generation and heating) from locally produced biomass. A great advantage of such cooperation is the independence from global commodity price fluctuations, and the provision of long-term stability through local contracts that regulate biomass supply and purchase. Resident initiators interested in the implementation of paludiculture play an essential role in the formation of such networks, as they can also encourage other local people to participate. The role of research partners is beneficial to integrate new technical developments and research findings into practical applications, whereas they receive feed-back relevant for their research from practitioners. With the integration of further stakeholders like land owners, energy suppliers as well as representatives of local population and decision making organs (compare Table 8.1), networks focused on paludiculture may establish pilot projects of exemplary prominence (Box 8.4).

Box 8.4: District heating plant Malchin – Added value and climate change mitigation

Tobias Dahms & Anke Nordt

A pioneering concept for the utilisation of biomass from rewetted peatlands is the close cooperation between the agricultural enterprise Voigt and the local energy supplier Agrotherm GmbH, who opened a heating plant to supply a district heating network for the town of Malchin, in the German federal state of Mecklenburg-West Pomerania.

As a result of a major fen restoration project ('Peenetal/Peene-Haff-Moor'), the fodder quality of the pastures used by farmer Voigt declined, which frustrated the continuation of grazing with suckler cows. In order to continue farming on the rewetted sites, a concept for the thermal utilisation of the biomass was developed since 2006.

An ideal opportunity was the short distance to the town of Malchin, which is situated 12 km from the farm and had a district heating network that solely operates with natural gas. It was possible to utilise regional biomass from sedges, rushes and reed canary grass by integrating a biomass boiler into the district heating network. The biomass of some 300 hectares are now annually used to produce 800–1000 t fuel, which equals 2,900–3,800 MWh or 290,000–380,000 l of heating oil. The plant currently provides district heating to 543 standard flats, two schools, and a day-care facility for children.

The biomass boiler is made by the Danish company LIN-KA/Danstoker and has a thermal capacity of 800 kW. The boiler has a water-cooled, movable step grate combustion system which prevents the formation of slag and ash deposits. Long residence times and the re-circulation of the exhaust gasses into the combustion chamber minimise the emission of harmful substances. Dust particles are removed by a multi-cyclone system coupled with a downstream fabric filter. The generated heat is transferred into a 24,000 l thermal storage tank, which is used to cover the base and medium load range. The local energy supplier guarantees the purchase of at least 4,000 MWh in the form of heat, which is equivalent to about 5,000 peak load operating hours per year. A gas boiler covers the peak load and the maintenance hours of the biomass boiler.

The required biomass is harvested by Voigt agricultural business during dry periods in summer using adapted machinery, then is pressed into round bales and transported to the boiler site. There, the bales are fed continuously into the boiler of the district heating plant with a conveyor belt, which can hold 24 bales of biomass. In winter, the boiler system is fed on a daily basis. The bales are shredded, and the loose stalks are transported via a double worm feeder and a screw stoker into the combustion chamber. The ash is automatically removed. Apart from biomass from wet fen peatlands, the boiler can also use straw as a fuel and – with a separate feeder, also wood chips.

The investment costs for a biomass boiler, particularly for herbaceous biomass, are significantly higher than for boilers operated on fossil fuels (FNR 2007). The net investment for the combustion plant (including fuel processing, flue gas filters, buffer store and installation) amounts to 630 EUR/kW of nominal heat output. Additionally, funding is needed for the building shell, chimney, storage and machinery. However, the great advantage compared to fossil fuel boilers is the much lower fuel cost. Agrotherm GmbH calculates fuel costs of 3.9 EUR per provided GJ of heat (Bork 2013, personal communication). Altogether, there is the advantage of reduced net heat production costs and long-term price stability. The energy concept of Malchin shows how local networks may contribute to the implementation of pilot projects and is an inspiring example for many other regions.

9 Sustainability and implementation of paludiculture

Land is becoming scarce. However, such scarcity is only one of the reasons for which a sustainable management of 'land' as a resource is needed. The use of land has also ecological, social and economic effects that often work beyond the utilised area itself (Foley et al. 2005). This should be evaluated when the sustainability of land use is assessed. Appropriate indicators and evaluation systems help to compare different land use options for peatlands, and to address aspects of sustainability when planning for paludiculture (Chapter 9.1). The availability of suitable land plays a significant role in the implementation of paludiculture. Accordingly, the potential for paludiculture may be identified using GIS data (Chapter 9.2). Whether an area is indeed suited or not has to be assessed in the field (Chapter 9.3). Based on the assessment, the implementation of paludiculture can be planned. Changes in water regulation and infrastructure may be required (Chapter 9.4). Thus, in Germany, the implementation of paludiculture is usually subjected to a plan approval assessment and permission procedure (Chapter 9.5).

9.1 Sustainable land use

Dieter Behrendt & Horst-Peter Neitzke

In the following, the term 'land use' is used in a broader sense than merely its cultivation. It rather refers to any allocation of land to meet human interests. Thus, if it is decided not to cultivate an area – for example to support conservation efforts, this is still considered land use. Land use may have considerable side-effects beyond the intended economic or social services. It either may release pollutants or greenhouse gases, or may actually help to hold them back. Land use may improve or reduce the recreational value of a landscape. Significantly, depending on the type of land use, it may support or impair the economic development of a region.

Decisions on land use almost always raise questions on prioritising needs and demands, specific interests, and societal objectives. The final question is, whether the available land, its use, and associated costs and benefits are distributed fairly. In this sense, fairness not only applies to the present, but also to future generations.

9.1.1 A concept to access sustainability of land use

Ideally, land use should simultaneously improve the ecological, social, and economic functions of the land, and adhere to the principles of intra- and inter-generational justice. In this sense, land use is considered sustainable if following conditions are met:

A. The ecological potential, particular for providing basic ecosystem services, is maintained or improved (MEA 2005: VI);
B. The quality of life of people who live on or near the used land is maintained or improved. Quality of life not only refers to the material living conditions, but also to the possibility to live one's life according to one's own ideals (see e.g. Kleinhückelkotten 2005, Nussbaum 2003);
C. Negative effects on the ecological potential of other areas are avoided or minimised, now and in the future, as well as negative effects on the quality of life of future generations or of people in other areas.

Following these principles:
- Demands on natural resources may be so high, that their ability to regenerate is not impaired. 'Natural resources' refers not only to raw materials and energy, but also to all ecosystem services, including the intangible values of nature, landscape and biodiversity.
- Concepts of weak sustainability are incompatible with this understanding of sustainable land use. Weak sustainability treats different forms of capital (natural, human, social, and physical capitals) as interchangeable, and merely requires that the total overall capital is maintained (see e.g. Klepper 2002).
- Economic capacity should be maintained or improved in the region subject to planning and decision making. In this way, it can be guaranteed that material conditions for the population do not deteriorate. The economic development in one region, however, must not be at the expense of other regions.
- Cultural diversity and the cultural heritage must be preserved.

The three principles – as presented above – correspond to an integrated understanding of sustainability. In contrast to the three-pillar model of sustainability (see e.g. Deutscher Bundestag 1998) we define the conservation or improvement of the ecological capacity of a region and the quality of the human life as fundamental principles, complemented by inter-generational and inter-geographical justice. Economic sustainability is derived from the demand for social sustainability. A sustainable economy should help to fulfil human needs and to improve living conditions – without violating the principle on conservation and improvement of ecological capacity. Economic aspects should, of course, be taken into account when assessing the sustainability of land use practices.

Land use that is oriented towards ´sustainable land managment´, needs to fulfil the following requirements at operational level:
- When assessing land use, all ecological, social and economic aspects – including their development over time if applicable, need to be considered;
- All people potentially affected by land use decisions have the opportunity to bring in their views and interests in the consideration process;
- The consideration process is based on sound knowledge on the ecological, social and economic consequences of the land use options under evaluation. Knowledge gaps are identified and closed where possible. The available knowledge is presented in a way so that all stakeholders can understand and use it. Uncertainties are specified;
- Criteria are developed and applied to assess sustainability (see Chapter 9.1.2);
- The evaluation process is transparent for all stakeholders.

9.1.2 Evaluation of sustainability

Decisions about the use of land are often based on short-term, cost-benefit considerations, current preferences, and knowledge. Ecological, social and (total) economic effects are usually not taken into account or at least not their long-term aspects. However, the concept of sustainability demands that precisely these effects are addressed in decision making. Planning tools are needed that enable an evaluation of the sustainability of different land use options by taking into acount specific local conditions, problems and objectives. Finally, once a specific land use practice is implemented, it should be monitored whether it indeed has the intended ecologic, social and economic effects.

A comparative assessment of the sustainability of different land use options could be made by looking at the respective contributions to the development objectives of a region (Behrendt & Neitzke 2013). These objectives should be legitimised by a broad participation of the local population, politicians and other stakeholders. Such consultations have been carried out in the past in the framework of setting up a local agenda on urban or (more rarely) regional development; thus far, the focus has rarely been on land use. Monitoring success in reaching the envisaged goals requires suitable criteria and tools. Experience with local agendas has learned that local stakeholders often find it difficult to establish appropriate tools. Therefore, in various projects, sustainability criteria or indicators have been developed that can be adapted to local or regional conditions. In part, these criteria have become common practice (Behrendt et al. 2010, Diefenbacher et al. 2005).

A set of criteria and indicators was developed in an interdisciplinary endeavour by scientists from several research groups associated with the funding programme 'Sustainable Land Management', as well as by local stakeholders from West-Pomerania (Behrendt & Neitzke 2013). First, the concept of 'sustainable land management' was operationalised by addressing ecological, social and economic aspects in 68 specific development targets. These targets were assigned to 10 ecological, four social, and three economic superior objectives (Table 9.1). Next, a weighting was allocated to each of the targets and objectives from the perspective of a) science and b) local stakeholders in West-Pomerania (Behrendt & Neitzke 2013). As shown below, the catalogue of weighted targets can be used in a comparative sustainability evaluation of the different land use options, whilst taking local priorities into account.

If land use is already implemented, it is possible to go one step further and actually measure the ecological, social and economic effects. Again, for this purpose, suitable indicators were defined by several research groups within the funding programme 'Sustainable Land Management' (Behrendt & Neitzke 2013). One example of an already established indicator is the proportion of the total area of a region that is taken up by settlements and roads. As an indicator of the ecological devaluation of land, the build-up area is regularly surveyed in the framework of reporting on the implementation of the German sustainability strategy, as adopted in 2002 (Statistisches Bundesamt 2012, 2013).

9.1.3 Assessing sustainability of fen peatlands use

As an example of a comparative assessment of the sustainability of different land use options, the current use of a 100 ha large fen peatland is compared with four alternatives (Table 9.1):
- Option 1 (current land use): Low intensity cattle grazing and fodder production (green fodder and silage);
- Option 2: Maize cultivation for biogas production;
- Option 3: Paludiculture – cultivating reed beds for bioenergy;

9.1 Sustainable land use

Table 9.1: Superior objectives and the result of evaluating the effects of five land use options to overarching ecological, social and economic aspects. For the evaluation of the superior objectives, in total 68 targets were requested. The weighting factors attributed to the superior objectives were established by stakeholders from West-Pomerania (Behrendt & Neitzke 2013).

Superior objectives	Weight	Land use options				
		1	2	3	4	5
		Grassland	Maize	Paludicultre: Common Reed	Short rotation coppice: Willow	Undisturbed fen peatland
Ecological aspects						
N1 Area: Conserve or restore areas	2.9	-2.2	-9.1	5.0	2.8	10.1
N2 Soil: Conserve or restore soil properties	2.8	-3.8	-10.2	7.4	4.1	9.8
N3 Groundwater: Conserve or restore	2.7	-5.8	-10.4	6.9	4.6	10.4
N4 Inland waters: Conserve or restore	2.7	-1.5	-6.8	4.1	2.7	7.0
N5 Coastal waters: Conserve or restore	2.8	-0.6	-8.6	4.2	2.0	8.3
N6 Global climate: Mitigate climate change	2.0	-3.2	-7.2	4.6	2.0	5.4
N7 Local climate: Conserve or restore	2.2	-0.7	-4.1	4.3	4.1	6.5
N8 Air quality: Conserve or restore	2.0	0.2	-3.6	3.2	3.5	5.2
N9 Biodiversity: Habitats, species, genetic resources	2.8	2.5	-7.5	4.2	3.1	7.4
N10 Landscape: Conserve or restore diversity and typical character	2.9	3.7	-6.9	6.6	3.7	3.3
Social aspects						
S1 Population: Promoting an optimal population development	2.1	1.3	-2.0	-1.3	0.7	-1.1
S2 Material living conditions	2.4	1.4	4.9	2.2	1.9	-3.0
S3 Health	2.4	6.2	-11.3	0.3	2.2	5.2
S4 Culture: Strengthen culture, education and local identity	2.4	10.2	-3.6	4.1	3.1	1.2
Economic aspects						
W1 Local economy/business: Improve or stabilise economic situation	2.5	-0.3	18.1	8.0	8.9	-14.0
W2 Public budget: Improve or stabilise financial capacity	2.3	-1.6	9.9	4.3	4.0	-11.4
W3 Resources: Guarantee availability	1.9	1.5	-1.8	11.0	10.6	-9.4
Ecological aspects		-11.4	-74.4	50.4	32.7	73.4
Social aspects		19.1	-12.0	5.3	7.9	2.3
Economic aspects		-0.4	26.2	23.3	23.5	-34.8
Total		7.3	-60.2	79.1	64.1	40.9

- Option 4: Short rotation coppice of willow for bioenergy;
- Option 5: Abandonment and restoration of hydrological conditions and development of near-natural fen vegetation.

Options 1 and 2 require active drainage to ensure sufficiently deep groundwater tables. Also option 3 and 4 require hydrological regulation to avoid deep inundation, or strong lowering of the water table (drainage and/or irrigation). Option 5 does not require active regulation of the water table, as long as the high water table does not negatively affect adjacent agricultural areas or settlements.

All five options were evaluated on basis on the catalogue of 68 targets. The evaluation was carried out by experts with scientific and practical backgrounds in soil science, agriculture, landscape ecology, nature conservation, and zoology. The experts assessed the ecological, social and economic consequences of each option and how it would contribute to long-term sustainable land use. For each of the 68 targets, the effect of the land use option was scored on a 7-point scale going from -3 (very negative) via 0 (no effect) to +3 (very positive).

The results were analysed as follows:
- The maximum number of points for each sector (ecological, societal, or economic) was set to 100. This maximum number would be attained if a land use option would score 'very positive' (+3) on all targets. Similarly, if all scores were 'very negative' (-3), the aspect would obtain -100 points. Although unachievable in practice, the highest possible score of a land use option would thus be 300 as a sum of the three sectors.
- A maximum number of points is defined for each superior objective assigned to the sectors by dividing the 100 points according to the weighting by the stakeholders (see column 'weight' in Table 9.1).
- This maximum number of points would be attained by an objective if all its inferior targets would score 'very positive' (+3), taking the weighting factors for the targets into account.
- The individual expert evaluations were combined to a mean score for each target, and then weighted.
- The scores of all single targets within one objective were added together to provide the score for each superior objective.
- The scores of each objective were added together to provide a score for each sector.

The result of the evaluation of the current land use and the four alternatives is shown in Table 9.1. For each objective, the mean weighting allocated by the local stakeholders is listed (see column 'weight'). The numbers in the third to seventh column show the scores achieved by each land use option for each objective. The final rows show the total scores of each land use option for the ecological, social and economic sectors, and also for all three sectors combined.

The highest scores were achieved by paludiculture (option 3, cultivation of reed beds and option 4, short-rotation willow coppice). Although the economic score for maize cultivation (option 2) was higher, this land use option had the lowest overall score, because it performed (very) negatively with respect to ecological and social objectives. Regarding social aspects, the current land use (option 1, grassland) achieved the best result, which is mainly because of the relatively high scores on health effects and regional identity compared with the other options. As would be expected, option 5 (abandonment and restoration) had a positive ecological score but a negative economic score. Apparently, the experts did not value a possible economic benefit from tourism as high as that of agricultural land use.

The above described method was also used to evaluate the sustainability of various land use scenarios in a 1.600 ha large part of the Thurbruch fen in the southern part of the island of Usedom, West-Pomerania (Kleinhückelkotten & Neitzke 2014). Also in this case, the highest score was reached by paludiculture.

9.2 Availability of suitable areas

Andreas Haberl, Philipp Schroeder & Christian Schröder

A detailed assessment of the regional potential for paludiculture helps finding specific sites for implementation, and can serve to facilitate funding assistance from investment programmes, agri-environmental schemes or agri-climate programmes (Chapter 7.3). An analysis of potential areas should consider site conditions, current land use, and current allocation of the land. Depending on the objectives, additional data (like size or accessibility) may be useful to localise suited sites. In the following sections, criteria for the localisation of potential paludiculture sites are presented and illustrated with help of two examples.

9.2.1 GIS-based search for suitable areas

In order to identify areas suitable for paludiculture, already existing spatial data can be combined in a GIS (Schroeder 2012). Suitable sites can be identified using a set of criteria. The quality of such a feasibility study depends on the age, accuracy and availability of data, and can vary between regions. To perform a feasibility study, a spatial delineation of peatland sites is indispensible. For most regions, soil maps of sufficient resolution are available. Yet, it must be assessed whether these maps suitably display the presence of organic soils (e.g. peat soils, organic content > 30%, peat thickness > 30 cm). Due to peat degradation and associated soil subsidence, older maps define areas as peatland that may no longer meet the required peat thickness criterion. However, as these areas were once peatlands, it should still be assessed whether they are suitable for paludiculture. The available GIS data used for area selection may range from information about size and habitats, to detailed information on the stratigraphy and condition of the peatlands.

9.2.2 Criteria for area selection

A wide range of criteria may be used to narrow down the search for potential paludiculture sites in a region (Table 9.2). Important criteria are the protection status and the use of an area of land. In principle, all peatland areas under agriculture or forestry may be suited for paludiculture. A further preselection of sites can be made if detailed peatland surveys are available, ideally including information on hydrology, trophic conditions, relief and topographic location in the landscape. Some of these parameters may be derived indirectly – e.g. from the hydrogenetic mire type, which provides information on the former water supply and thus on the rewetting potential (Table 9.3). Information on existing drainage infrastructure, polders, and groundwater catchment areas, helps to assess the effects of rewetting and to avoid potential conflicts.

The current vegetation of an area can serve as a proxy for the water table. It furthermore helps to assess the effort needed to establish target vegetation. If target vegetation is already present, no costs arise for its establishment and use may start sooner (Wichmann & Wichmann 2009). Such areas may be identified using habitat surveys or via remote sensing (Steffenhagen et al. 2010, Frick et al. 2011).

Potential competition for land can be identified with land use models (Kuhlman et al. 2013). For instance, on areas where only low income is achieved, the willingness to change current land use will be higher than on profitable arable land. Competition is stronger for more profitable land and therefore a change in land use is unlikely on the short term. Implementation of paludiculture on such land is largely controlled by legislative and agro-political framework conditions.

In addition, competing development objectives may determine the availability of land areas. In this respect, nature conservation may restrict the establishment of paludicultures, but may also stimulate it (Table 9.4, Holsten et al. 2012, Chapter 7.1). Strictly protected areas – like the core zone of a national park, are in many cases exempt from land use. Therefore,

Table 9.2: Example set of criteria to identify prospective paludiculture sites.

Criterion	Aim	Comments/example criterion
Soil type	Organic soils only	Peat layer of at least 30 cm depth.
Current land use	Restriction to agricultural land	Suitable: Arable land, grassland and fallow land.
Habitat type/vegetation type	Restriction to suitable habitats and vegetation	Unsuitable: Forest, urban areas, water bodies. Suitable: Reed beds, sedge stands, humid forb communities.
Conservation status	Conservation objectives	Excluded: Core zones of national parks and biosphere reserves, pristine mires. Conservation management: Maintainance of short grassland as breeding habitat for birds; avoidance of shrub encroachment.
Size of area	Economically viable land use	Minimise logistic effort: e.g. minimum area >15 ha and >100 ha within a radius of 10 km.
Hydrogenetic mire type	Assessment of water and nutrient availability	Rewettability (Table 9.3) Suitability for different paludicultures.
Degree of drainage	Assessment of the current status	Necessary rewetting measures.
Catchment	Assessment of rewettability	Assessment of impacts on neighbouring areas; assessment of water supply.

9 Sustainability and implementation of paludiculture

Table 9.3: Suitability of different hydrogenetic fen mire types (after Succow & Joosten 2001, Joosten & Clarke 2002) for paludiculture: +++ = very good, ++ = good, + = intermediate, ~ = conditionally, − = unsuitable.

Hydrogenetic mire type	Horizontal peatland				Sloping peatland			
	Schwing moor terrestrialisation mire	Immersion terrestrialisation mire	Water rise mire	Kettle-hole mire	Flood mire	Percolation mire	Surface flow mire	Spring mire
Water supply (pre-drainage)	continuous	mostly continuous	small	periodic/ continuous	periodic	continuous	continuous	continuous
Slope	none	none	none	none	none/ medium	medium	medium-steep	medium-steep
Internal water storage (pre-drainage)	large	mostly large	medium	medium - large	small	large	small	small
Rewettability	+	++	+++	++	+++	++	~	~
Suitability for paludiculture	−	+	+++	−	++	++	−	−

protected areas should not be included unreservedly when compiling a pool of potential areas. It should be verified whether paludiculture is compatible with conservation aims. Consequently, synergies with existing conservation management can be sought and utilised. Existing plans that already define targets for development, like peatland protection programmes or landscape planning frameworks, may be useful as well.
Further criteria for the identification of suitable areas include accessibility of sites, transport connections, and proximity to potential users of the biomass. Information on these criteria is only partially evaluable via GIS data, and should rather be considered when assessing a specific site in detail (Chapter 9.3).

Table 9.4: Suitablility for paludiculture in relation to the conservation status of an area: + = suitable, (+) =conservation has priority, possibly no land use allowed, − = no use allowed (including paludiculture).

Status	Suitability
Agricultural area	+
Biosphere reserve, nature park, landscape conservation area	+
National park, nature reserve, protected habitats	(+)
Natura 2000 site (EU Habitats Directive); EU bird sanctuarie; Special Protected Area (SPA); Ramsar site	(+)
Core zones of protected areas	−

9.2.3 Potential areas in Mecklenburg-West Pomerania

Christian Schröder & Philipp Schroeder

A pool of fen peatlands that are potentially suited for paludiculture was compiled for Mecklenburg-West Pomerania. The aim was to identify and map the regional distribution of all areas that are suitable in theory, thus providing a basis for site specific studies into the practical possibilities of implementation (Chapter 9.3). The analysis was based on the current official peatland map of the German federal state of Mecklenburg-West Pomerania (soil map, scale 1:25,000 – organic soil types; Schiefelbein et al. 2011). This map was combined with other spatially explicit data (Table 9.5). The analysis focused on agricultural areas, including fallow land; peatland forests and water bodies were excluded. The hydro-genetic mire types 'raised bog', 'spring mire' and 'kettle-hole mire' were excluded as well, because of their high conservational value (Table 9.3 hydrogenetic mire types). Nationally or internationally protected areas or biotopes were considered to be only available for paludiculture conditionally (see Chapter 9.2.2).

The resulting map presents a traffic light rating of the sites: suitable areas are displayed in green, unsuitable areas in red, and areas where paludiculture would need to accommodate nature conservation in yellow (Colour pictures 90 and 91). The potential areas were also mapped on a 10 x 10 km grid (Figure 9.1). This kind of presentation supports the search for paludiculture pilot areas, where producers and users need to come together.

9.2 Availability of suitable areas

Table 9.5: GIS data used to identify potential areas for paludiculture in Mecklenburg-West Pomerania.

Criterion	Data based on	Source
Basic map	GDI-MV – Topographic maps	State administration for internal afairs (LAiV), last updated, 2013
Peat soils	Soil map 1:25.000 (KBK25) –organic soils	State administration for the environment, nature conservation and geology (LUNG), last updated, 2011
Hydrogenetic mire type	Peatland survey 1:50.000	State administration for the environment, nature conservation and geology (LUNG), last updated, 1998
Forest area	Map of forest cover 1:5.000	State forestry of Mecklenburg-West Pomerania, last updated, 2011
Water bodies	Network of water bodies in M-WP (DLM25W) 1:25.000	State administration for internal afairs (LAiV), last updated, 2003
Conservation status	Nature reserves 1:50.000	State administration for the environment, nature conservation and geology (LUNG) Online portal: http://www.umweltkarten.mv-regierung.de/script/, last updated, 2012
	Protected habitats 1:10.000	State administration for the environment, nature conservation and geology (LUNG), Online portal: http://www.umweltkarten.mv-regierung.de/script/, last updated, 2011
	National parks 1:10.000	State administration for the environment, nature conservation and geology (LUNG), Online portal: http://www.umweltkarten.mv-regierung.de/script/, last updated, 2012

Table 9.6: Potential area for paludiculture in Mecklenburg-West Pomerania (Data sources Table 9.5).

Categorie	Area in [ha]
Peatlands in Mecklenburg-West Pomerania	280,800
Unsuitable:	76,000
Forest	67,900*
Water bodies	4,100*
Hydrogenetic mire types: spring mire, kettle-hole mire, raised bog	8,600*
Suitability to be reviewed:	102,600
Protected habitat	63,200*
Nature reserve	36,500*
National park	8,000*
Natura 2000 site and SPA	138,600*
Area suitable for paludiculture:	102.200

* area of land cover types partially overlap, so totals are smaller than the sum of parts.

The analysis revealed that of the 280,800 ha of peatland of Mecklenburg-West Pomerania, 36.4% (102,000 ha) may be suitable for paludiculture. An additional 36.5% (102,600 ha) may or may not be suitable, depending on whether paludiculture can be carried out in line with nature conservation. Finally, 27.1% (76,000 ha) is not suitable for paludiculture (Table 9.6). Although all woodlands on peat soils were deemed unsuitable, a separate analysis should look at the possibilities to convert drained forests to wet forests – e.g. for Alder cultivation.

9 Sustainability and implementation of paludiculture

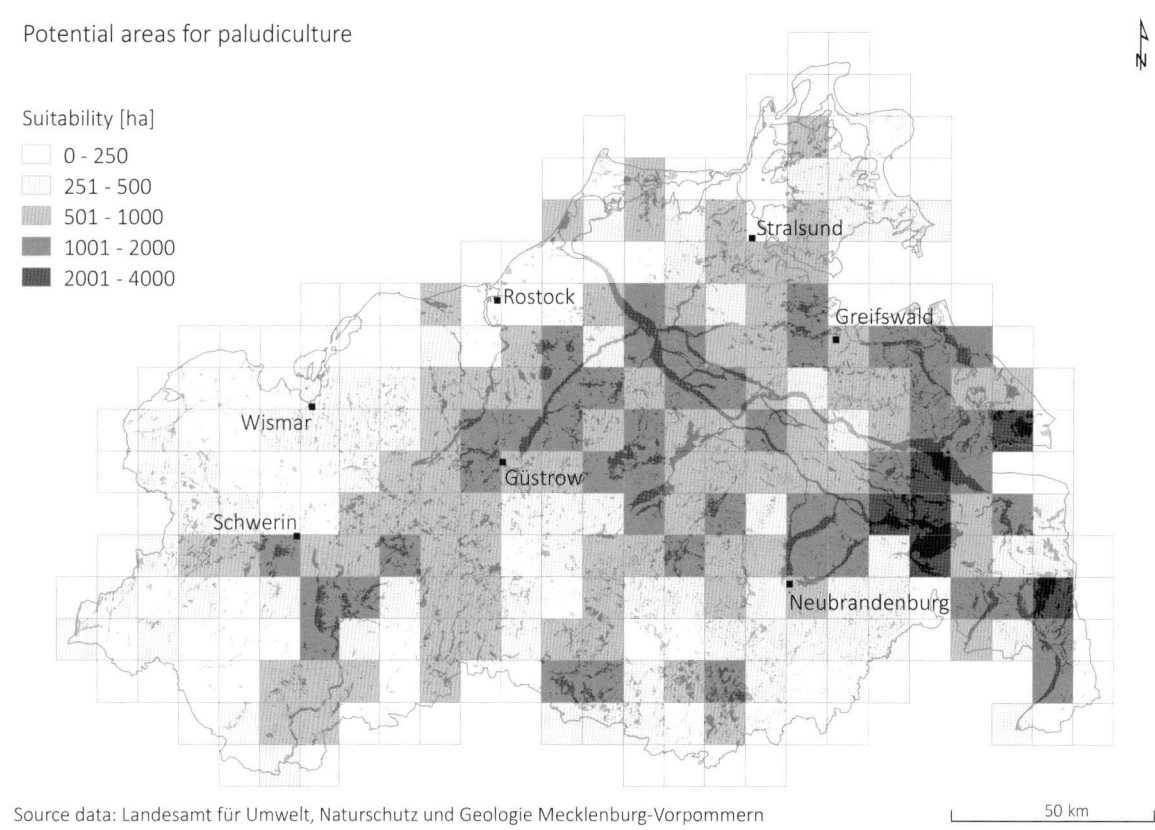

Figure 9.1: Potentially suitable areas for paludiculture in Mecklenburg-West Pomerania in a 10 x 10 km grid (Map: Philipp Schroeder).

9.2.4 Finding a paludiculture pilot area in the German Federal State of Brandenburg

Wendelin Wichtmann & Bernhard Hasch

The Ministry of Environment, Health and Consumer Protection of the German Federal State of Brandenburg commissioned a feasibility study to identify peatland areas suitable for a short-term implementation of paludiculture pilot projects. A total of 71 (partial) catchments with a large proportion of peatland area were evaluated using a points system (Table 9.7, Hasch et al. 2012).

The analysis was based on a digital land registry (field block cadastre) that includes all agriculturally used areas. The proportion of peatlands of this agriculturally used area was identified by combining the digital land registry with the peatland distribution map of Brandenburg (Bauriegel 2013). Areas with less than 500 ha of agricultural land on peat were excluded from further analysis, because the potentially suitable area was likely too small for a pilot operation. As a result, the number of partial catchments was reduced to 48.

The larger the proportion of peatlands in a partial catchment, the larger the potential of finding sufficient sites for paludiculture projects. A large proportion of peatland within polders or designated flooding areas produced a high score, because of a high natural tendency for rewetting and an associated larger willingness to adopt wet land use. In contrast, a large proportion of peatlands in protected areas meant a low score, because of strongly competing nature conservation targets. For example, the planting of Common Reed or Cattail could clash with the aim of protecting meadow breeding birds. Similarly, a large proportion of arable land use on peat resulted in a low score. Although conversion from arable land to paludiculture would be sensible in many ways, resistance to change will prevent a shift to paludiculture on the short term. A large proportion of low intensity peat grasslands was rated with an additional point, because conversion to paludiculture is more likely in those areas. Peat grasslands are used with very low intensity only, and profitability strongly

9.2 Availability of suitable areas

Table 9.7: Points awarded to (partial) catchments to evaluate the possibilities of implementing paludiculture on the short term. Areas with at least 5 points were studied in more detail (Hasch et al. 2012).

Feature	Criteria	Points
Agricultural land on peatlands within the considered catchment (based on land register)	500-900 ha	1
	901-1,500 ha	2
	> 1,500 ha	6
Thereof		
Proportion of polders (with grassland proportion > 70%) or Proportion of flood retention area	> 50%	3
Proportion of protected Natura 2000-habitats	> 20%	-1
Proportion of arable land	≥ 30%	-1
Average proportion of low intensity grassland in relation to the total area of each field	> 33%	1

depends on current subsidies. Moreover, future agri-environmental measures with a focus on climate and peatland protection will more likely meet with acceptance here.

Particularly, the large and unfragmented fen areas of the Rhinluch, the Havelluch, the Nuthe floodplain and the Baruth Valley, as well as the Spreewald, Ucker Valley and Randow Valley, appeared to be well suited for paludiculture (Figure 9.2). The set-up of a pilot operation was assessed in detail for the most suitable areas against additional criteria, including water availability, mire type, required constructions, local potential for biomass utilisation, potential project management, willingness of farmers, and stance of local authorities. Besides the decisive criterion of the presence of a suitable farm, important criteria were the absence of major opposition in the region and the possibility to avoid a plan approval procedure by the authorities. Eventually, the choice fell on two peatland areas (together about 300 ha) at Lake Blindow and Lake Möllen in the Ucker Valley close to the town of Prenzlau. Most of the area is already covered by wet meadows and reed beds. As the area is largely the property of the city of Prenzlau, the local nature conservation administration was intensively involved at the planning stage. Extensive peatland areas are found nearby, which are currently under high intensity grassland use or covered with reed beds. These areas could become available to expand paludiculture on the long term. A decisive factor in the project was the willingness of a farmer to cooperate. Thus, a farm-level pilot project was prepared together with all identified partners (Box 9.1).

Box 9.1: Improving infrastructure near Lake Blindow

Wendelin Wichtmann & Bernhard Hasch

In the framework of a project on integrated rural development (IRD), a paludiculture pilot was planned in cooperation with the town of Prenzlau (Ucker Valley, German federal state of Brandenburg) (Chapter 9.2.4). Besides construction measures to optimise the water table, additional measures were planned to improve the infrastructure of the area and make it more accessible. For example, a dead-end track that leads to a former bridge over the river had to be paved to enable transport of the biomass – by tractor and trailer, to the end user or to a transfer point. An additional storage and turning place has to be built at the end of the road. The road tracks and the storage facility will be paved with 50 cm of gravel, partly in combination with geotextiles, to ensure an adequate load-bearing capacity. To facilitate crossing the river, a mobile pontoon bridge (10 x 6 m) – with a load-bearing capacity of 700 kg m^{-2} and an overall capacity of 40 t, is initially planned. The bridge can be put in place for a short period during harvest, which allows the river to be crossed not only at the location of the former bridge, but also elsewhere. In addition, paved transfer points will be built for reloading the biomass to lorries.

9 Sustainability and implementation of paludiculture

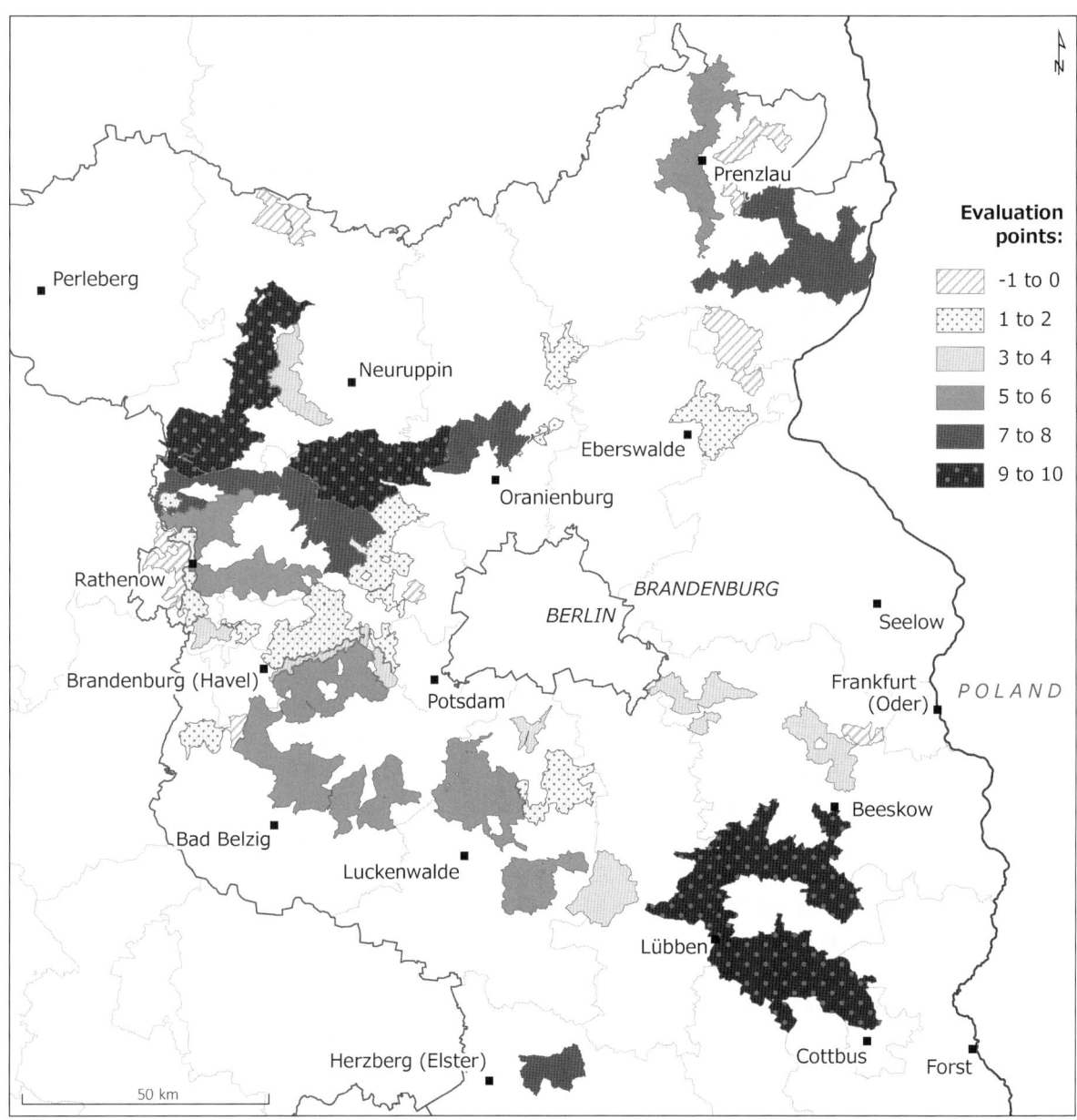

Figure 9.2: Suitability for paludiculture of (partial) catchments in the German federal state of Brandenburg, assessed using a points system. Catchments that received many points contain many areas suitable for paludiculture (Hasch et al. 2012).

9.3 The decision-support tool TORBOS

Paul Schulze, Claudia Schröder, Vera Luthardt & Jutta Zeitz

A change in land use on peatlands requires a good understanding of the site conditions and associated land use options. The complexity and wide range of possibilities make it hard for a farmer to keep an overview of the management requirements, which moreover must be tailored to each individual site. The decision support system 'DSS-TORBOS' (Decision Support System for a peat preserving cultivation of organic soils, German: Decision Support System *zur TORferhaltenden Bewirtschaftung OrganiScher Böden*) is an online tool that offers planning support to farmers, land use planners, and decision makers, in the conversion to a peat preserving land use practice on a particular site. All decisions are left to the user and are transparently documented by the system. The general structure of the tool is presented in the following sections.

9.3.1 The Decision Support System (DSS) as a guidance tool

Decision support systems (DSS) are information systems that provide support to users in complex decision-making processes and also help to find solutions. For this purpose, information is gathered using targeted questions and all data relevant to decision-making are processed. The result of the structured query is a list of possible solutions. The final decision is always left to the user, but the DSS serves to consolidate and document this decision.

A DSS usually consists of a graphical user interface, a database and logically linked models (Figure 9.3) (Turban 1995, Cox 1996, Shim et al. 2002, Janakiraman & Sarukesi 2004). Various types of models can be included, ranging from a simple decision tree (Box 9.2) to elaborate numerical simulation models. The models are central to the DSS, as they process the entries of the user into a decision supporting result. They are connected to the database, and combine and filter relevant data from internal and external sources (Jacob 2012). The graphical interface guides the user through the system, and requests all data and information necessary to solve the problem. After all queries of the DSS have been successfully processed, one or several suitably weighted results are produced (Bortfeldt et al. 2008).

Over the past years, decision support systems have been developed to address problems in agriculture and forestry, as well as nature conservation. Often, simulation models were involved and linked to a GIS to make predictions (e.g. calculation of water and matter dynamics). Examples of such DSS are ELANUS (landscape based analysis of land use strategies, German: *Landschaftsanalyse und Bewertung alternativer Landnutzungsstrategien*, Lutze et al. 2000); DSS-WAMOS (peatland forest management, German: *Waldmoormanagement* Hasch et al. 2008), a DSS for the management of degraded and abandoned peatlands in Russia (Abel et al. 2011), and one for the prediction of changes of peatland functions (Knieß 2007).

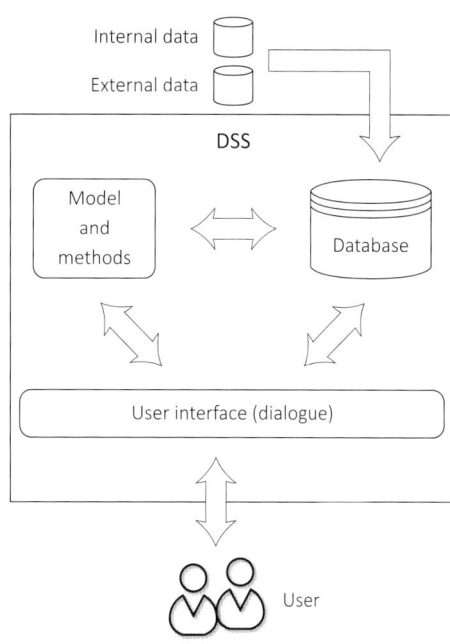

Figure 9.3: Important elements of a decision support system (DSS; after Jacob 2012).

Box 9.2: What is a decision tree?

Paul Schulze & Claudia Schröder

A decision tree is a hierarchical series of questions to guide a decision concerning an object or a process. Each question concerns an attribute of the object or a process, and prompts an answer that affects the further path through the tree. Going through the tree, attributes with the largest information content are addressed first, followed by the next most relevant, and so on. Attributes can be expressed either on a categorical scale or a metric scale. Through its hierarchical structure and by reducing decisions to a single attribute at a time, a decision support tree allows its user to make a quick, substantiated decision (Petersohn 2005, Schwaiger & Meyer 2011).

9.3.2 The structure of the DSS-TORBOS

The DSS-TORBOS can be accessed via a web page (www.dss-torbos.de). It offers planning support for a change to peat preserving or degradation reducing land use on a specific, agriculturally used fen peatland. Fen peatlands are here defined as areas with at least 30 cm of peat that has originated under groundwater-fed conditions (Sponagel 2005). The DSS-TORBOS addresses farmers, farming consultants, and decision makers. Users are guided through the system independently of their professional expertise. The system invokes the expert knowledge of the user but provides relevant information (e.g. explanatory texts and links to maps or graphic materials).

To keep the system transparent and easy to use, unwieldy simulations and complex models were omitted in favour of dichotomous decision trees (i.e. trees that require 'yes' and 'no' decisions; Box 9.2). This approach allows the user to return to the previous question and correct a decision if needed. The query addresses a concrete area, assessing as little data as possible but as much as necessary to help decide on the adoption of land-use.

The DSS is divided into two parts. The first part provides possible land use options for the area of interest. The second part specifies the measures required for the conversion to the new type of management. Related topics are arranged in query modules (Figure 9.4), which increases clarity, reduces complexity, and lowers error proneness (Heindorf 2010). The decision trees within each module provide intermediate results. An example of an intermediate result of module 'water availability' is that 'topographic location and site conditions indicate that it is possible to rewet the peatland'. The intermediate results are evaluated by using a dichotomous decision key, and potential land use options are indicated. After finishing the DSS, a number of land use options are suggested to the user.

As a final result, the system produces a printable summary of all queries and the final choice of land use options. In addition, the user receives recommendations for the required measures before the new land use can be implemented. The summary also includes information on the consequences of retaining the current land use practice.

9.3.3 Modules of the DSS-TORBOS

The first module, called 'general input information', serves to gather site parameters such as peat thickness, average soil moisture conditions, and current land use. These variables help assess the consequences if current land use is maintained.

The module 'water availability' looks at the water supply in the catchment of the peatland, and assesses whether sufficient water is available to raise the water table. The DSS-WAMOS on peatland forest management (Hasch et al. 2008) showed that simple descriptive proxies are as useful as technically more elaborate methods in assessing the rewetting potential of a site. For instance, the potential for rewetting may be restricted by a lack of inflow, or by water losses from the peatland (e.g. by evaporation, seepage) that can not be sufficiently compensated by water retention. In contrast, a high potential for rewetting may be assumed if, for example, existing drainage infrastructure can be dismantled. If the species composition of forests in the catchment has a negative impact on water availability, forest conversion can support rewetting (Müller 2009). The module has three possible outcomes with respect to rewetting potential: 'Already rewetted', 'possible', and 'impossible'.

The module 'restrictions' assesses resistance against rewetting and land use change, including whether rewetting might affect protected areas like nature reserves, Natura 2000 habitats, or other legally protected habitats (Hölzel 2009). The risk of a negative impact on nearby water bodies by increased phosphorous loads is assessed by a separate query. The module aims to shift the focus of rewetting and agricultural usage away from endangered habitats and protected species.

The intermediate results of all modules such as 'topography indicates that the site can be rewetted' (module water availability) and 'the peatland is not protected' (module restrictions) are linked, and the DSS lists one or more possible land use alternatives for the area.

Finally, the module 'water management' presents site specific measures to carry out rewetting, based on user input in other modules and the consequent land use options chosen. The presented catalogue gives information about different rewetting measures such as fixed weirs with rock ramp, movable weirs or (partially) backfilling of ditches.

9.3.4 Result: recommended land use options

Paludiculture includes land use practices that preserve the peat (Chapter 1). The rate of peat mineralisation increases with lower groundwater tables (Mundel 1969, Luthardt 1987, Chapter 5.1). Peat preserving land use is therefore only possible if the groundwater table does not fall below 20 cm below the ground surface, which correlates with the soil moisture class 4+ (wet to very humid) (Couwenberg et al. 2011, Couwenberg et al. 2011, Box 5.3). Under such conditions, land use options in fen peatlands include the cultivation of Alder (Chapter 3.1.5), the utilisation of reedbeds and stands of Cattail (Chapter 3.1) as well as grazing by Water Buffalo (Chapter 3.3.2, Box 5.6).

9.3 The decision-support tool TORBOS

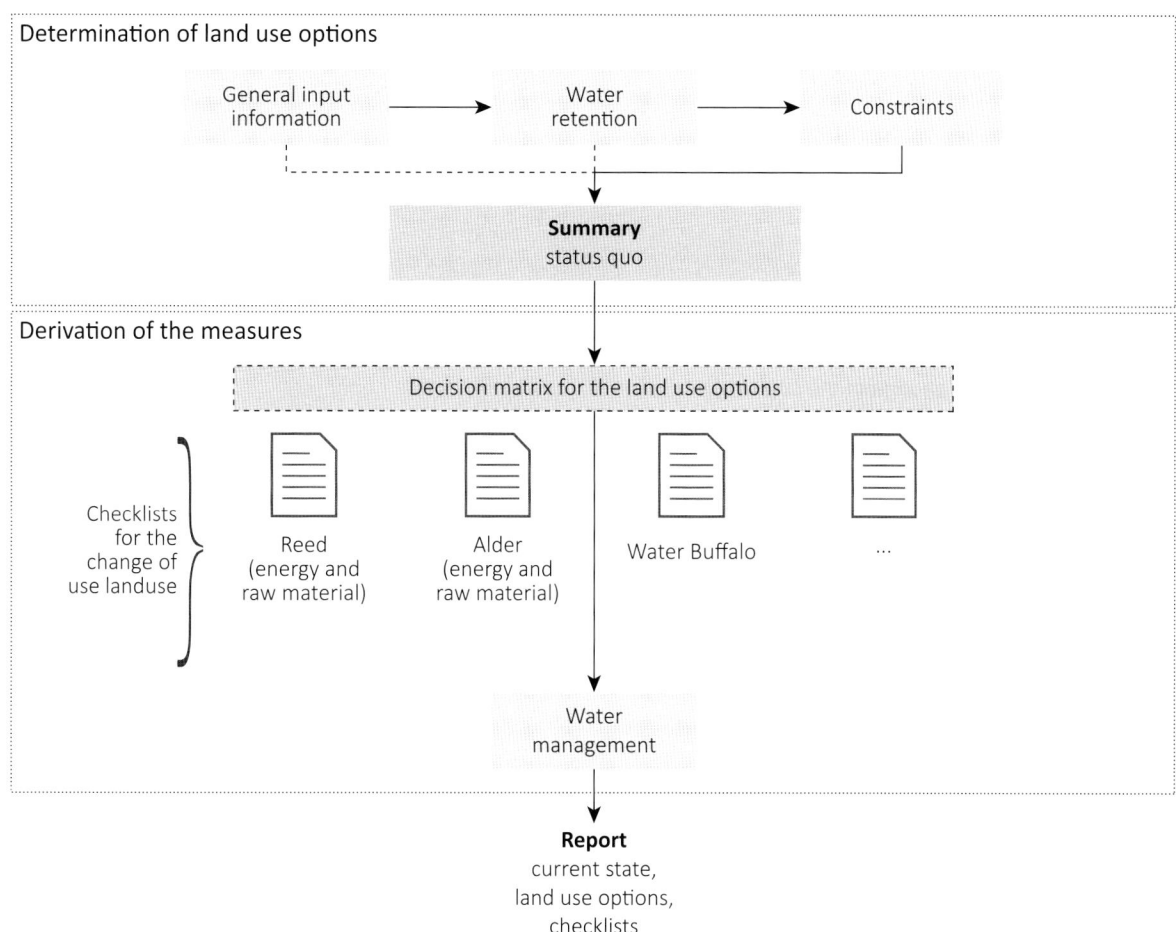

Figure 9.4: Basic structure of the decision support system TORBOS (the acronym stands for peat preserving cultivation of organic soils; German: Decision Support System zur TORferhaltenden Bewirtschaftung OrganiScher Böden).

The possible land use options are presented in fact sheets, which are included in the DSS but can also be used separately. The fact sheets give the user quick information about the essential facts and issues for implementing the respective land use. Also, land use options for not fully rewettable sites are presented.

Alder (*Alnus glutinosa*) cultivation is for example possible along water bodies and on peatlands with a wide water level amplitude. Nonetheless, the production of timber whilst conserving the peat is only possible under very humid conditions (soil moisture class 4+; Schäfer & Joosten 2005). Alternatively, short rotation coppice with a rotation of 5–8 years is an option. The disadvantage of this option is that hitherto no technique for harvesting is available. Sedge vegetation and reedbeds require high water levels and also tolerate inundation (i.e. reed). On wet and very humid grassland (soil moisture classes 5+ and 4+) Water Buffalo (*Bubalus bubalis*) is superior to other breeds of cattle. For areas with limited water supply, land use options are recommended that reduce peat degradation as much as possible, including low intensity grazing and the production of energy biomass at the highest possible water levels. Also, short rotation choppice with Alder or Willow must be taken into account.

The DSS and the fact sheets are available under www.dss-torbos.de.

9.4 Technical measures for implementing paludiculture

Wendelin Wichtmann & Christian Schröder

The assumption that peatlands can easily be rewetted and then used for paludiculture is rarely appropriate. Establishing paludiculture is fundamentally different from simple rewetting for peatland restoration. Rewetting for nature and climate protection often results in considerable spatial variation in water level. To reduce losses of carbon and nutrients, preserve the peat body, and provide optimal growth conditions to wetland plants in paludiculture, the water level needs to be near the surface during the whole year. A stable water table may require regulation and appropriate technical measures. A mosaic of different crops and land use practices can help to respond to differences in site conditions (Box 9.3). This may be more challenging with respect to site management, but it adds to operational diversification, regional added value and biodiversity. Moreover, infrastructure is needed for on site management and for transport of the harvested biomass (Chapter 4.4). These logistic challenges can equally be met by technical measures.

Table 9.8: Water management measures for paludiculture areas.

Water management	Examples	Characteristics
Water retention (passive water management)	• Ground sills • Blocking ditches • Filling-in ditches • Dams • Fixed weirs • Linking to receiving water • Spillway gates	• Water retention • No or low maintenance costs • Natural dynamics • Frequent strong fluctuations • No regulation possible • No influence on water table during extreme weather events
Water regulation (active water management)	• Adjustable weirs • Pumping facilities • Mobile pumps	• Active regulation of water table • Stable water tables possible • Possible to adapt to weather conditions • Maintenance costs • Potentially improved flood protection

---

Box 9.3: Land use mosaic after rewetting

Christian Schröder & Wendelin Wichtmann

In many cases, the rewetting of a peatland area creates a mosaic of very different site conditions. Particularly in shallow peatlands, peat subsidence has often resulted in a surface structure that reflects the relief of the mineral substrate below the peatland (Zeitz 1992). In peatlands with a thick peat layer, the rate of subsidence can differ considerably between sites; in sloping peatlands, it is hardly possible to restore a sloped groundwater table. Consequently, areas with different degree of wetness are formed after rewetting; in extreme cases, they are either deeply inundated or remain too dry. If peat conservation should be achieved over a maximum area, some parts of the peatland will be inundated too deeply for implementing paludiculture. For this reason, a difference between the gross rewetted peatland area and the net area that can be harvested can arise. Differences in surface relief can be reduced by levelling the area before rewetting or by dividing the area into different subsections, with different water levels. This makes rewetting measures for paludiculture complex. It may thus make sense to implement different kinds of uses within one area. In deeply inundated areas, water reeds could be established, which need another management than moist meadows at the rim of the rewetted area (Figure 9.5). Differences in quantity and quality of the feeding water can result in different management options as well. For example, if nutrient rich water is available, it can be used for irrigating fields of Cattail, which needs a large nutrient supply. However, in many cases only a small area may profit from this extra nutrient supply because of the high nutrient uptake by the plants. Sites that are not close to the inlets can be cultivated with crops with less nutrient demand or used to produce special qualities. For example, Common Reed from rather nutrient poor sites has good quality for thaching (i.e. thin stems with shorter internodes). Sites fed by spring or oxygen rich water are suitable for timber production with Alder. Studies have shown that peat accumulation in Alder stands also takes place under rather dry conditions (soil moisture class 4+, mean water level 0–20 cm below surface) (Schäfer & Joosten 2005). Thus, Alder is an option for somewhat less wet sites. Less accessible sites can also be planted with Alder trees if it is accepted that harvest can only take place occasionally after severe frost.

The planning of infrastructure for paludiculture has to take into account the requirements of the desired crops regarding harvesting time and frequency, and biomass transport (Chapter 4.3). A mosaic of different land use practices is more challenging for management, but offers added value with regard to operational diversification and regional development. The implementation of different management options offers perspectives for species diversity as well (Chapter 5.2).

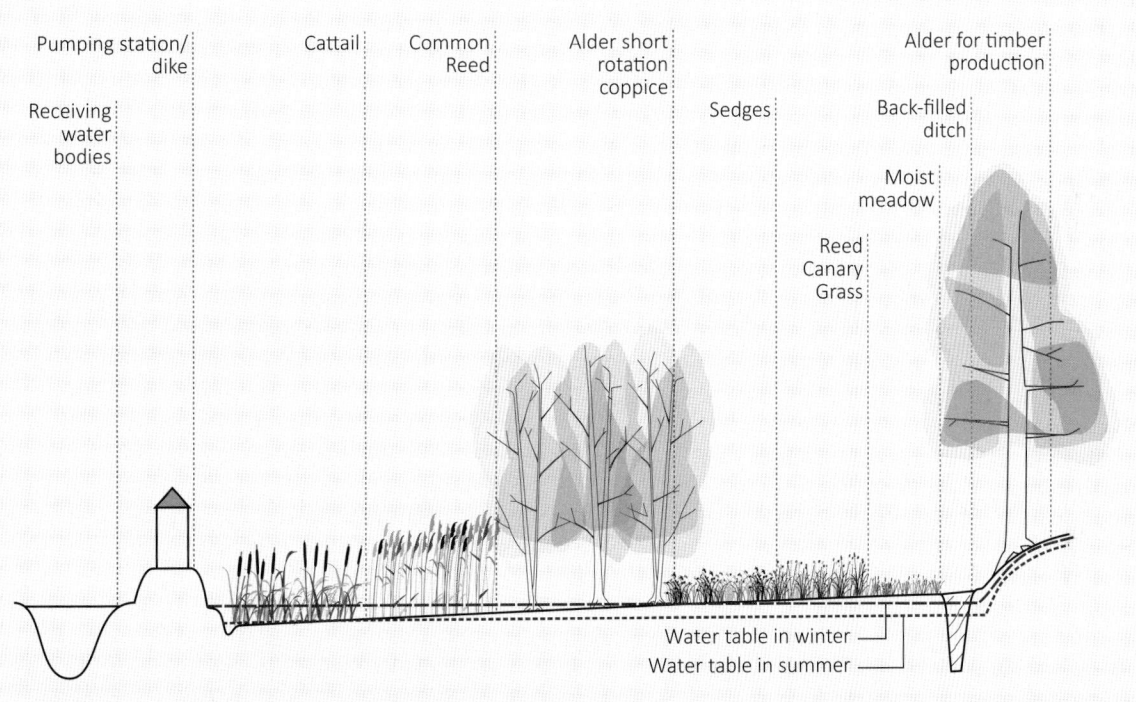

Figure 9.5: Mix of different land use options after rewetting.

9.4.1 Water management

Rewetting is a precondition for paludiculture on degraded peatland areas. Depending on the condition of the peatland and its supply of ground and surface water, different actions are needed. Two types of measures can be distinguished to restore high water tables: those that require a single intervention (one-off measures) and those that allow for active regulation of the water table (Table 9.8).

One-off measures include ground sills, infilling of ditches, weirs, and dams, all of which have proven useful in nature conservation projects. Moreover, costs for their installation and maintenance are low. Particular technical solutions have been described in detail in the pertinent literature (Wheeler & Shaw 1995, Hawke & José 1996, Brooks & Stoneman 1997, Kratz & Pfadenhauer 2001, Wagner & Wagner 2003, LUA Brandenburg 2004, Grosvernier & Staubli 2009, Kozulin et al. 2010). These one-off measures serve foremost to keep the water in the peatland, and hardly allow for targeted regulation of the water table. As drainage will have caused considerable changes to the peat body (e.g. reduced pore volume), water table fluctuations will be larger than in undrained mires. A fixed top water level may then result in inundation in winter alternated with deep water tables in summer, depending on water availability in summer, the input from the below- and aboveground catchment, and evapotranspiration (Box 9.4). To prevent the sites from becoming too dry in summer, the retention level must be set higher. Consequently, in most cases, optimum water tables for paludiculture can hardly be achieved with one-off measures. Active regulation of the retention level or even irrigation with additional water may be necessary (Box 9.4). The more additional water is available, the easier it is to balance seasonal demands and keep water tables constant. Additional water is easily available if the formerly drained peat surface has subsided below the level of the receiving waters. Thus, low lying polders are suitable for paludiculture, but the existing water regulation infrastructure must be adaped for water level regulation in both directions (Figure 9.6, Table 9.8).

Conventional high performance machinery can only be used if the water table is brought to about 40 cm below the surface. Accordingly, it is sometimes discussed to lower the water table during harvest to reduce the investment costs for adapted harvesting machinery. To limit negative impact, water table should be lowered for as short a period as possible. However, a quick lowering of the water table requires a drainage system that is much more efficient than a conventional one (higher gradient, denser network of ditches). Apart from the high costs of such drainage system, lowering the water table in summer bears the risk that water supply is not sufficient to raise the water table back to the surface, and that additional water is needed. Moreover, particularly in summer, even a short period of low water tables can induce a significant increase in greenhouse gas emissions. Such situation is inconsistent with the principles of paludiculture. Moreover, the many ditches needed for drainage and irrigation would present an obstacle during harvest, and could render the removal of the harvested biomass from the site impossible. Thus, draining a site for a short period during harvest can, at best, be an option for special crops.

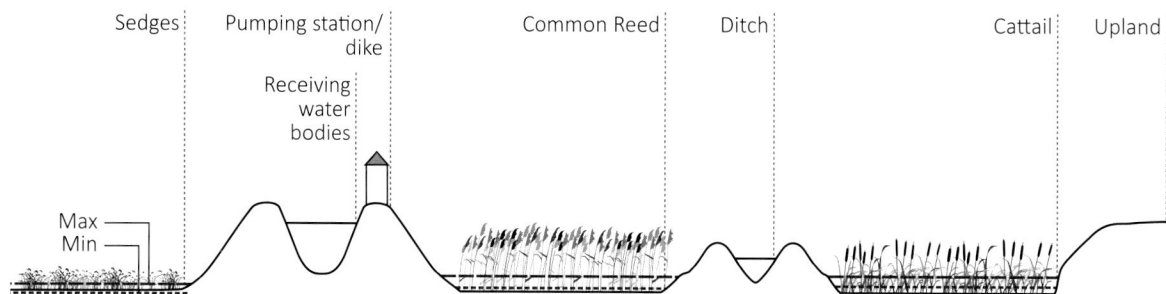

Figure 9.6: The water management can be designed differently depending on the requirements of the culture. When peatlands are already below the level of the receiving water, such as in many polders, a sufficient supply with additional water is usually possible.

9.4 Technical measures for implementing paludiculture

9.4.2 Adjusting infrastructure

To limit damage to the site and the harvested vegetation, harvest and removal of biomasss require adapted infrastructure (Chapter 4.4). If such infrastructure does not exist, its construction should be addressed at the planning stage. In many cases, the existing infrastructure needs to be adjusted to the special requirements of management under wet conditions (Box 9.1). Insufficient drainage of paved roads can result in damages due to frost, and often there are merely dirt 'summer' roads, which can be used only when the subsoil is dry. Some land use options of wet peatlands, however, require also a trafficability during winter (e.g. harvesting thathing reed). To improve transport infrastructure, the following measures may be neccesary:
- Construction of paved transport routes and biomass transfer sites;
- Adaptation of existing transport routes to high water levels;
- Improving accessibility by filling in ditches and realignment of fields;
- Establishment of additional access points;
- Regular maintenance of infrastructure;
- The use of mobile boards.

Nevertheless, even outside the harvest areas, an optimisation of infrastructure is often required. The biomass transfer points must be accessible via paved roads that are suitable for roadworthy transport vehicles. For this purpose, an upgrading of the existing roads may be needed. Sometimes, the construction of new bridges may be necessary to connect harvested areas with existing infrastructure. The use of mobile pontoon bridges for biomas removal (Colour pictures 83 and 84) is only recommended in exceptional cases because of risk of accidents.

The cost for the implementation of infrastructure measures should be beared by the general public if it implies a deconstruction of infrastructure that has been financed in the past with public funds. Grants can be provided through regional development programmes and agricultural structural programmes (Chapter 6.2; Chapter 7.3).

Box 9.4: Water demand of paludiculture, illustrated for northeast Gemany

Andreas Wahren

The constant availability of water normally makes evaporation from peatlands higher than from mineral soils. As a result, paludiculture on fen peatlands in northeastern Germany would need additional water to cover the evaporative losses in summer. In undisturbed peatlands, the catchment provides (above or below ground) either the necessary water, or internal retention mechanisms constrain water table fluctuations (mire oscillation = change in pore volume). However, paludiculture on hydrologically disturbed peatlands (particularly on soils with a low storage capacity) requires systematic water management. The water deficit during summer can either be covered by retaining water from the wet season (November to April) or by irrigation. Example hydrographs for different target water tables were (theoretically) derived from the water balance of a peatland in northeast Germany (Figure 9.7). This simple example was calculated using average data (1997–2011) from the climate station Angermünde (NE Germany). Evaporation was calculated using Dalton's law for water levels above the surface (DVWK 1996), or following Romanov for water tables below the surface (Edom et al. 2010) with an assumed 15% of macropores for the upper peat layer. In this example, a water table near the surface (mean annual water table +/- 0 cm) can only be maintained if the site is inundated by 12 cm in March, at the end of the winter period. If the water level is just 5 cm above the surface in March, the water table will already drop to 46 cm below the surface in June (mean annual water table ca. -10 cm). If the water table in March is reduced to 12 cm below the surface, June water tables would drop 54 cm below the surface and the water table would remain below surface most of the year (mean annual water table ca. -20 cm). The example underlines the importance of retaining winter excess water instead of draining it away.

Should inundation and associated water retention not be possible, water must be added during summer to avoid deep water tables. To prevent the water table to drop below -20 cm, a mean annual irrigation of 110 mm or 110 m³ ha^{-1} is needed between April and September (Figure 9.8).

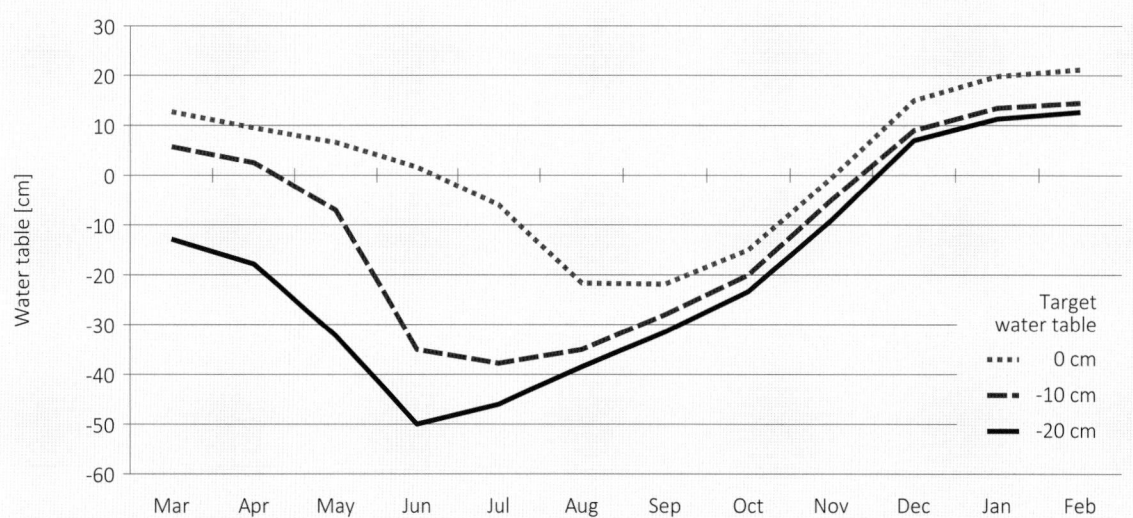

Figure 9.7: Modelled hydrographs for a peatland in northeast Gemany for mean annual target water tables of +/- 0, -10 and -20 cm. Hydrographs were derived from the water balance (precipitation minus evapotranspiration), not taking in- and outflow and infiltration into account (climate data for Angermünde: 1997–2011). The upper graph corresponds to soil moisture class 6+/5+; the two lower graphs to 5+/3+.

9.4 Technical measures for implementing paludiculture

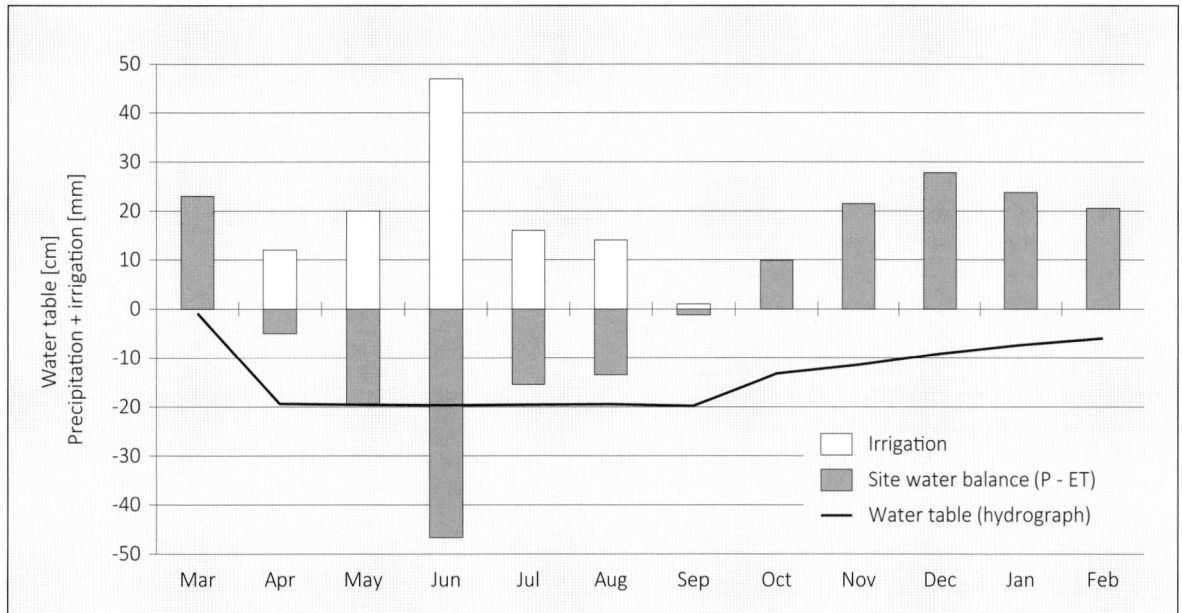

Figure 9.8: Mean monthly amount of irrigation (white) required during summer to prevent the water table from falling more than 20 cm below the surface. Water balance calculated as precipitation (P) minus evapotranspiration (ET) without taking in- and outflow and infiltration into account, and using climate data (1997–2011) from the Angermünde station. The hydrograph corresponds to soil moisture class 4+, with a mean annual water table of -14 cm.

9.5 Implementation and administrative approval in Germany

Bernhard Hasch

The requirements for the administrative approval of paludiculture projects differ depending on the actual project component and the necessary construction works. Hydrologic engineering, claims on water rights, improvements in infrastructure (roads, storage areas, and so on), changes in land use, and aspects of nature conservation must all be reviewed according to their respective sectoral requirements. The obligation to obtain a permit follows from the expected effects of the planned measures. These effects must be presented in the course of a plan approval procedure. Additional expert reports or an environmental impact assessment may be necessary.

9.5.1 Planning steps for the implementation of a paludiculture project

Paludiculture projects are generally large-scale projects that need to fulfill complex technical, operational and legislative prerequisites. Therefore, it is recommended to first draw up a general development plan (Chapter 9.2 and 9.3) that gives a rough presentation of:
- Potential area;
- Potential land use practices (crop, harvest period, and so on);
- Potential operators (farms) and machinery;
- Rewetting potential/water supply in the catchment;
- Necessary constructions (hydrologic engineering, infrastructure, storage areas, and so on);
- Availability of areas (ideally based on communication with owners and farmers);
- Possible impacts on environment and landscape;
- Potential project managers;
- Availability of funding;
- Potential processing facilities.

On this basis, coordination should be sought at an early stage with the responsible authorities for agriculture and nature conservation, and with the water board. They can provide guidance, assess whether permits are possible in principle or whether restrictions apply, and advise on the necessary documentation (see below).

In addition, it may be useful to organise public participation at an early stage. Public participation is essential if substantial reservations towards the planned project are expected (see Chapter 8.2, Box 8.3).

9.5.2 The approval procedure

If the early communication with the authorities and the public does not indicate any insurmountable obstacles, the formal plan approval procedure can start. To this end, the project management typically has to submit the following documents:
- Comprehensive project layout, including a cost estimate;
- Evidence of area availability;
- Presentation of the potentially affected area;
- Environmental impact minimisation and compensation plan;
- Expert opinion on potential impact on protected species (if applicable);
- Study on compatibility with the EU Habitats Directive (if applicable);
- An environmental impact assessment (if applicable);
- Hydrological assessment covering implications for flood protection and the European Water Framework Directive (if applicable).

The plan approval procedure includes a formal hearing that involves the authorities and the public; the complete planning documentation must be presented at this hearing. After the (amended) plan has been approved, a concrete plan of execution can be made to implement the required measures and start paludiculture.

Legally, the plan approval procedure replaces all other required permits, including those regarding conservation law. The plan approval stipulates that all stakeholders, as well as the public are involved (Regierungspräsidium Gießen 2011).

Under some circumstances, a simple permission by the authorities may be sought instead of going through the plan approval process. According to §68 of the German Water Resources Act (German: Wasserhaushaltsgesetz, WHG) this simplified procedure is possible if:
- The hydrologic engineering does not require an environmental impact assessment;
- The rights of third parties are not affected or affected parties have agreed in writing to claims on their property.

The obligation for an environmental impact assessment (EIA) is regulated by law (German: *Gesetz zur Umweltverträglichkeitsprüfung*). Annex 1 of the German EIA law stipulates the need for a pre-assessment for each individual case, (e. g. for hydraulic engineering projects in agriculture). The use of a water volume of 5,000 to 100,000 m³ per year requires a site-specific pre-assessment of the impact on local hydrology. From 100,000 m³ per year upwards, a general pre-assessment is required. If the pre-assessment reveals potential negative impacts (as defined by Annex 2 of the EIA law), an environmental impact assessment must be carried out.

If a Natura 2000 Habitat (according to §32 of the Federal Law on Nature Conservation, – German: *Bundesnaturschutzgesetz*) might be affected by the paludiculture project, a risk assessment has to be conducted (according to §34 of the Federal Law on Nature Conservation).

9.5.3 The legal approval procedure with regard to water

The extent and range of hydrological effects are considerable, and may well extend beyond the paludiculture area itself. Therefore, the implementation of paludiculture on formerly drained peatlands can hardly avoid an elaborate water law procedure. Also, in case of a paludiculture project in a spontaneously rewetted area, a review against water law requirements is recommended to enhance planning security.

In paludiculture projects, the groundwater table is usually raised by damming-up or filling in ditches and canals. Also, supply of water from external sources may be necessary. Damming-up of water constitutes utilisation of a water body according to §9 of the German Water Resources Act. Therefore, the building of a dam requires a legal permit according to §8. If it causes large scale rewetting (which is actually the intent), damming is subject to plan approval (§68), because it may affect river banks and floodplains (Regierungspräsidium Gießen 2011).

Any restructuring or closing of surface waters requires a plan approval, because it is a significant alteration of a water body according to §67, section 2 of the German Water Resources Act. Legally, ditches and canals are considered water courses as well and included under this Act (LUA Brandenburg 2004, BLU 2005, Regierungspräsidium Gießen 2011). Moreover, requirements of flood protection need to be met. Projects situated inside designated flood risk areas may not negatively affect flood retention potentials (e.g. if retention areas are cultivated) or increase flooding risks by reducing run-off.

Water law furthermore demands to check whether the project may violate the ban on deterioration stipulated in §27 of the German Water Resources Act (environmental objectives of surface water according to the European Water Framework Directive). A violation may occur if a new dam disturbs the connectivity along the flow path of a stream (§34) or if damming-up or abstraction of water for irrigation would lower stream water flow below the minimum threshold, as set out in §33.

9.5.4 Land use change and intervention regulations

All alterations of the shape and use of land are subject to intervention regulation according to §14, section 1 of the German Federal Law on Nature Conservation (German: *Bundesnaturschutzgesetz*).

In certain cases, a change in land use practice to paludiculture may require permission by the nature conservation authorities. If agricultural land use follows 'best practice', it is considered without impact on the environment according to §14, section 2 of the German Federal Law on Nature Conservation. However, this rule only applies if the area was agriculturally used before paludiculture was implemented, and if the paludiculture practice is officially recognised as agriculture (cf. cultivation of reed; Chapter 7.2).

If paludiculture is established on previously unused (or abandoned) areas, the authorities need to be convinced that functions and services of the environment are not substantially impaired. Otherwise, the project implies an impact on the environment that has to be stopped or compensated.

If agricultural practice is changed in a substantial way, the laws on habitat protection (according to §30, section 2 of the German Federal Law on Nature Conservation) and the ban on ploughing up fen grasslands (according to §5, section 2, No. 5 of the German Federal Law on Nature Conservation) must be respected. Habitat protection (according to §30, section 3 of the German Federal Law on Nature Conservation) includes an exemption clause that entails compensation of negative impacts. If, for example, paludiculture is implemented on legally protected wet meadows, an exemption is required. In case of a considerable negative impact on the protected habitat, losses will have to be compensated.

It may be necessary to plough or slit the sward to supress competition and facilitate a quick establishment of reed beds, for example. It is not yet clear, whether this form of grassland renewal or change in the sward is in line with the Federal Law on Nature Conservation. Thus, permission for establishing new crops should be sought from the appropriate authorities. In addition, conservation status and associated decrees must be observed.

Paludicultures usually require changes in the infrastructure as well (Chapter 4.4, Chapter 9.4.2). Changes in the road network will need the appropriate permits from the authorities, if construction goes beyond repairs and involves building new road stretches (or widening and paving exisiting ones). The road works can be dealt with in the framework of the plan approval procedure, under the water act.

10 Paludiculture in a global context

Globally, usable land is becoming an increasingly scarce resource. In many regions of the World, the pressure on land is continuously rising, because of continuous population growth and their need for shelter, infrastructure, raw materials, and importantly, food. In recent years, usable land has been lost to overexploitation and climate change. New forms of land use – like growing crops for energy and raw materials, are additional strong drivers of the growing demand for land. Additionally, peatland areas get lost for the production of comestibles and raw material, because of rewetting of degraded peatlands driven by nature conservation and environment protection.

Supported by international framework conditions and obligations (Chapter 10.1) there is a huge potential for the rewetting of degraded peatlands and their conversion to paludiculture worldwide (Chapter 10.2). Consequently, many possibilities for its implementation arise, supported by international framework conditions and obligations.

Until now, rewetting of degraded peatlands with the explicit aim to implement paludiculture has only been carried out in pilot projects. Nevertheless, paludiculture is already practised around the world (e.g. Table 10.1). The uses of wet peatland sites may range from traditional production of raw materials to the targeted promotion of specific ecosystem services.

In Northeast Germany (Chapter 10.3), the use of peatlands for the protection of nature and the climate receives a lot of attention, whereas the traditional use

Table 10.1: Examples of paludiculture around the world.

Country (Region)	Germany (Mecklenburg-West Pomerania) Chapter 10.3	Belarus (Grodno Region) Chapter 10.4	Poland (Biebrza, Podlasie) Chapter 10.5	Indonesia (Kalimantan) Chapter 10.6	China (Heilongjiang, Jilin) Chapter 10.7	Canada (Manitoba) Chapter 10.8
Peatland type/ soil type	Fen (degraded valley mires, polders)	Raised bog and fen (cutover peatlands)	Fen (semi-natural)	Raised bog (degraded, deeply drained former peat swamp forest)	Lake edges (lake sediments, terrestrialisation mires)	Fen (flood mires, flood plains)
Land use so far	Drained grassland	Drainage and superficial peat cutting, then abandoned	Haymaking, later succession of shrubs and trees	Rice cultivation, abandonment	Traditional land use	Recreation, hunting, trapping, agriculture
Objectives	Climate protection, sustainable land use, biomass utilisation	Climate protection, species conservation, biomass utilisation	Habitat management, biomass utilisation	Prevention of peat fires, sustainable land use, climate protection	Biomass production (as raw material for pulp production)	Nutrient retention, biomass utilisation
Plant species	Common Reed, Reed Canary Grass and sedges	Common Reed, Reed Canary Grass and sedges	Mainly sedges	Multiple crops	Common Reed	Cattail
Harvest/ Transport	Converted snow groomers (Pistenbully)	Adapted tractors, converted snow groomers (Pistenbully)	Converted snow groomers ('Ratraks')	Manual	During frost, manual, common agricultural machinery	Adapted tractors
Utilisation of biomass	Energy and material use	Energy use: briquettes, pellets	Energy use: pellets	Food, timber, pulp	Paper, wicker work, forage	Energy use: loose biomass, bales, briquettes, pellets

of Common Reed from near-natural peatlands has been declining for years. The use of biomass from re-wetted peatlands for energy still is in its initial stages, and is largely restricted to biomass from conservation management.

In Belarus, replacement of peat briquettes by biomass grown on rewetted cut-over peatlands is being tested (Chapter 10.4). Using biomass from rewetted peatlands for energy reduces greenhouse gas (GHG) emissions in two ways: Firstly, peat – which is a fossil source of energy, is replaced by sustainably produced biomass, and secondly, the emissions from the degradation of drained peatlands are avoided. Moreover, land use switches to wet, semi-natural peatlands, where biomass harvest supports habitat development for species protection (Tanneberger & Wichtmann 2011).

The Polish paludiculture activities presented in Chapter 10.5 are mainly concerned with habitat management for the protection of species. However, there is a targeted effort to use agri-environmental funding to achieve large scale implementation. The use of the biomass is currently being studied in pilot projects.

The Indonesian example (Chapter 10.6) illustrates the many options offered by paludiculture. Rewetting and paludiculture will not only combat GHG emissions, but also social problems, and importantly, the devastating peat fires.

China is one of the few countries where the use of wetlands focusses on biomass production. Although overall declining, the production of reed-based paper still occurs at a large scale (Chapter 10.7).

In Canada (Chapter 10.8),the large scale cultivation of Cattail is being tested with the objective to reduce nutrient loads to receiving waters, as well as to provide raw materials to biomass industries.

The diverse approaches to the use of wet peatlands not only demonstrate the diverse ways of profitable use of biomass, but also the diverse synergies that paludiculture may provide.

10.1 Global demands and international commitments

Hans Joosten

The current developments are only the beginning. The global population will grow to 9 billion people by 2030 and will need 50% more grain, more than 35–60% more clean water, and 45% more fuel (Foresight 2011). The pressure on peatlands will increase because of population growth, which is unstoppable for the time being, because of the justified demand of the many poor people for more prosperity, and because of the need to replace fossil raw materials (such as oil, gas, ores and minerals) by products of agriculture and forestry, as the geological resources are depleting or we cannot allow their utilisation for climate-related reasons any more.

This implies that the experiences made with the destruction of European and Southeast Asian peatlands will be repeated in other parts of the world, if no adequate alternatives are found. We will only be able to protect peatlands by clearly demonstrating their crucial importance for humankind: as global climate regulators, providers of drinking water, and because of the significant role they play in securing inhabitable and productive land, in protecting our physical and mental health and in preserving our value system. We need learn to keep the impact on peatlands to a minimum, to avoid serious risks, to appreciate functions of wet peatlands and to seek synergies (Joosten & Clarke 2002)

And we need to find ways to avert the pressure of the expanding land use from the ever more decreasing wilderness areas and to direct it to the degraded areas to make them productive again, in order to preserve the remaining wilderness as a mirror of our culture. If we want to protect peatland wilderness we must rehabilitate the degraded peatlands.

Both rewetting of peatlands and subsequent paludiculture are consistent with and supported by a wide variety of international policy agreements (Barthelmes et al. 2015). The Ramsar Convention on Wetlands has called upon its member states to take urgent action … 'to reduce the degradation, promote restoration, improve management practices of peatlands … and to encourage expansion of demonstration sites on peatland restoration and wise use management in relation to climate change mitigation and adaptation activities' (Resolution X.24, 2008). Resolution X.25 (2008) explicitly encouraged Contracting Parties 'to consider the cultivation of biomass on rewetted peatlands (paludiculture)'.

Recently, the UN Framework Convention on Climate Change (UNFCCC) has substantially improved the perspectives for peatland rewetting. Whereas the Kyoto Protocol did not provide real incentives for peatland rewetting, in Durban 2012 the UNFCC has adopted the new activity 'Wetland drainage and rewetting'. The activity has deliberately been created to allow a 'hotspot approach' for peatland rewetting, as long as no complete wall-to-wall land-based accounting has been achieved, or not all land use activities have to be accounted mandatorily.

The Intergovernmental Panel on Climate Change (IPCC) has meanwhile produced extensive guidance on how to report and account for these peatland emissions (IPCC 2014a, 2014b). For the longer term, the architecture of the post-2020 climate treaty will be decisive. The disproportionally large emissions from drained peatland offer an excellent opportunity to implement a hot spot approach for peatlands as a first step toward the full coverage of land. Furthermore, peatland rewetting is – certainly in combination with paludiculture, a good example of the integration of climate change mitigation and adaptation. The Convention on Biological

10.1 Global demands and international commitments

Diversity (CBD) recognised in 2008 'the importance of the conservation and sustainable use of the biodiversity of wetlands and, in particular, peatlands in addressing climate change'. Whereas not specifically focusing on peatlands, the Strategic Goals and Aichi Targets adopted by CBD's Decision X/2 (2010) are of special importance, specifically Target 15: 'By 2020, ecosystem resilience and the contribution of biodiversity to carbon stocks have been enhanced, through conservation and restoration, including restoration of at least 15 per cent of degraded ecosystems, thereby contributing to climate change mitigation and adaptation and to combating desertification.'

Since 2011, the Food and Agricultural Organisation of the United Nations FAO pays much attention to peatlands in the framework of its MICCA (Mitigation of Climate Change in Agriculture) programme. The FAO recognises that agriculture and forestry are the main drivers of peatland drainage worldwide. However, it also acknowledges that, in turn, peatland drainage – through huge GHG emissions and subsidence associated land loss is frustrating the aims of a sustainable provision of food, fodder, fibre and fuel. In 2012 FAO produced the report 'Peatlands – guidance for climate change mitigation through conservation, rehabilitation and sustainable use' (Joosten et al. 2012), which identified 10 elements of strategic action, including '5. Restore degraded peatlands by rewetting, reforestation (in the tropics) and subsequent conservation and/or paludiculture.' A FAO report on the technical options for climate responsible peatland utilisation – published in 2014 (Biancalani & Avagyan 2014), gives best practice examples with recommendations for paludiculture.

Following the embrace of the new Kyoto Protocol activity Wetland Drainage and Rewetting, the European Union aligned its own accounting rules with the new standard and adopted Decision No. 529/2013/EU in May 2013, which effectively imposes mandatory accounting for most peatland rewetting activities. In Annex IV of the Decision, 'incentivising sustainable paludicultural practices' is explicitly mentioned as a measure 'to improve the management of agricultural organic soils, in particular, peat lands'. With respect to peatlands, the recitals of the decision furthermore state, that 'the Union should endeavour to advance the issue at the international level with a view to reaching an agreement within the bodies of the UNFCCC or of the Kyoto Protocol on the obligation to prepare and maintain annual accounts' for Wetland Drainage and Rewetting, 'with a view to including this obligation in the global climate agreement to be concluded no later than 2015.'

The EU Habitats Directive (together with the Birds Directive) forms the cornerstone of Europe's nature conservation policy, and protects a wide variety of mire and peatland types (Chapter 7.1). The EU Biodiversity Strategy to 2020 sets as midterm goals: 'To halt the deterioration in the status of all species and habitats covered by EU nature legislation and achieve a significant and measurable improvement in their status … by 2020'. The Strategy also includes a 'restoration subtarget' to restore at least 15% of degraded ecosystems.

The EU Water Framework Directive (WFD) has as an aim to establish a framework for the protection of inland surface waters, transitional waters, coastal waters and groundwater and, with regard to their water needs, terrestrial ecosystems and wetlands directly depending on the aquatic ecosystems (Box 5.7). Through Article 1 mires and peatlands are thus protected by the Water Framework Directive against further deterioration. Moreover, wetland restoration is explicitly suggested as a measure. In various regions of the EU, the nitrogen and phosphorus pollution into water courses, lakes and coastal waters has to be reduced in order to reach the target of the Directive. Wet mires act often as nutrient sinks, whereas drained peatlands have short flow paths and a decomposing peat body, which increases nutrient pollution of surface waters. In this way, rewetting of drained peatlands contributes to the implementation of European targets in water pollution control.

Similar goals have also been formulated in the HELCOM Baltic Sea Action Plan to restore the good ecological status of the Baltic marine environment by 2021. Peatland rewetting is one of the options that have been adopted by various member states to reduce the emission of nitrogen and phosphorus to the Baltic Sea.

These examples show that the international legal and policy framework opens ample opportunities to realize important synergies by the implementation of paludiculture worldwide.

10.2 The global potential and perspectives for paludiculture

Alexandra Barthelmes

The global peatland area is estimated at 4,5 million km², most of which is found in the boreal and subarctic zones (Gorham 1991), and in tropical climate zones (Lappalainen 1996, Joosten 2009). Although there are many global peatland inventories (e.g. Kivinen & Pakarinen 1980, 1981, Gore 1983, Lappalainen 1996, Joosten & Clarke 2002), most of the data on location, extent, and particularly, on the status of peatlands (natural or degraded) are based on estimates.

About 15% of the global peatland area is drained, 80% of which for agriculture or forestry (Joosten & Clarke 2002, Joosten 2009). The current global hotspot of drainage and degradation lies in Southeast-Asia. In Western Indonesia alone, about 1,400 km² are deforested and drained each year (cf. Miettinen et al. 2012a, cf. Joosten et al. 2012). Between 1960 and 1980, natural peatlands were drained at a rate of about 5,500 km² per year (cf. Immirzi et al. 1992). In light of the developments in Southeast-Asia, the rate now is most likely just as high (Colour picture 89).

Most of the degrading peatland areas are located in the temperate climate zone and in Southeast-Asia – i.e. in densely populated areas with a climate that is favourable for agriculture (Joosten & Clarke 2002, Barthelmes et al. 2012). In Europe, there are about 286,000 km² of degraded peatlands. In 19 European countries the area of heavily damaged peatlands exceeds 1,000 km². Thus, of all continents, Europe has the largest potential for rewetting and paludiculture (Barthelmes et al. 2012), followed by East Asia. In China alone, where almost all peatlands are disturbed to some degree, there are more than 20,000 km² of degraded peatland (Yang 2000). Furthermore, there are about 125,000 km² of degrading peatland in Indonesia and about 8,400 in Malaysia (Joosten et al. 2012; see Table 10.2).

Although peatlands cover only 3% of the global land area, they store as much carbon as all terrestrial biomass (cf. Houghton et al. 1990, Joosten & Couwenberg 2008). Only 15% of the global peatland area has been drained for land use (about 0.4% of the global terrestrial area). This relatively small area, however, is responsible for almost 5% of the global anthropogenic CO_2 emissions (Joosten 2009, Joosten et al. 2012). Efforts to rewetting have to focus on these hotspots of peatland degradation (Table 10.3). This could help curtail greenhouse gas emissions and mitigate climate change.

The same negative effects of drainage and degradation as observed in Central Europe are seen all over the world. Yet, due to faster degradation processes in tropical climates, drainage often has a more severe impact on the environment, the economy and living conditions in such areas (Box 10.1).

Table 10.2: Worldwide regions with large potential areas for paludiculture. Current dominating land use mainly agriculture.

Country/region	Degrading peatland area km²
Europe	
Northern Baltic[1]	109,192
East[2]	110,660
West[3]	45,193
China	20,020
Mongolia	15,000
Southeast Asia	
Indonesia	124,900
Malaysia	8,380
Papua-New Guinea	10,015
Africa	
Lake Victoria region[4]	1,815
Zambia	1,620
Angola	1,010
USA[5]	10,516
Brazil	3,000

[1] Includes Finland (64,930 km²), Sweden (15,458 km²), Latvia (6,067 km²), Estonia (5,900 km²), Lithuania (4,679 km²), Norway (6,600 km²), Iceland (3,665 km²) and Denmark (1,893 km²). Currently dominating land use is forestry and agriculture.
[2] Includes Russia (90,600 km²), Belarus (16,970 km²) and Poland (10,090 km²). Currently dominating land use is agriculture and forestry.
[3] Includes Germany (12,550 km²), UK (18,089 km²), Ireland (11,971km²) and The Netherlands (2,583 km²).
[4] Includes Uganda (204 km²), Burundi (630 km²) and Rwanda (552 km²), Kenya (208 km²) and Tanzania (221 km²).
[5] Includes Alaska, Hawaii, American Samoa and Guam.

One of the key problems of the agricultural use of drained peatlands and the associated degradation is the loss of the productive function of the land, because of subsidence (by peat oxidation, shrinkage and compaction), desertification, and acidification (Joosten & Clarke 2002, Joosten et al. 2012). In many parts of Asia and sub-Saharan Africa, peatlands are among the most productive land for agriculture. Many of these peatlands are currently overexploited: they are deeply drained, regularly ploughed, or overgrazed. In conjunction with climate change and human induced fires, their overuse has led to large losses in productivity and of productive land. In Mongolia, for example, there are large stretches of degraded, dry peatlands without ve-

10.2 The global potential and perspectives for paludiculture

Table 10.3: Global Hot Spots (top 50) of peatland related Greenhouse Gas Emissions (CO_2, CH_4, N_2O) (excl. fires), sorted for 'Total degrading peatland area'.

Country	Current peatland area [km²]	Peat carbon stock [Mton C]	Degrading peatland area [km²]				Total annual emissions [Mton CO_2]
			Total	Agriculture	Forestry	Peat extraction	
Indonesia	265,500	54,016	124,900	62,096	62,804		687.7
Russian Federation	1,411,796	137,554	90,600	51,000	35,000	4,600	198.6
Finland	90,000	9,000	64,930	4,116	59,691	1,123	85.8
China	33,499	3,477	20,020	20,000		20	71.0
United Kingdom	26,838	2,684	18,089	15,497	2,523	69	19.6
Belarus	25,600	2,560	16,970	13,627	2,975	368	44.7
Sweden	66,450	6,680	15,458	2,060	13,300	98	22.8
Mongolia	26,291	750	15,000	15,000			47.4
Germany	12,800	2,018	12,550	10,434	1,918	199	36.0
Ireland	14,664	1,466	11,971	8,091	3,000	880	12.4
USA	223,846	29,167	10,522	10,292		230	34.6
Canada	1,700,000	170,000	10,252	2,000	6,972	1,280	10.6
Poland	12,110	876	10,090	7,542	2,515	33	27.2
Papua New Guinea	59,921	5,982	10,015	10,015			46.3
Malaysia	26,684	5,431	8,380	560	7,820		49.9
Ukraine	12,000	1,200	7,000	4,769	1,926	305	16.5
Norway	45,700	4,570	6,600	2,055	4,250	295	11.8
Rumänien	7,690	769	6,262	6,262			20.0
Latvia	9,231	923	6,067	1,264	4,337	466	10.0
Estonia	9,150	919	5,900	2,180	3,177	543	11.3
Lithuania	6,460	646	4,679	2,834	1,622	223	11.2
Iceland	5,777	650	3,665	3,665			9.8
Brazil	54,729	5,440	3,003	3,000		3	14.4
The Netherlands	2,733	273	2,583	2,451	132		7.0
Japan	2,323	225	2,170	2,170			7.6

10 Paludiculture in a global context

Country	Current peatland area [km²]	Peat carbon stock [Mton C]	Degrading peatland area [km²]				Peat extraction	Total annual emissions [Mton CO₂]
			Total	Agriculture	Forestry			
New Zealand	2,587	259	2,103	1,857	196		50	5.4
France	2,875	287	2,089	2,013	56		20	6.5
Guinea-Bissau	2,500	250	2,067	2,067				10.3
Bangladesh	2,500	250	2,010	2,000			10	10.0
Denmark	2,029	203	1,893	1,486	391		16	3.0
Australia	10,828	567	1,870	1,850			20	6.6
Myanmar	1,910	130	1,700	1,700				7.0
Zambia	15,410	780	1,508	1,508				6.4
Vietnam	2,382	224	1,310	1,300			10	6.5
India	955	90	1,171	1,171				6.5
Austria	1,200	120	1,025	1,025				2.7
North Korea	1,209	117	1,010	700	300		10	2.6
Guyana	7,910	780	1,000	1,000				5.0
Mozambique	3,000	300	1,000	1,000				5.0
Mexico	9,910	1,483	1,000	1,000				4.8
Ethiopia	2,191	218	1,000	1,000				3.2
Burundi	656	66	928	898			30	2.2
Madagascar	1,854	180	926	925			1	3.7
Venezuela	7,928	1,984	901	500	400			4.6
Cuba	2,500	250	644	223	400		21	3.6
D. R. Congo	11,955	1,190	600	500	100			2.7
Thailand	630	57	540	350	190			2.9
Guinea	955	40	500	500				2.5
Colombia	9,999	1,000	427	417			10	2.3
Sri Lanka	1,000	100	409	408				2.4
Sum			517,307					

getation cover that are eroding quickly, thus accelerating desertification (Joosten et al. 2012). The Lac Alaotra peatlands of Madagascar are another example of large-scale loss of productive land. Once the most productive rice cultivation area of Madagascar, these peatlands have lost about 40% of their original productivity during the past years. The loss was mainly caused by the sedimentation of thick layers of unfertile laterites and sands, which were washed in from the surrounding mountains after these had been cleared of their forest cover (Andrianandrasana et al. 2005). The prevailing use of slash and burn (Bakoariniaina et al. 2006) and decades of drainage have additionally degraded the land. As many parts of the Lac Alaotra peatlands can no longer be used, new areas continue to be drained (Mutschler 2004).

Water retention and water flow regulation are important ecosystem services provided by undisturbed peatlands (Chapter 2.4). These services are severely restricted by drainage. In mountainous areas, peatlands are often at the source of large rivers and play an important role in water supply. For instance, the Ruoergai peatlands in Tibet (China) are part of the source catchment of the Huanghe River (also called the Yellow River). Millions of people depend on a regular water flow in this river, which the Ruoergai peatlands help sustain in times of fluctuating supply by irregular rainfall and snow melt (cf. Joosten et al. 2012). However, by the end of the 20th century, most of the Ruoergai peatlands were damaged by drainage and overgrazing, greatly increasing the risk of flooding and erosion (Wetlands International 2005, Joosten et al. 2008). During the past decade, rewetting efforts by the government and NGO's, together with a ban on peat extraction, drainage, and agriculture, helped partly restore the most important ecosystem services in the Ruoergai peatlands.

In parts of the Lac Alaotra peatlands of Madagascar, the surface had subsided to two metres below its original height already by the mid 20th century (Straka 1960). As a result, these parts of Alaotra peatlands had also lost their ability to retain and regulate water. During the past decades, extreme weather events could not be buffered anymore, leading to catastrophic droughts and floods (FAO 2000). The exploitation of drained peatlands may be profitable on the short term anywhere in the world. Yet, in the long term, such unsustainable use will have serious socio-economic consequences (Joosten et al. 2012)

The total area of about 530,000 km² of degraded peatlands worldwide and the associated ecological, economic, and social problems emphasise the global potential of paludiculture as an alternative land use option. Paludicultures are currently concentrated in the temperate climate zone, but the same needs, challenges and possibilities exist for a sustainable use of peatlands anywhere in the world. There is a wide range of wetland plants that can be used in paludiculture, as (former) traditional use reveals (see e.g. Box 3.1).

Box 10.1: Health hazards posed by peat fires

Alexandra Barthelmes

The large scale drainage and degradation of peat swamp forests in Southeast Asia has led to repeated catastrophic peat fires (e.g. van der Werf et al. 2004). These peat fires caused a significant increase in respiratory diseases and exposed the population to alarmingly high levels of carcinogenic metal ions such as cadmium and nickel (Betha et al. 2013). Between September and November 1997, during one of the most desastrous peat fire episodes, close to 1.5 million cases of respiratory infections were reported in Indonesia compared to a total population of 12 million people of the eight affected islands (Cochrane 2009, Colour picture 5). Also, in other parts of the world, local populations have been exposed to health hazards caused by large peat fires. The 2008 fires in North Carolina (USA) caused a significant increase in respiratory disease and acute cardiac problems (Rappold et al. 2011). During the massive 2010 peat fires in Russia, the mortality rate in Moscow alone almost doubled (Goldammer 2010); it is likely that the resulting significant air pollution caused hundreds of deaths (van Donkelaar et al. 2011).

Peat fires pose a particular risk to people with cardiac problems and chronic respiratory diseases, as well as to very young and very old people in general. Besides the emission of large amounts of fine particles, the incomplete combustion of organic material causes the emissions of carbon monoxide, hydrocarbons, organic acids, nitrogen and sulphur compounds, and so on (Goldammer 2010).

10.3 Germany – Rewetting and paludiculture in Mecklenburg-West Pomerania

Christian Schröder

Mecklenburg-West Pomerania has 280,800 ha of peatlands, which makes it one of the most peatland rich areas in Germany. About 12% of the federal state are covered with peatlands, 60% of which (168,485 ha) are under agricultural use and 24% (67,883 ha) under forestry (Schroeder 2012). In 1990, less than 3% of the peatland area had not been drained (MLUV 2009). Peatland drainage had led to many problems (loss of biodiversity, soil degradation, GHG emissions, peat subsidence and more, Chapter 2.2). Therefore, a 'Plan for the conservation and development of peatlands in Mecklenburg-West Pomerania' was initiated in the year 2000 (German: 'Konzept zum Bestand und zur Entwicklung der Moore in Mecklenburg-Vorpommern'; Lenschow 1997) and updated in 2009 (MLUV 2009). The original 2000 programme already emphasised the importance of peatlands in relation to climate. In 2009, the updated programme linked two relevant elements: the mitigation of GHG emissions by rewetting and the production of biomass to replace fossil fuels (MLUV 2009).

The strong focus on climate aspects follows from the important role of peatlands in the total GHG balance of the federal state. Peatland emissions amount to some 6 million tonnes of CO_2 equivalents (t CO_2e), which is 27% of the total GHG emissions of Mecklenburg-West Pomerania (Figure 10.1, Jensen et al. 2012). In the framework of the peatland conservation programme, 29,764 ha of drained peatlands have been rewetted (10.6% of the total peatland area, MLUV 2009), reducing GHG emissions by about 300,000 t of CO_2e. The aim is to rewet 100,000 ha (around 30%) of peatlands in Mecklenburg-West Pomerania, taking future aspects of land use into account (personal communication Till Backhaus, Minister for agriculture, environment and consumer protection MWP, 17 April 2013).

The peatland conservation programme is based on voluntariness. Land owners and users have to agree to the planned rewetting measures. This means that restriction for management or abandonment is compensated financially. The land is purchased from the owner in order to retain high water levels. Alternatively, the new water levels are stipulated in the land register. In many cases, the programme has been used to facilitate retreat from areas that are difficult to cultivate. Initially, the conservation programme concentrated its rewetting activities on deeply drained polders, where drainage caused high costs and had become technically challenging. Now it is important to implement higher water levels on a wider scale. Changes in agricultural policy, like the assignment of EU direct payments for grassland to match payments for arable land, have led to a reduced willingness among land users to participate in rewetting projects. Thus, it is important to develop economically viable paludicultures. Such paludicultures allow combining agricultural production interests with climate change mitigation to meet demands of peatland conservation.

Particularly, biomass production for a decentralised energy supply offers new prospects, both for agriculture and for regional development (Box 8.4). If paludiculture were implemented on 20% of the peatland area of Mecklenburg-West Pomerania (56,160 ha), 280,800 t of biomass with a net calorific value of about 4,352 TJ could be produced each year (Table 10.4). This amount would be sufficient to operate 280 decentralised heating plants, each with a thermal output of 800 kW. These numbers illustrate the enormous potential of paludiculture for regional development. By replacing fossil fuels, the emission of almost 320,000 t CO_2e would be prevented. In addition, the rewetting

Table 10.4: Potential GHG emissions reduction in Mecklenburg-West Pomerania by implementing paludiculture on 20, 50 or 100% of the peatland area and thermal utilisation of the produced biomass. Assumptions for the calculation: Productivity: 5 t dry matter $ha^{-1}a^{-1}$; Net Calorific Value: 15.5 MJ kg^{-1} (water content: 10%; Box. 3.5); Emission reduction by rewetting: 15 t CO_2e $ha^{-1}a^{-1}$; Emission reduction by replacing heating oil: 73.6 t CO_2e TJ^{-1}; Emissions due to biomass handling 10 t CO_2e TJ^{-1}).

	Scenario 20%	Scenario 50%	Scenario 100%
Paludiculture area [ha]	56,160	140,400	280,800
Biomass [t TM a^{-1}]	280,800	702,000	1,404,000
Net Calorific Value [TJ a^{-1}]	4,352	10,881	21,762
Emissionsland from use change [Mt CO_2e a^{-1}]	-0.84	-2.11	-4.21
Emissions from heating oil replacement [Mt CO_2e a^{-1}]	-0.32	-0.80	-1.60
Emissions from biomass handling [Mt CO_2e a^{-1}]	0.04	0.11	0.22
CO_2-Balance [Mt CO_2e a^{-1}]	-1.12	-2.80	-5.60

of peatland would avoid soil emissions from land use of about 840,000 t of CO_2e. In total, thermal utilisation of biomass from rewetted peatlands would reduce emission by some 1 million t CO_2e per year. If paludiculture is implemented on the majority of agriculturally used peatlands in Mecklenburg-West Pomerania (50% of the peatland area), emission reduction would even amount to 2.8 million t CO_2e per year (Table 10.4). Thus, paludiculture can contribute substantially to reaching the emission reduction targets of the federal state, providing an example for other regions in Germany, Europe, and the world.

Apart from small areas traditionally managed for harvesting reed thatch, paludiculture in rewetted peatlands has not yet been implemented on a large scale in Mecklenburg-West Pomerania. The first heating plant operating with biomass from wet peatlands was only realised in 2014 (Box 8.4). Although various options for the use of biomass from wet peatlands have been identified, their implementation will need more time and, above all, successful demonstration projects. In addition, agricultural policies need to be adjusted so that farmers can change their land use practices.

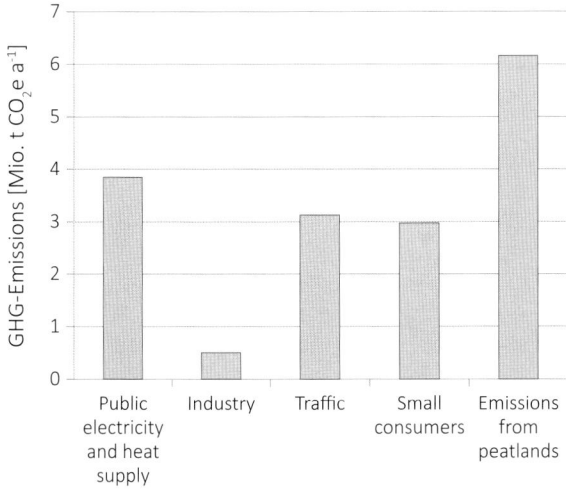

Figure 10.1: Emissions from peatlands in comparison to other relevant sources in Mecklenburg-West-Pomerania (Moorschutzkonzept MV 2009).

10.4 Belarus – Biomass from rewetted peatlands as a substitute for peat and for promoting biodiversity

Wendelin Wichtmann, Vladimir Kapitsa, Franziska Tanneberger & Nina Tanovitskaja

Belarus has a very high proportion of peatlands, which cover about 2,560,500 ha or 11% of its total surface area. About 34% (863,000 ha) of peatlands are still undisturbed hydrologically, whereas the remaining 66% (1,697,500 ha) mainly consist of drained fens (Nikiforov et al. 2013). About 12% (299,100 ha) are partially exploited, under peat extraction or have been abandoned after peat excavation. Extraction (mainly for the production of peat briquettes) is still ongoing on 17,600 ha (Bambalov et al. 2016). Most of the drained peatlands are used for agriculture or forestry. The associated GHG emissions are drastically increased by frequent peat fires, which burn especially in cutover peatlands. The peatlands are often abandoned for a long time and are characterized by two different traits: bare peat or covered with spontaneously developed stands of birch and pine trees.

Low intensity grasslands on moderately to shallowly drained fens have been abandoned on a large scale over the past 25 years. In the past, these fens were (more or less) regularly mown for the production of hay or litter or were grazed by cattle in dry years. Sometimes the fens were burnt to remove excess biomass and litter, to stimulate growth, and to keep the landscape open. Data on the extent of these abandoned sites are lacking – estimates mention about 100,000 ha. Vegetation succession introduces shrubs and trees to these grasslands; thus they lose their typical open landscape character together with their function as habitat for rare species.

Since the beginning of the 21st century, large scale peatland rewetting projects have been carried out by non governmental organisations (NGOs), as well as by the Ministry for the Environment – often financed by international foundations and sponsors. Since 2009, the rewetting of cutover peatlands and other degraded peatlands is legally regulated (Kozulin 2011). In total, more than 51,000 ha peatlands have been rewetted in the framework of various projects (Nikiforov et al. 2013). Rewetting was mainly carried out to reduce GHG emissions and to restore habitats for important target species like the Greater Spotted Eagle (*Aquila clanga*) and Aquatic Warbler (*Acrocephalus paludicola*) (Box 10.2, Tanneberger & Wichtmann 2011).

After rewetting, projects aiming to promote certain target species require vegetation management that is specifically tailored to these species and often entails regular mowing. Several pilot projects were initiated for mowing and processing the biomass (Wichtmann et al. 2012).

At present, peat briquettes for combustion are still being produced in 26 factories in Belarus. The aim of the project 'Wetland Energy' (funded by the EuropeAid-programme, 2011–2015), is to test whether it is

> **Box 10.2: Biomass production and habitat management in Sporava, Belarus**
>
> Wendelin Wichtmann, Uladzimir Malashevich & Franziska Tanneberger
>
> The harvest and utilisation of biomass from wet fen peatlands was demonstrated as part of the project 'Rewetting and sustainable peatland management in Belarus – a climate protection project with benefits for the economy and biodiversity', funded by the German Ministry of the Environment (2008–2011, Tanneberger & Wichtmann 2011).The mid-term objective is to restart mowing on more than 10.000 ha of abandoned fen peatlands in the Yaselda River valley (Region of Brest). Previous litter and fodder production on the extensive flood- and percolation mires had largely been abandoned during the past 20 years. Conservation management was implemented on some sites using a converted snow groomer (Colour picture 58) and will be expanded in the framework of follow-up projects. The most important reason for resuming land use is safeguarding the habitat of the Aquatic Warbler (*Acrocephalus paludicola*). This bird is the only globally threatened passerine song bird species of Continental Europe, and Sporava is one of the four most important breeding areas in the world. The quality of the habitat had declined after management had been abandoned (Flade & Lachmann 2008).
>
> The vegetation in Sporava is mainly characterised by Reed Canary Grass stands (*Phalaris arundinacea*), several sedge species (especially *Carex elata*), Purple Small-Reed (*Calamagrostis canescens*), and Common Reed (*Phragmites australis*). Representative areas showed average yields of 9 tonnes of dry matter per ha for Reed and Reed Canary Grass, and 7 tonnes of dry matter per ha for sedges (Wichtmann & Tanneberger 2009).
>
> Harvest is between September and March/April. The converted snow groomer used is equipped with a cutter-harvester and a trailer that is also mounted on the chassis of a snow groomer (Colour picture 58). The biomass is taken up by suction and blown onto the trailer. Then, the biomass is transported by the harvesting machine to the field margin and unloaded using the drag chain conveyor of the trailer. The subsequent processing of briquettes and pellets is planned.
>
> The production of biomass fuels offers new perspectives for the use of the abandoned fen peatlands. Apart from keeping the sites in productive use, targeted habitat management can help protect endangered species. To prevent conflicts between biomass production and habitat management, management plans have been drawn up.

possible to switch to biomass for briquette production. The potential for growing biomass on rewetted cutover peatlands is being assessed, together with the possibilities of making briquettes (or pellets) from the biomass produced. The idea is to replace fossil peat with renewable biomass and create a sustainable production of energy (Wichtmann & Tanneberger 2009).

The pilot sites of the 'Wetland Energy' project are located in the Dokudovskoye peatland (Grodno oblast, Lida rayon) in western Belarus. The total area of peatland comprises 7,811 ha, about 3,500 ha of which have been cut over. Bare peat covers about 30–40 % of the area. Currently, the average water table of the excavated sites is 1.2–1.5 m below the surface, which causes frequent peat fires to occur (Wichtmann & Tanneberger 2009, Wichtmann et al. 2012). A part of the peatland is protected ('zakaznik') and is separated hydrologically by dams from the peat extraction sites. In some parts, fen peat extraction for briquette production is ongoing. After peat extraction – which on average leaves a residual peat layer of 50 cm, all sites shall be rewetted and, if possible, taken into paludiculture.

Over several hundred hectares, reed beds have developed spontaneously after extraction ceased. Here, the water level is on average about 0.5 m above the surface. The reed grows to a height between 1.8 and 2.7 m. At the margins, Greater Pond Sedge (*Carex riparia*) is dominant. In addition, Tufted Sedge (*Carex elata*) and Cattail (*Typha angustifolia*) are found. Willows (*Salix* sp.) developed at the margins. The average potential biomass yield of the reed beds amounts to 9 t DM per hectare (Wichtmann et al. 2014). Whereas the mean yield was 11.7 t DMa^{-1} (standard deviation (5.9) in 2009, it was only 4.6 t DM a^{-1} (standard deviation 2.9) in 2010 – which is probably due to a lack of older stalks. The late sampling date (end of March/beginning April) in both years results in low chlorine, sulfur and nitrogen content, which means that combustion in heating plants should not pose any problems (Wichtmann et al. 2013; Chapter 3.5.3).

In cooperation with the peat factory in Lida, different mixing ratios of peat and biomass are tested in briquette and pellet production. For this a briquetting and a pelleting plants were built up within the 'Wetland Energy' project (Colour picture 92). In addition, different harvesting and transport practices are being tried. Next to tractors with wide, low-pressure tyres, special adapted harvesting machinery is being developed. Furthermore, the project will consider if the existing railway for extracted peat can be used to transport biomass from the paludiculture fields to the peat factory. Societal acceptance, economic efficiency, and relevance of paludiculture for Belarus are assessed as well.

10.5 Poland – Paludiculture for biodiversity and peatland protection

Franziska Tanneberger, Dariusz Gatkowski & Jarosław Krogulec

Poland has a total peatland area of 1,495,000 ha (just under 5% of the territory). Fens cover 93%, transitional mires 3%, and raised bogs 4% of this area (Kotowski et al. 2016). Peat formation occurs in only 16% (201,938 ha) of the area, as the remaining 84% is degraded (Kotowski & Piórkowski 2003). The 23.5 million tonnes of annual CO_2 emissions from peatlands places Poland in the 10th position in the list of countries with the highest emissions (Joosten 2009). Most Polish peatlands are used as hay meadows and pastures (70%); forests cover 12% of the peatland area whilst 4% is excavated and 0.5% is used as arable land (Kotowski et al. 2016).

Since the year 2000, Poland has been gained considerable experience in developing integrated management concepts that combine peatland and species conservation with rural development. A major stimulus for the development of these concepts was provided by projects on the protection of birds, particularly the globally threatened Aquatic Warbler *(Acrocephalus paludicola)* (Colour picture 93). In 2004, specific EU agri-environmental programmes were introduced to support the sustainable use of peatlands, while fulfilling the requirements of species and habitat protection in fens, bogs, and wet meadows. These programmes were well received by land users (Dommain et al. 2012).

The extensive fen peatlands of the Biebrza National Park and the Lublin region are a core area in the distribution of the Aquatic Warbler, comprising 25% of the global population of 10,200–14,200 singing males (Flade & Lachmann 2008). Until the early 20th century, the Aquatic Warbler was common in temperate open wetlands between the North Sea and Siberia (Schulze-Hagen 1991). Due to a large-scale destruction of its habitat, the global population has decreased massively. In large parts of its former distribution range, the Aquatic Warbler is now extinct. The bird's natural habitat consists of undisturbed mesotrophic fen mires (Schulze-Hagen 1991), which have all but disappeared in Europe. The remaining areas are threatened by encroachment of dense reed stands, shrubs and trees, as well as by continued drainage and eutrophication (Kloskowski & Krogulec 1999, Flade & Lachmann 2008). The quality of all remaining Polish breeding habitats depends on management, such as low intensity mowing and grazing.

After traditional mowing by scythe was abandoned around 1970 in Eastern Poland, meadows became increasingly overgrown with shrubs and bushes. By 1999, woody vegetation had developed on more than 15,000 ha of formerly open fens. Therefore, various management activities were established to maintain open fens, particularly as a habitat for Aquatic Warbler.

Between 2005 and 2011, the EU LIFE project 'Conserving *Acrocephalus paludicola* in Poland and Germany' (managed by OTOP BirdLife Poland) initiated large scale restoration of habitats and their sustainable management. Since 2007, mowing equipment has been tested on the sensitive peat soils, and since 2009, converted snow groomers (so-called 'ratraks') have been in use (Lachmann et al. 2010). After specific agri-environmental measures were introduced in 2007 to stimulate mowing in peatlands, the Biebrza National Park and protected areas in Lublin region currently lease about 10,000 ha of public land to be mown by farmers (Colour picture 94). The follow-up EU LIFE project 'Aquatic Warbler and Biomass Use' (2010–2015) has been testing the use of biomass for energy, while trying to optimise harvest and enlarge the mown areas (Cris et al. 2014).

In both LIFE projects, options to utilise biomass harvested in late summer were evaluated (Gańko et al. 2008). The quality of the biomass is too low to use it as fodder. Three other options were explored in a feasibility study: biogas production, compost production, and combustion.

As areas are only mown after the Aquatic Warbler breeding season ends (i. e. starting in August), fermentation of the biomass will only produce low gas yields. In other words, the potential for biogas production is only low. Although the biomass can successfully be composted, there is no market for compost in Eastern Poland. High investment costs would not be outweighed by the low added value. The third option of combustion offers two possibilities: direct combustion of bales and the production of briquettes/pellets. Direct combustion of bales requires biomass with a low moisture content (< 20%). Only a small portion of the biomass fulfils this criterion. Cost effective production of briquettes and pellets can also be based on moister raw material, once it is dried. Overall, the feasibility study showed that briquettes and pellets are the best option for using the biomass, because there is a local market and the material processing allows for higher selling prices.

In 2012, some 3,000 ha of fen peatland were mown in the framework of agri-environmental programmes in the Biebrza Valley National Parc alone. Although yields on the near-natural sites are only low (1.0–1.5 t of dry mass per ha), some 3,000–4,500 t of biomass become available each year. Until recently, its high moisture content (about 30%) precluded utilisation of the biomass. Following the advice of a feasibility study carried out by the EU LIFE+ Nature project 'Aquatic Warbler and Biomass Use' (2010–2015), OTOP built a pellet manufacturing facility in Trzcianne, which has been operational since February 2013.The target capacity of this pelleting plant lies at 4,500 t of dry biomass per year. Thus, the entire amount of biomass from the Biebrza National Park could be utilised.The land is managed by OTOP itself or by other land leasers. Biomass from the land of other leasers is either collected free of charge or, if it has been mown and dried, against payment. The manufactured pellets are sold mainly to wholesalers or industrial users (power plants), as well as to retailers for domestic heating. Because of their moderate ash content (5.9 ± 0.3%) the pellets, with a heating value of 16.9 ± 0.1 MJ kg^{-1}, can effectively substitute wood pellets. The profits of OTOP are reinvested in the conservation work in the Biebrza Valley, but are foreseeably too low to cover all of the harvesting costs. Although the problem of biomass utilisation seems to be solved, additional financial support from agri-environmental programmes is still needed to cover the high costs of the specialised harvesting equipment.

Since 2010, the wet fens of the Lublin Region are being mown again, thanks to the EU LIFE+ follow-up project, agri-environmental programmes, and the new management plans of the Poleski National Park. Biomass is harvested in three peatland complexes, where each year part of the area (20–50%) is mown: 130 ha in Ciesacin Fen, an EU Special Protected Area for birds (SPA); 960 ha in the SPA Chełm Calcareous Fens; and 130 ha in the Bubnów Peatland, a SPA within the Poleski National Park. Most of the land is state-owned and some of it is private. The National Park leases state-owned land to farmers for 5-year periods to be managed in the framework of agri-environmental programmes and the Park's management plan. Ciesacin Fen is entirely privately owned and since 2010, 50% of the area is mown each year as part of an agri-environmental programme. Farmers use the biomass as fodder and bedding for Charolais cattle. In the Chełm area, 520 ha of state-owned *Cladium mariscus* fens are managed. In 2011, an area of 155 ha was cleared of old entangled *Cladium* biomass and scattered shrubs. As the area was mown rather high above the surface, only a small amount of biomass was harvested (about 1 t ha^{-1}). In 2012, a 5 year agri-environmental programme started, in which each year 20% of a 490 ha large area is mown. In the first year, 1.9 t of biomass were harvested per hectare on 98 ha. Close to the Chełm Fens there are two pelleting factories and a cement plant that is equipped to use biomass as alternative fuel. However, the biomass market is not very stable yet, and prices fluctuate strongly (between 22.5–30 Euro per tonne in 2011 and 40–47.5 Euro in 2012).

Count data revealed that the Biebrza and Lublin populations of the Aquatic Warbler have remained stable between 1993 and 2010. A positive effect on the birds' population size has been observed on subsites with re-introduced mowing to halt shrub encroachment, particularly in the second year after mowing (Bellebaum personal comm.). A similarly positive effect on the bird's breeding success has been described by Kubacka et al. (2014). Apparently, Aquatic Warblers profit from regular – but not annual, mowing. However, early results from vegetation studies indicated that the new harvesting technique using converted snow groomers can have unintended negative effects on the surface microtopography of the fen peatlands, and consequently, on plant diversity (Kotowski et al. 2013). Further research is needed to guide the temporal and spatial planning of mowing, particularly in the Biebrza National Park.

The future of large-scale mowing in near-natural fens in Eastern Poland mainly depends on whether it is accepted by the administration of protected areas (especially the National Parks). If large-scale mowing is included in management plans, it could be secured for the next decades. Another challenge lies in adequate agri-environmental programmes for the funding period 2014–2020. Appropriate remuneration and well tailored, flexible programmes will help keep farmers interested in mowing their own – or leased, land.

Colour pictures 87–102

Colour picture 87: The painting 'Wiesen bei Greifswald' ('Meadows near Greifswald') by Caspar David Friedrich (between 1820 and 1822) shows the beauty of the fen landscape north of the Hanseatic town of Greifswald.

Colour picture 88: 'Herbst im Moor' ('Autumn in the bog'), Otto Modersohn, 1895, Kunsthalle Bremen, VG Bild-Kunst Bonn 2012.

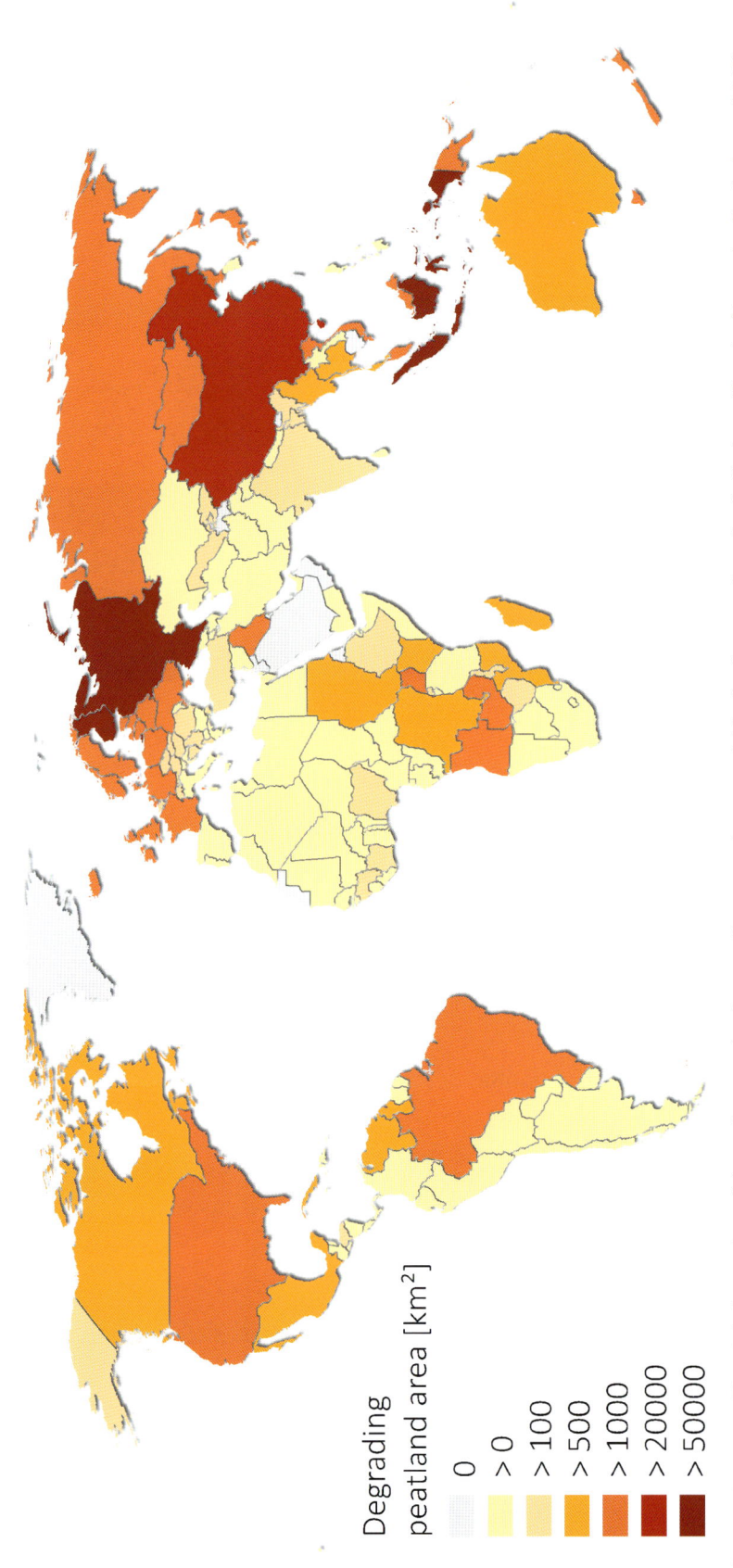

Colour picture 89: The global overview of degrading peatland areas illustrates the potential for implementation of paludiculture (Graph: Alexandra Barthelmes, 2013).

Colour pictures 87–102

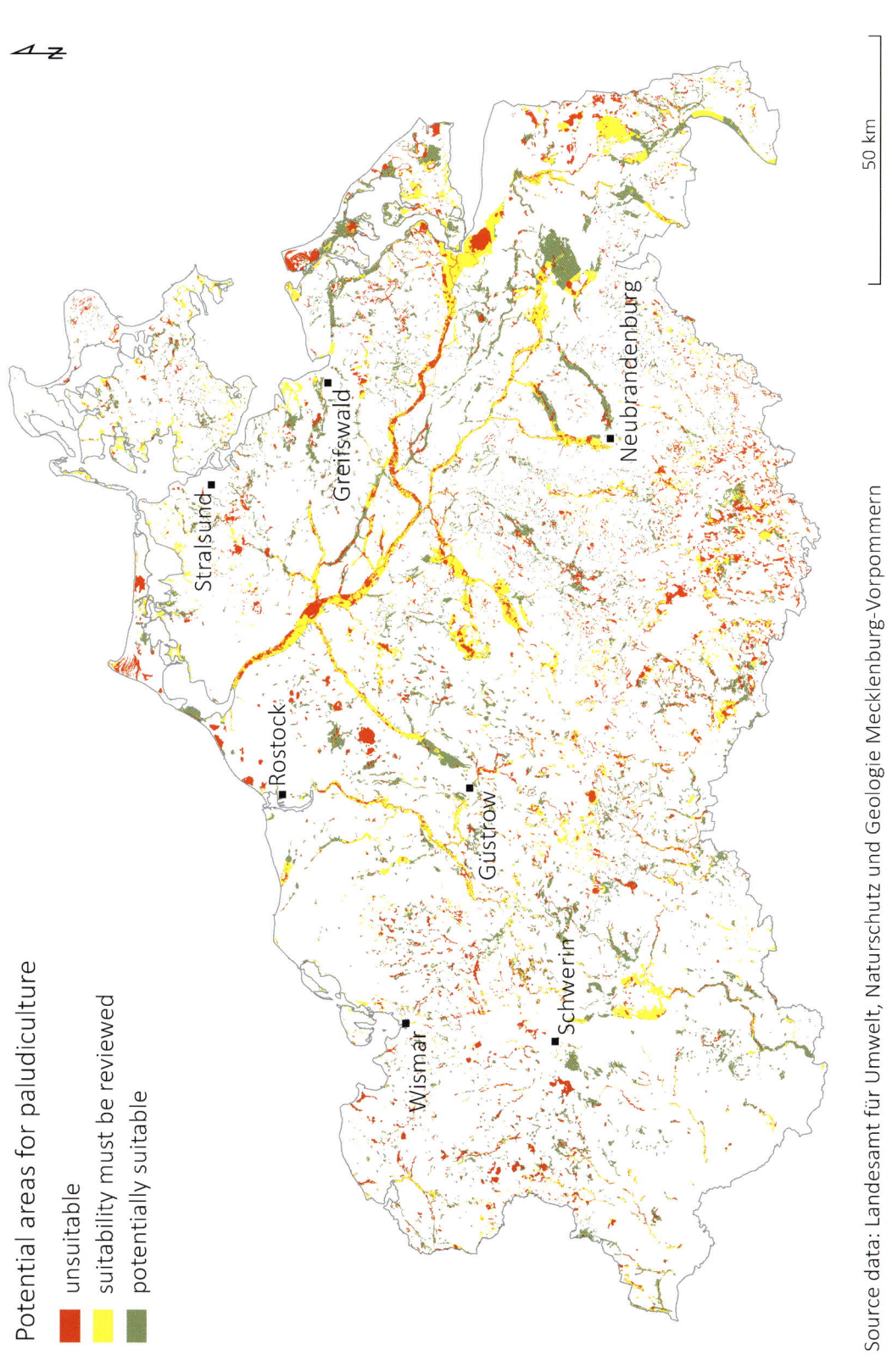

Colour picture 90: Potential areas for paludiculture in Mecklenburg-West Pomerania. Red: unsuitable areas (forests, water bodies, spring mires, kettle hole mires, raised bogs); yellow: suitability must be reviewed because of conservation status (legally protected habitats, nature reserves, Natura 2000 sites and SPAs, national parks); green: potentially suitable, agriculturally used peatlands (Map: Philipp Schroeder).

Source data: Landesamt für Umwelt, Naturschutz und Geologie Mecklenburg-Vorpommern

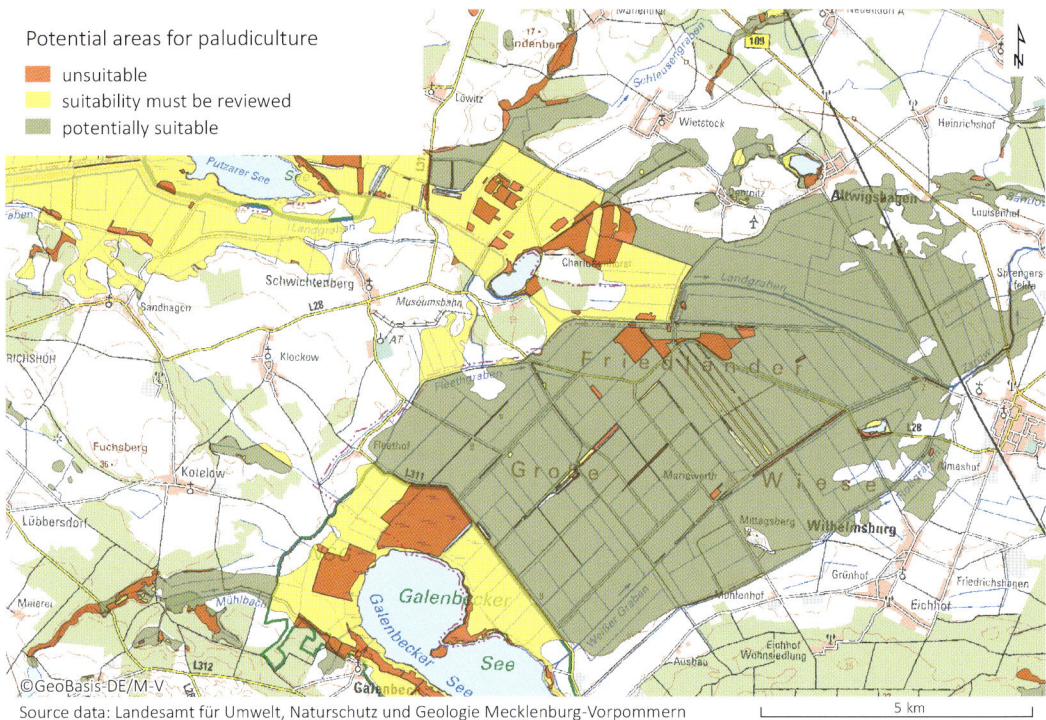

Colour picture 91: Potential areas for paludiculture in an exemplary part of Mecklenburg-West Pomerania. Red: unsuitable areas (forests, water bodies, spring mires, kettle hole mires, raised bogs); yellow: suitability must be reviewed because of conservation status (legally protected habitats, nature reserves, Natura 2000 sites and SPAs, national parks); green: potentially suitable, agriculturally used peatlands (Map: Philipp Schroeder).

Colour picture 92: Facility for the production of briquettes from peat and biomass in Lida, Belarus. In the framework of the EU-Aid funded project "Wetland Energy" the peat is substituted by biomass from rewetted peatlands in a stepwise manner (Lida, Belarus; Wendelin Wichtmann, March 2014).

Colour picture 93: The Aquatic Warbler (*Acrocephalus paludicola*) is globally threatened by extinction (Nemunas Delta, Lithuania; Zymantas Morkvenas, June 2011).

Colour picture 94: Harvesting sedge in the lower Biebrza valley (Bagno Ławki, Polen; Paweł Świątkiewicz, December 2012).

Colour picture 95: Cooperative tree nursery in Tumbang Nusa with seedlings of Pulai (*Alstonia* spp.), Blangeran (*Shorea belangeran*), Durian Hutan (*Durio carinatus*) and Jelutung (*Dyera lowii*) for planting in rehabilitated peatland areas of Block C ex-Mega Rice Project area (Central Kalimantan, Indonesia; René Dommain, March 2011).

Colour picture 96: Rewetted area in Sebangau National Park with various wetness adapted tree species (Indonesia; Hans Joosten, April 2007).

Colour picture 97: Rewetting of logged and fire-damaged peat swamp forest in Sebangau National Park with large dams. A tree nursery for production of planting material for afforestation is visible on the left side (Central Kalimantan, Indonesia; René Dommain, November 2008).

Colour picture 98: Loom for the traditional production of reed mats (Inner Mongolia, China; Jan Felix Köbbing, October 2011).

Colour picture 99: Reed bales are loaded on to a lorry using a conveyor belt (Inner Mongolia; China, Jan Felix Köbbing, October 2011).

Colour picture 100: Transport of reed bales to a paper factory (Inner Mongolia, China; Jan Felix Köbbing, October 2011).

Colour picture 101: Cattail harvest and swathing (Libau Marshes, Winnipeg, Canada; Richard Grosshans, September 2012).

Colour picture 102: Cattail harvest and baling of square bales (Libau Marshes, Winnipeg, Canada, Richard Grosshans, October 2012).

10.6 Indonesia – Paludiculture as sustainable land use

René Dommain

The largest peatland area of the tropics is found in Indonesia. The Southeast Asian island state has more than 20 million ha of peatlands (Page et al. 2011), which is more than 12 times the peatland area of Germany. After Russia, Canada and the USA, Indonesia is the fourth most peatland rich country in the world (Chapter 10.2). The peatlands of Indonesia are covered by tropical rain forests known as peat swamp forests.

Due to their huge size and the inaccessibility of the swamp forests, one might assume negligible forests losses and the general absence of high-intensity land use in those peatlands. In reality, however, the same mistakes are being made in Indonesia that led to the wide-scale destruction of peatlands in Europe. There has been large-scale deforestation and drainage, especially during the past 20 years, mainly for industrial plantations of oil palm and acacia. In addition, devastating peat fires burn almost every year in association with the drainage and cultivation of peatlands. As a consequence of this inappropriate land use, Indonesia has the highest CO_2 emissions from human induced peat degradation in the world (Joosten 2009).

The largest peatland areas of Indonesia are located on the island of Sumatra (7.2 million ha); in Kalimantan, the Indonesian part of the island of Borneo (5.77 million ha), and in Papua, the Indonesian part of the island of New Guinea (7.98 million ha) (Wahyunto et al. 2003, 2004, 2006). The total carbon pool of Indonesian peatlands is estimated to exceed 55×10^9 t (Jaenicke et al. 2008), whereas current carbon losses amount to more than 200×10^6 t per year (Couwenberg et al. 2010).

The equatorial peatlands, located between 5° north and 9° south, are subject to a humid to perhumid tropical climate with constant high temperatures of around 27°C and annual precipitation of 2,000–4,000 mm (Walter & Lieth 1960–67). The southern part of the country experiences a dry season from July to September, which can be more pronounced during 'El Niño' events (Aldrian & Susanto 2003). Peat fires mostly occur during this dry season. The peatlands of Southeast Asia are generally forested, oligotrophic-acidic raised bogs (Brünig 1990). More than 100 tree species grow in the Indonesian peat swamp forests, where trees are up to 35 m high (Shepherd et al. 1997), and the woody peat deposits up to 14 m deep (Hope et al. 2005). In inland Kalimantan, there are huge raised bogs that are more than 50 km in diameter and cover a total area of some 2 million ha. However, by far the largest peatland areas of Indonesia are found on the coastal plains of Sumatra, Borneo and Papua. Due to their high peat accumulation rates, these peatlands re-

present globally important carbon sinks (Dommain et al. 2011, 2014).

The dense population of Indonesia (the fourth largest in the world, with more than 241 million inhabitants) determines the current land use. The current population growth is 1.6 % (ADB 2012) and the poverty rate was 18.1 % in 2010, so more than 43 million people were forced to live on less than $ 1.25 USD per day (ADB 2012). Consequently, national policies focus on economic growth, food security, and the alleviation of poverty (Coordinating Ministry for Economic Affairs 2011).

In the 1930s, the Dutch colonial government initiated the transmigration programme (Indonesian: transmigrasi) to relieve population pressure on the main island of Java and to stimulate economic development of the poorer outer islands of Indonesia. To this day, people are still being relocated through this programme from Java and Bali to the less populated islands of Sumatra, Borneo, Sulawesi and Papua. Once at those places, these people are allocated uncultivated and often forested land for agricultural use (Bowen et al. 2001). Initially, transmigration settlements were established on mineral soil, so that traditional rice cultivation could be continued. Since the 1960s and 1970s, many swamps and peatlands have been targeted because they were very sparsely populated and provided enormous areas for land reclamation. The objective of the Indonesian government was to increase national rice production in order to become independent from imports (UNEP 2005). However, the productivity of rice on acidic raised bogs steadily declines after initially high yields, eventually resulting in complete crop failure, because rice is not adapted to the nutrient-poor acidic peat soils (Abdoellah 1996). Drainage, fertilisation and liming are required to avoid these drawbacks, but at the same time, such measures cause considerable peat subsidence. Under the tropical climate conditions, subsidence rates can reach up to 5 cm per year (Box 2.1), which makes permanent land use exceedingly difficult. As a result, fields are abandoned and new land is deforested and drained.

Originally, indigenous people never cultivated rice permanently on peatlands because of the low soil fertility. Land use on peat soils was marginal and largely restricted to fishing, hunting and gathering of non-timber forest products. In floodplains, Sago Palm (*Metroxylon sagu*) was harvested, which is still the case in remaining swamp forests.

Modern drainage-based land use practices result in soil acidification, loss of soil fertility, and peat subsidence (Box 2.1), which create a poverty trap and contribute to widespread poverty in peatland settlements, such as transmigration villages (van Eijk & Kumar 2009). As a result, many settlers leave their villages. This scenario was exemplified by the Mega Rice Project in Central Kalimantan, which started in 1996 and was the largest peatland reclamation project in Indonesian history. Of over 15,000 transmigrant families relocated to Block A of the Mega Rice Project area, more than half had already left within a decade (Euroconsult Mott MacDonald/Deltares/Delft Hydraulics 2008). Before it failed completely, the Mega Rice Project was set up to cultivate rice on 1.4 million ha of land with 900,000 ha of peat swamp forest. More than 4,000 km of drainage canals were excavated and large areas were deforested (Böhm & Siegert 2001). During the extreme 'El Niño' dry season of 1997/98, the widespread practice of burning crop residues led to out-of-control peat fires in the deeply drained Mega Rice Project area. As the peatlands are unsuitable for rice cultivation, many drained areas were abandoned and still burn regularly to this day.

A particular challenge with implementing alternative land use concepts in inhabited peatland areas is to ensure that local populations have access to a sufficient food supply. The problem of food security will only increase as the population of Indonesia is predicted to grow to more than 300 million people by 2050 (World Population Prospects 2010), which in turn will increase pressure to develop peatland areas. Yet, the most important driver for current and future peatland cultivation is the rapid expansion of the drainage-based plantation industry in Southeast Asia (Miettinen et al. 2012a). Demand for palm oil on the world market is rising steadily, mainly for food and, to a lesser extent, for biodiesel. Between 2007 and 2011, the increasing demand led to a global rise in palm oil production by 10 million tonnes to a total of more than 50 million tonnes. Together, Indonesia and Malaysia are responsible for almost 90% of the world production of palm oil (FAS/USDA 2011). Palm oil production and export is highly important to the Indonesian economy, and the government plans to increase production to 40 million tonnes by 2020 (Koh & Ghazoul 2010). According to the Indonesian Ministry of Forestry, 5 million ha of peat swamp forest are available for agricultural conversion in Sumatra and Kalimantan, as well as large areas in Papua (MoF 2010, Miettinen et al. 2012b). The rate of plantation expansion has increased continuously since 1990 and from 2007 onwards, more than 100,000 ha of new plantations are added each year (Miettinen et al. 2012b). Should this trend continue unrestrained, about 30% of Indonesian peatlands will be cultivated with oil palms by 2030 (Table 10.5).

The Presidential Decree No. 32/1990 prohibits agricultural use of peatlands where peat is more than 3 m deep, which in theory would restrict the expansion of plantations. This legislation is rarely complied with, however (Silvius & Suryardiputra 2005), which is evident from ongoing oil palm cultivation on deep peat in the Ex-Mega Rice Project area. Until now, Sumatra has been, by far, more affected by the plantation industry than Kalimantan, as almost all the Indonesian acacia plantations are located on Sumatra (Table 10.5). The cultivation of exotic acacia species (*Acacia crassicarpa*, *A. mangium*) for pulp production is expanding because of low costs and extremely rapid plant growth, despite the trees requiring deep drainage. Thus far, the large pulp mills on Sumatra operate below their full capacities and therefore rely on secondary wood supply from existing natural forests (Barr 2004).

In 2010, more than 2.2 million ha of peatland were managed as industrial plantations in Kalimantan and Sumatra (Miettinen et al. 2012a). Together with more than 2.4 million ha of peatland used by smallholders (Miettienen & Liew 2010), a total area of 4.6 million ha of peatland is under agricultural use in western Indonesia – which is 35% of the total peatland area. In Sumatra, 50% of all peatlands are currently used for agriculture.

The expansion of agriculture on peat soils is the main reason for the continuing reduction in Indonesian peat swamp forests. Currently, only 32% of the peatlands in western Indonesia are still covered with forest (Table 10.5). At the present rate of deforestation, all remaining peat swamp forests will be disappeared by 2030 (Miettinen et al. 2012c). Associated with this loss, an invaluable wealth of knowledge on traditional use of peat swamp forest plants would be irretrievably lost as well.

In Indonesia, paludiculture on deforested and burned peatlands not only requires rewetting, but also afforestation (Colour pictures 95–97). The forest vegetation plays an essential role in the hydrological self-regulation of the raised bogs – without swamp forest, high water levels cannot be maintained (Dommain et al 2010). In Indonesia, paludiculture should therefore focus on the use of tree species from peat swamp forests. Management options range from enrichment plantings in selectively used natural peat swamp forests to plantations on completely deforested areas (Muuß 1996). Paludiculture can strengthen and restore important ecosystem services, particularly climate change mitigation, carbon storage, fire reduction, provision of food and forest products, and consequently, reduce poverty. In addition, paludiculture will have a positive impact on the biodiversity of peat swamp forests, which is extremely threatened. Objectives for future implementation of paludiculture include:

- Securing the food supply in populated areas by identification and cultivation of food plants that can grow on wet peat soils;
- Making substitutions for products of commercial crops that depend on drainage;
- Commercialising/marketing local products (e.g. medicinal plants);
- Providing alternatives to agricultural clearance fires.

The tropical peatlands of Southeast Asia are very species-rich. More than 1,300 plant species have been described, with 720 of them being tree and shrub species (Giesen 2013). More than 530 species have known uses, including 222 for timber, 221 for medicinal purposes, and 165 for food (Giesen 2013) (Table 10.6, Box 3.1). Half of the 222 timber species provide high quality timber to existing markets (Giesen 2013). Of particular interest are species providing non-timber

Table 10.5: Used peatland area in western Indonesia (Sumatra and Kalimantan), including a projection for the area under oil palm plantation in 2030 (after Miettinen et al. 2012b, c).

Land use	Region	Peatland area	1990		2000		2010		2030 Projections[3]	
		ha	ha	%	ha	%	ha	%	ha	%
Oil palm plantation[1]	Sumatra	7,234,069	17,985	< 1	512,341	7	1,026,922	14	2,431,722	34
	Kalimantan	5,764,645	0	0	15,982	< 1	258,299	4	1,237,534	21
	West-Indonesia	12,998,714	17,985	< 1	528,323	4	1,285,221	10	3,669,256	28
Acacia plantation[1]	Sumatra	7,234,069	306	< 1	80,176	1	874,921	12		
	Kalimantan	5,764,645	0	0	250	< 1	22,797	< 1		
	West-Indonesia	12,998,714	306	< 1	80,426	1	897,718	7		
Total industrial plantations[1]	Sumatra	7,234,069	18,291	< 1	604,995	8	1,936,436	27		
	Kalimantan	5,764,645	0	0	16,567	< 1	306,968	5		
	West-Indonesia	12,998,714	18,291	< 1	621,562	5	2,243,404	17		
Natural peat swamp forest[2]	Sumatra	7,234,069	4,921,600	68	3,078,500	43	1,806,900	25		
	Kalimantan	5,764,645	3,857,000	67	2,869,200	50	2,403,500	42		
	West-Indonesia	12,998,714	8,778,600	68	5,947,700	46	4,210,400	32		

[1] Data from Miettinen et al. 2012b.
[2] Data from Miettinen et al. 2012c.
[3] future expansion of oil palm plantations is based on a linear expansion rate of 117,910 ha a^{-1} calculated between 2007 and 2010 (see Miettinen et al. 2012b).

forest products (NTFP). The range of products in this group is considerable and includes fruit trees such as Durian (*Durio carinatus*) and Mango (*Mangifera* spp.) plus many medicinal plants such as the White Samet (*Melaleuca cajuputi*), which has long been cultivated in Southeast Asia. This plant not only provides wood for construction, but is best known for its aromatic cajeput oil (Wibisono et al. 2005). The high diversity of useful plants means that the prospects for paludiculture in Indonesia are considerable.

Thus, paludiculture could be implemented in the following ways:
1) Small holder farming of food products in orchards and agroforest land in populated peatland areas (e.g. transmigration villages, Ex-Mega Rice Project area);
2) Community forestry and agroforestry in buffer zones of peat swamp forests that are still undisturbed, protected or being restored;
3) (Large-scale) plantations with native peat swamp forest trees for commercial use.

The first large scale paludiculture projects must aim at securing food for the local populations. In this respect, targeted cultivation should be tested for the 165 peat swamp plant species that are (locally) used for food (Giesen 2013). These peatland plants can be cultivated together with food plants that are less tolerant of swamp conditions, by planting them on artificial ridges or mounds. On shallow peat soils, the cultivation of Sago Palm (*Metroxylon sagu*) should be promoted. Sago does not require drainage, and its yield is ten times higher than rice. In addition to being a starchy food source, it also provides material for roofing, basket making, fodder, and so on (Rijksen & Persoon 1991).

Wet community forests should be established primarily in cultivated or degraded margins of still (largely) undisturbed peat swamp forests. The goal should be to restore forests in the rewetted peatlands to provide timber and additional non-timber forest products. Such an approach could be implemented in so-called 'Hutan Desa' (community forest) concessions, which are allotted to village communities and comprise 10,000 ha at the most. A spatial division in areas under protection, forestry, or agroforestry would be possible. This zonation of land use could prevent the (illegal) exploitation of peat swamp forest (both the one that is already there and the one that is regenerating) by providing renewable forest resources.

The plantation approach involves cultivating commercial hardwoods, as well as species which produce non-timber products. This approach achieves two

Table 10.6: Promising plant species and their potential for paludiculture (outside Europe); Plant type: c = climber; g = grass; h = herbs/tall herbs; s = shrub; t = tree; * Phloem = living tissue below the bark of trees. Information taken from the data base of potential paludiculture plants (DPPP; Box 3.1).

Scientific name (common/ local name)	Plant type	Current use						Used parts of the plant								
		Raw material (industry)	Energy	Food	Forage	Medicine	Jewellery	Wood	Timber	Above ground biomass	Flower	Fruit	Leaves	Roots	Phloem juice*	Seeds
TROPICAL AFRICA																
Cyperus papyrus (Papyrus)	g	■	■					■		■						
Entandrophragma palustre	t	■						■	■							
Erythrina excelsa	t	■						■								
Heritiera utilis (Niangón)	t					■			■							■
Mitragyna ledermannii oder *M. stipulosa* (Abura)	t	■						■	■							
Pentadesma butyracea (Kanya)	t	■						■				■				
Phoenix reclinata (Senegal date palm)	t	■		■				■				■	■			
Raphia hookeri, R. farinifera, R. palma-pinus (Raffia Palm)	t	■		■				■					■		■	
Saccharum spontaneum (Kans Grass)	g			■						■						
SOUTHEAST ASIA																
Casuarina equisetifolia (Common Ironwood)	t	■						■								
Durio carinatus (Durian paya)	t	■						■				■				
Dyera polyphylla (Jelutong)	t	■		■		■		■					■			
Fibraurea tinctoria	c	■				■										
Gonystylus bancanus (Ramin)	t	■						■	■							
Ilex cymosa	t	■						■					■	■		
Korthalsia flagellaris (Purun, Ant Rattan)	t	■														
Lepironia articulata (Grey Sedge)	g	■								■						
Mangifera havilandii	t							■				■				
Melaleuca cajuputi (Tea Tree, Honey Myrtles)	t	■				■		■					■			
Metroxylon sagu (True Sago Palm)	t	■		■				■					■			
Nypa fruticans (Nipa Palm)	t	■		■				■					■		■	
Pandanus tectorius (Screwpine, Pandanus)	t	■		■				■				■	■			
Shorea spp. (Meranti, Alan)	t	■							■							
CHINA																
Euryale ferox (Fox Nut)	h	■		■						■						
Glyptostrobus pensilis (Chinese Swamp Cypress)	t	■						■	■							
Metasequoia glyptostroboides (Dawn Redwood)	t	■						■	■							
Miscanthus sacchariflorus (Amur Silver-Grass)	g	■	■							■						
Oenanthe javanica (Japanese Parsley, Chinese Celery)	h			■		■							■			
Vetiveria zizanioides (Khas Kha, Khas or Khus Grass)	g	■				■				■				■		■
Zizania latifolia (Wild Rice)	g	■		■						■						

10.6 Indonesia – Paludiculture as sustainable land use

Scientific name (common/ local name)	Plant type	Current use						Used parts of the plant								
		Raw material (industry)	Energy	Food	Forage	Medicine	Jewellery	Wood	Timber	Above ground biomass	Flower	Fruit	Leaves	Roots	Phloem juice*	Seeds
SOUTH AMERICA (AMAZONAS)																
Carapa guianensis (Andiroba, Crab Wood)	t	▨						▨								▨
Euterpe oleracea (Açaí Palm)	t	▨		▨								▨				
Grias peruviana (Sachamangua)	t			▨								▨				
Myrciaria dubia (Camu Camu, Cacari or Camocamo)	s			▨		▨						▨				
Spondias mombin (Yellow Mombin)	t			▨								▨				
PANTROPICAL																
Acmella oleracea (Jambu)	h			▨		▨							▨			
Bacopa monnieri (Waterhyssop, Brahmi, Thyme-leafed Gratiola, Herb of Grace, Indian Pennywort)	h					▨				▨						
Coix lacryma-jobi (Job's Tears)	g			▨												▨
Echinochloa colona (Jungle Rice, Awnless Barnyard Grass)	g				▨					▨						
Echinochloa crus-galli (Cockspur, Common Barnyard Grass)	g				▨					▨						
Ipomoea aquatica (Water Spinach, Kangkong)	h			▨									▨			
Limnocharis flava (Yellow Sawah Lettuce, Genjer, Phak Khan Chong)	h			▨									▨			
Nasturtium officinale (Watercress)	h			▨									▨			
Nelumbo nucifera (Lotus)	h			▨							▨			▨		
Neptunia oleracea (Water Mimosa)	h			▨									▨			
Talinum fruticosum (Waterleaf, Cariru, Ceylon Spinach)	h			▨									▨			
Trapa natans (Water Chestnut)	h			▨												▨
Typha domingensis (Southern Cattail, Cumbungi, Berdi)	g	▨	▨							▨						
Typha latifolia (Broadleaf Cattail)	g	▨	▨							▨						

goals: cultivation of degraded and abandoned peatlands for peatland revitalisation, and economic development and provision of alternatives for the palm oil and acacia pulp industry.

Many tree species of Southeast Asian peat swamp forests are used commercially and are partially endangered due to overexploitation – e.g. Ramin (*Gonystylus bancanus*). The large scale cultivation of these species would ensure market availability of their products and contribute to species conservation. Besides offering *in situ* conservation, the plantations would also ease the pressure on remaining natural populations. In addition to Ramin, important tropical hardwoods include Borneo Kauri (*Agathis borneensis*), Meranti (*Shorea* spp.), Belangeran (*Shorea balangeran*) and Kempas (*Koompassia malaccensis*). In strongly degraded peatlands where subsidence has created a secondary surface relief, planting on ridges perpendicular to the slope helps reduce surface runoff, increase water retention, and thus reduce the risk of fire. In areas with periodical deep inundation, seedlings must be planted on peat mounds (Wibisono et al. 2005). The optimum harvesting method would focus on removing single trees that are transported using the traditional 'kuda-kuda' system, in which trunks are loaded onto sledges and are pulled out of the forest on wooden tracks. In this way the peat soil remains undamaged (Brünig 1996).

There is a variety of fast-growing, native peat swamp forest species that can serve as an alternative to the drainage-based cultivation of Acacia (*Acacia crassicarpa*, *A. mangium*), including *Alstonia pneumatophora*, *Combretocarpus rotundatus* and *Macaranga pruinosa*. These tree species do not require drainage, are partially resistant to fire and flooding, and often form pioneer forests after deforestation or fire (van Eijk & Leenman 2004, Giesen & van der Meer 2009). Trials to test their suitability as a viable alternative to Acacia are urgently required in collaboration with the pulp industry.

Globally, no alternative exists for the African Oil Palm (*Elaeis guineensis*) in terms of productivity, but tengkawang trees create a product of comparable quality and do not have a harmful effect on climate (Box 10.3). The cultivation of oil palm on peatlands can only be restricted through strict legislation. However, at present, land use concessions on peatlands are still being granted to the palm oil industry, and the prospects for a large scale implementation of paludiculture in Indonesia are bleak. Few areas seem to be remaining for paludiculture, because land concessions have already been allocated.

One opportunity to test paludiculture would be within the framework of REDD+ projects (Reducing Emissions from Deforestation and Forest Degradation). REDD+, an international climate protection mechanism, not only aims to protect forests to reduce carbon emissions, but also supports participation of local communities in emission reductions and carbon sequestration through alternative forest management. Paludiculture greatly improves access of local commu-

---

Box 10.3: Prospects of poverty alleviation and climate protection with non-timber forest products from paludiculture in Indonesia

René Dommain

Smallholder cultivation of peat swamp forest species in mixed orchards or plantations would make an important contribution to the alleviation of poverty and to the restoration of degraded peatlands. Particularly, the climate effect of peatlands could be improved in several ways: Firstly, rewetting will result in CO_2 emission reductions of up to 70 t per hectare and year; secondly, belowground tree biomass may form new peat; and finally, water tables of less than 40 cm below the surface considerably reduce the probability of fires (Colour picture 97). Cultivation of non-timber forest products of high value supports long-term preservation of tree stands, which in turn fosters peat formation and tree carbon sequestration, as well as a cool and wet micro-climate. The cultivation of mixed tree stands provides a range of products and helps secure income, even if single crops fail or specific market demands decline. Experiences in the production of high-quality non-timber forest products exist for Jelutong (*Dyera polyphylla*), Tengkawang nut trees (*Shorea* spp.) and for Rattan palm trees (*Calamus* spp.). Jelutong, a species limited to peat swamp forest, produces high quality latex that is obtained by tapping the stem, and is used for manufacturing chewing gum. During the 1990s, Jelutong was commercially cultivated for the first time in a 2,000 ha large plantation in Sumatra (Muub 1996). Between 2005 and 2008, Jelutong was planted on 250 ha of rewetted peatland in Block A of the Ex-Mega Rice Project area by Wetlands International, in collaboration with the local communities (CKPP 2008). It is still being cultivated in the province of Jambi in Sumatra (Giesen 2013). From 1997 to 1998, Indonesia exported 2.79 million tonnes of jelutong latex. In general, the production has collapsed because natural tree populations have often been overexploited (Giesen 2013). The cultivation of Tengkawang nuts (produced by several species of the genus *Shorea*, which grow naturally on mineral soil) also could be promising. Planting trials in West Kalimantan have shown that these trees also thrive on peat soils, are tolerant to flooding, and have similar yields as on mineral soil (Giesen & van der Meer 2009). The nuts yield a buttery fat that is used as a surrogate for cocoa butter (Blicha-Mathiesen 1994), and as a cooking oil that can substitute palm oil (Giesen 2013). In Kalimantan, the cultivation of Gemor (*Alseodaphne coriacea*) should be considered. The bark of this tree is traditionally used as a mosquito repellent. To this day, the bark is still an important source of income in Central Kalimantan, but current tree stocks are overexploited (Suyanto et al. 2009). Mixed orchards can also be cultivated with Rattan palm, which provides an important raw material for furniture but is also used for weaving baskets, mats and bags (Wibisono et al. 2005). These climbing plants could be cultivated together with native peat swamp forest trees that support their growth. Such mixed rattan plantations did exist in peatlands of Central Kalimantan during the 1960s, when the demand for rattan was high (Suyanto et al. 2009). Critical for promoting the establishment of non-timber forest product plantations is the directive of the Indonesian Ministry for Forestry (No. P.19/Menhut-II/2009) on the National Strategy for Developing Non-Timber Forest Products (see Giesen 2013). Collaboration with the palm oil industry would be advantageous in establishing alternatives to palm oil – which also started as a non-timber forest product.

nities to peatland resources and valuable plant species, and thus would be an important step in reducing poverty in Indonesian peatlands (Box 10.3). As paludiculture reduces both losses of carbon and of biodiversity, it qualifies for the Climate, Community and Biodiversity Standard (CCB) for projects of the voluntary carbon market. It is critical to start immediately with demonstration projects to show the feasibility and benefits of paludiculture to local and national stakeholders. A rapid implementation would help achieve the ambitious emission reduction targets that the country has set (Presidential Decree No. 61 2011). The plan is to reduce national CO_2 emissions by 26% until 2020. The Indonesian government has specifically listed peatland protection and restoration as key measures to fulfil this plan (Wibisono et al. 2011).

10.7 China – Paper from the water

Jan Felix Köbbing

China has a vast wetland area of 38.48 million ha (2008), which consists of lakes, marshes, coastal areas and river valleys (MEP 2008). In 2004, Common Reed (*Phragmites australis*, Chin. lúwěi) covered about 1 million ha outside of protected areas (Table 10.7). Large reed beds are found in the eastern coastal regions and along the big rivers like the Yangtze and the Huang He (Yellow River), as well as in the provinces of Inner Mongolia and Xinjiang (Pöyry 2006a). In 2004, the Chinese production of Common Reed was led by Hunan Province with 650,000 tonnes, followed by Liaoning Province with 470,000 tonnes (incl. 15–18% water content, Pöyry 2006a).

Reed yields vary considerably between cultivation sites (Table 10.7) because of differences in site conditions (temperature, precipitation/water availability, soil fertility, solar radiation) and in management practices. The average yield is 5.5 t DM per hectare. The data presented in Table 10.7 are conservative compared to other sources that show significantly higher yields. For example, Xiao & Li (2004) present annual yields of 496,800 tonnes for the Liaohe Delta, in the province Lioaning. According to Köbbing et al. (2014), 200,000 tonnes of reed were harvested annually at Wuliangsuhai Lake in Inner Mongolia alone. In contrast, Pöyry (2006a) mentions a total reed harvest for Inner Mongolia of only 100,000 tonnes.

In some Chinese provinces, yields can be increased considerably with appropriate investments and management. Brix et al. (2014), for example, report on a sophisticated pumping and drainage system in the Liaohe Delta, which enables ideal water table regulation for growth and harvesting. The water management is strictly regulated to maximize yields (Box 10.4). Other ecosystem services, like habitat protection or greenhouse gas emissions, are not taken into account.

No specific data are available about soil properties, but it seems that the natural debris of reed has formed a considerable layer of dead biomass – provided the site is consistently flooded. In some cases, *Phragmites* peat may be present. For the Wuliangsuhai Lake in Inner Mongolia, it could be shown that a gyttja layer of 12 cm had accumulated over the past 20 years (Liu et al. 2007).

In China, the majority of the harvested reed (95% in 2004) is used for paper production (Pöyry 2006a). Other applications include construction materials, mats (Colour picture 98) and panels, fuel, fodder, and protection against erosion and desertification (e.g. straw checker boards).

Traditionally, reed is harvested in winter, particularly in the northern provinces where a reliable ice cover enables an easy harvest. If it is used for fodder, reed is cut in spring or summer using boats or it is directly grazed by livestock (cows, goats, and sheep). Winter harvest is mainly carried out manually by local farmers with the help of additional hired workers. The reed is bundled or pressed into bales and loaded onto lorries to be transported to the paper mills (Colour pictures 99 and 100).

Depending on the location, various regulations exist concerning property and harvesting rights of reed beds. Pöyry (2006a) distinguishes: a) state-owned enterprises; b) harvesting rights leased or auctioned to farmers; c) privately owned enterprises; and d) collectively owned enterprises. In general, there is a strong trend away from traditional collective land use towards privatisation. Private enterprises bring the necessary capital to invest in management and infrastructure.

Already for decades there has been a high interest in Common Reed as a raw material for the paper industry, because of its high cellulose content (up to 49%). Whereas the reed-based paper industry is still large in China, until a few decades ago reed paper mills also existed in Turkey, Sweden and Egypt (Sainty 1985), as well as in Romania, Iraq, Iran, Italy, the GDR and the USSR (Wayman 1973, see also Chivu 1968a, 1968b). In Europe, reed fell into disuse because of a lack of sufficient raw material (partly because of the drainage of peatlands), as well as high harvesting costs, and greater availability of substitutes (Wichmann & Köbbing 2015).

Due to its limited domestic wood resources, China has traditionally produced a high proportion of its paper from non-wood fibres. Since the 1950s, Common Reed has been used for paper production, together with agricultural residues like straw or baggase. Its high cellulose content and long fibres make reed particularly suitable for the production of strong paper and cardboard. During the time of the 'Great Leap Forward' and the 'Cultural Revolution', many collectively owned paper mills were opened, each one with a production capacity of only a few thousands tonnes. Since the 1990s, stricter environmental regulations forced many small paper mills to shut down, as most of them did not have wastewater treatment plants. These closures had a

Table 10.7: Reedbed areas and yields in different provinces of China 2004 (Pöyry 2006a).

Province	Yield [t ha^{-1}]*	Reedbed area [ha]	Total yield [t]*	Reed price [CNY t^{-1}]$^{a)}$	Reed price at the end user [CNY t^{-1}]
Heilongjang	1.3	160,000	210,000	200–300	500–600
Hunan	10.0	63,000	650,000	–	–
Liaoning	7.5	66,000	470,000	400	–
Xinjiang	9.0	43,000	400,000	250–300	–
Hubei	9.7	36,000	350,000	500	–
Inner Mongolia	4.0	26,000	100,000	–	300
Jiangsu	0.7	113,000	80,000	500	–
Jilin	0.2	470,000	110,000	–	–
Shandong	7.5	13,000	100,000	300	–
Anhui	5.3	15,000	80,000	–	–
Total	5.5	1,005,000	2,550,000	343	466

* 15–18% moisture content; $^{a)}$ price after harvest.

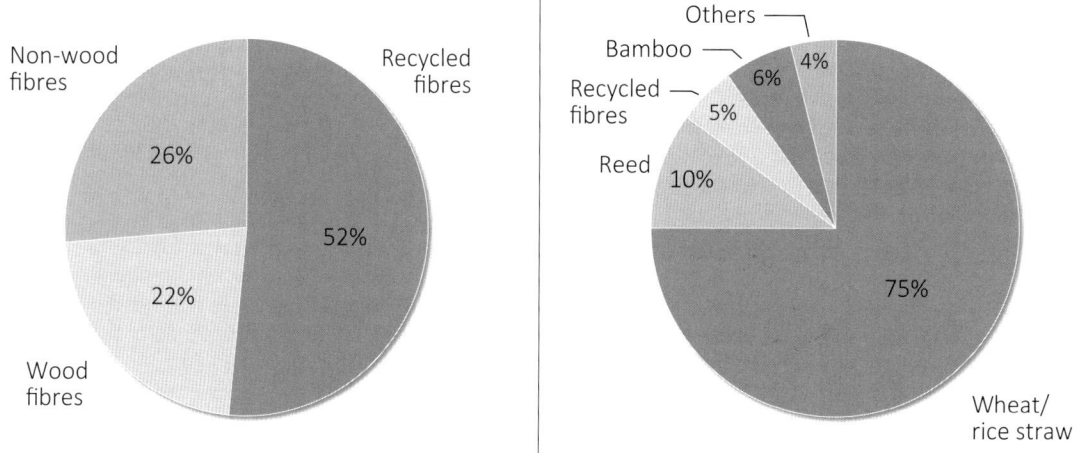

Figure 10.2: Raw materials for pulp production in China 2004. Left: total production; right: non-wood fibres (Pöyry 2006b).

strong effect on rural areas of China, where in 2002 approximately 8 million jobs depended on selling non-wood fibres (mainly straw) for the production of paper (Savcor Indufor Oy 2006).

In 2004, 2.6–2.7 million tonnes of reed were harvested in China, of which 95% were used for the production of paper (Pöyry 2006a), covering 10% of the non-wood based paper (Pöyry 2006a; Savcor Indufor Oy 2006, Figure 10.2). The proportion of non-wood pulp to the total market is decreasing, even if the absolute amount is still increasing. In China, paper production has been, and still is, increasingly concentrating on wood-based pulp factories that are usually situated near the coast. It is likely that more of the small non-wood paper mills will close in the coming years due to stricter environmental regulations. Between 2006 and 2010, non-wood paper mills with a total production capacity of more than 10 million tonnes were closed (Leponiemi 2011). These closures will further aggravate the crisis on the raw materials market. Besides, higher labour costs and designation of reed beds as conservation areas pose additional challenges. Finally, reed is considered an outdated construction material that people want to have replaced by steel or concrete. Currently, there is a sales crisis for reed biomass, and prices for reed are falling (Box 10.4).

In the coming decades, the extremely large demand for energy and raw materials offers a new opportunity for reed as a valuable resource in China. Cultivation of reed does not directly compete with food production and there are ample opportunities for its development – for example in energy production (incl. liquid biofuel).

However, a higher degree of mechanisation and professionalization in the whole reed processing chain is needed. The current harvesting techniques are simple and of limited performance, so there is great potential for improvement. Further modernisation of the paper industry should lower the environmental impact, particularly considering wastewater, and thus make the use of reed more attractive again. Last but not least, reed cultivation should take into account additional ecosystem services provided by reedbeds, such as provision of bird habitats and climate change mitigation.

Box 10.4: Paper production from reed in the Liaohe Delta, China

Jan Felix Köbbing

The 90,000-100,000 ha large reed area in the Liaohe Delta (Panjing) in Northeast China (121°10'–122°30'E, 40°30'–41°30' N), in the province of Liaoning, is considered to be the largest in the world (Ji et al. 2009; Xiao & Li 2004). Over the past decades, high intensity management of artificial and natural reed beds has resulted in strong increases in harvested area and yields. Between 1984 and 2006, the reed area grew by 600 ha on average per year (Ji et al. 2009). Several measures helped achieve these results, like disease control, suppression of other plants by flooding, burning of old reed stands, seasonal flooding and draining of the soil. These measures also served other purposes, like promoting soil thaw, preventing frost damage to the shoots, flushing salts, and soil mineralisation (Ji et al. 2009; Ye et al. 2013).

The reed is harvested from areas drained for the occasion, using conventional tractors with a side mower. Reed is bundled manually and transported to local paper mills by lorries or small trains. In 2011, 450,000 tonnes of dry mass were harvested (Ye et al. 2013). The majority of the reed is processed by local paper mills of the Jincheng group (Yingkou Paper and Jin Cheng Paper) and by paper mills in Yingkou and Dandong (Pöyry 2006a). The gross price for raw reed was 400 CNY t^{-1} (1 EUR = 8 CNY in 2004) in 2004 (Pöyry 2006a). The paper mills in Jincheng and Yinkou have a production capacity of 150,000 and 100,000 tonnes of paper fibres respectively, which corresponds to a total reed consumption of 575,000 tonnes dry mass (reed to fibre ratio of 2.3:1) (Pöyry 2006b; Ye et al. 2013). In 2011, there was a shortage of reed biomass at the two Jincheng paper mills of 140,000 tonnes (Ye et al. 2013).

10.8 Canada – Harvesting Typha spp. for nutrient capture and bioeconomy at Lake Winnipeg

Richard Grosshans

There are different potential functions and use potentials of *Typha* which opens up opportunities for innovations regarding nutrient recovery and bioenergy. In North America, *Typha latifolia* (Broadleaf Cattail), the European introduced *Typha angustifolia* (Narrow-leaf Cattail), and their hybrid *Typha x glauca* are all commonly occurring (Stevens & Hoag 2006). They are very tolerant of water table fluctuations and extremely competitive – spreading clonally or by seed (Grace & Wetzel 1981). Like other highly productive plants, *Typha* sequesters large amounts of CO_2 from the atmosphere and plays a significant role in the mass balance of wetlands by assimilating nutrients in its biomass. In this way, it effectively removes nutrients and toxins that cause eutrophication in aquatic systems, and therefore *Typha* is commonly used in engineered constructed wetlands for wastewater treatment (Kadlec & Knight 1996). *Typha* can produce significant biomass within a single growing season. Stands are often cut and mown to control the highly invasive plants. Harvesting *Typha* captures nutrients such as nitrogen and phosphorus, which are often the cause of eutrophication in aquatic environments, and removes them from the nutrient cycle. Integrating watershed management with *Typha* harvesting provides an innovative solution to reduce nutrient loading downstream to rivers and lakes (Box 10.5).

Lake Winnipeg, in Manitoba, Canada, is the 10th largest freshwater lake in the world. It is also considered one of the most eutrophic and suffers from excessive loading of nutrients (i. e. nitrogen and phosphorus) from its 100 million ha catchment (Figure 10.3). High levels of phosphorus in Lake Winnipeg are causing algal blooms of increasing intensity and frequency that consume oxygen and can release dangerous toxins (Lake Winnipeg Stewardship Board 2005). New concepts are needed to solve the problem of eutrophication, which may also promote new developments in the field of bioeconomy.

In Winnipeg, the International Institute for Sustainable Development (IISD) and its partners from the University of Manitoba and Ducks Unlimited Canada demonstrated, within a bioeconomy proof of concept, that plants like *Typha* spp. (Cattail) can be a key driver for a regional bioeconomy. These plants soak up nutrients that would otherwise flow into waterways and cause eutrophication and large-scale algal blooms (Grosshans et al. 2011a). It was demonstrated that harvesting *Typha* and other novel biomass crops can reduce nutrient loading in catchments of aquatic systems while also providing raw plant materials for the biomass industry and recovering valuable nutrient resources (Grosshans et al. 2011a, IISD 2013). Phos-

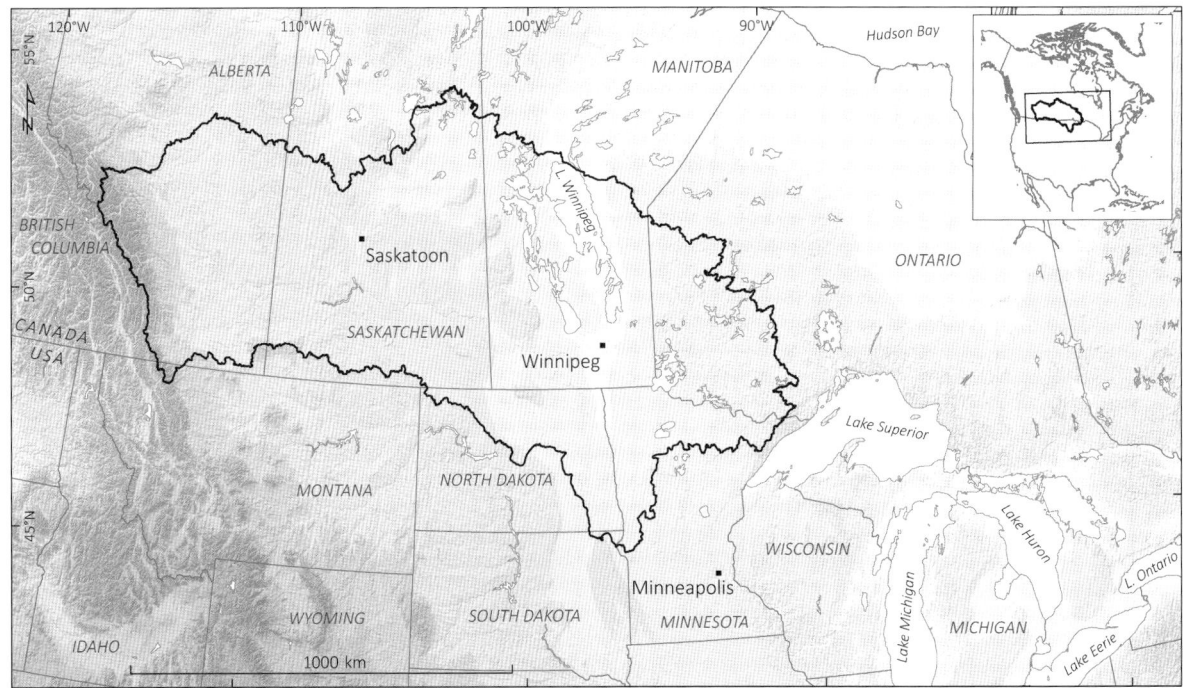

Figure 10.3: Lake Winnipeg, Manitoba, Canada with its catchment area outlined.

10.8 Harvesting Typha spp. for nutrient capture

phorus, the noxious pollutant fouling Lake Winnipeg, is an important natural resource for plant growth, and critical for agriculture and global food security (Ullrich et al. 2009).

Typha spp. (hereafter *Typha*) is characteristic of wet environments in North America. It represents an under-utilized renewable and sustainable source of biomass that can be integrated into solid and liquid bioenergy systems to help meet increasing sustainable energy demands.

Although the current research in Manitoba focuses on naturally occurring stands of *Typha* and harvesting for catchment scale nutrient capture and bioenergy, the rapid clonal growth and reproduction of *Typha* can be useful in restoration or for constructed wetlands. Cost-effective means for establishing plants on larger areas are available. Also, *Typha* plants or roots can be transplanted from other sites. Seeds collected from seedheads (in autumn, when they are dry or slightly immature) will germinate on moist soils, thus providing a cost-effective means for establishing stands on larger areas (Stevens & Hoag 2006).

In several countries, *Typha* is an important economic resource harvested for traditional products such as building material (Özesmi 2003), and similarly to Common Reed (*Phragmites australis*), it is used as a solid fuel for cooking and heating (Ba et al. 2009). *Typha* was explored as a potential alternative bioenergy crop in North America as early as the 1970s (Lakshman 1984, Dubbe et al. 1988), but the economic feasibility

Table 10.8: *Typha* biomass yield and phosphorus retention in a pilot harvest in the Netley Libau Wetlands, Manitoba (harvested in August).

	Ditches	Stormwater retention area
Calculated biomass yield per area [t ha^{-1}]	13	13
Total harvested area [ha]	17.1	54.1
Total calculated biomass yield at site [t DM]	222	703
Average total phosphorus [%]	0.11	0.10
Total phosphorus in *Typha* biomass [kg]	244	703
Total average phosphorus per area [kg ha^{-1}]	14.3	12.99
For comparison results for harvest in autumn: Total phosphorus [%]	0.097	0.087

Box 10.5: Prospects of the use of Typha biomass – The Netley-Libau Nutrient-Bioenergy Proof of Concept

Richard Grosshans

By applying modern economic and environmental valuations, Grosshans et al. (2014, 2011a, b) demonstrated that *Typha* is an alternative bioenergy crop with major environmental and economic co-benefits. In Manitoba, *Typha* reaches maturity in less than 90 days. Late summer/early autumn harvests yield an average of 14–20 tonnes of dry mass per ha, removing up to 30 kg of phosphorus per ha and year. Once harvested, nutrients locked in plant tissue are prevented from being released into the environment via natural decomposition.

Utilizing the harvested biomass for energy provides a further benefit by displacing fossil fuels, generating valuable carbon offsets. Combustion trials have shown a gross calorific value of 17–20 MJ kg^{-1} DM, which is comparable to commercial wood pellets (17 MJ kg^{-1}). The average ash content is 5–6%, and no major concerns were identified regarding combustion emissions and ash (Table 10.9, Dubbe et al. 1988, Grosshans 2014). The average potential energy yield of *Typha* is almost 300 GJ per ha. Up to 88% of total phosphorus was recovered in ash following combustion in solid fuel burners, which could be recycled for fertiliser.

Table 10.9: General characteristics of Cattail biomass.

Characteristic	Average
Biomass yield [t ha^{-1}]	14-20
Moisture [% DM]	6-15
Carbon [% DM]	38.8-43
Gross Calorific Value [MJ kg^{-1}]	17.1-19.2
Ash content [% DM]	5.5-7.5
Ash fusion temperature [°C]	1,378
Phosphorus capture [kg ha^{-1}]	20-60

of this type of use was not demonstrated until recently (Box 10.5).

Commercial-scale harvesting of wetland biomass has not been demonstrated in Canada or North America.

Harvesting in wetlands presents serious challenges because of the trafficability of soils. Within several research and pilot projects, diverse harvesting approaches were tested, including a complete harvester for chopped biomass, as well as baling with adaped machinery (Colour pictures 101 and 102). Furthermore, the logistics of baling, transportation, storage and processing were assessed. In total, almost 800 bales of *Typha* were harvested from a pilot area of ca. 17 ha for a total of 220 t DM; (Colour picture 102), and almost 250 kg of phosphorus were removed from the site (Zubrycki et al. 2012). Currently research is being done to evaluate *Typha* biomass for higher value uses, like biochar, third generation ligno-cellulosic ethanol, fibers and composite materials, as well as the recovery of high value elements. Nutrients can be reclaimed and reused directly from biomass or from ash following heat production (Grosshans et al. 2011, Grosshans 2014). Estimates of yields and phosphorus captured are given in Table 10.8. Due to the late harvest and beginning of plants senesce, the biomass contained less phosphorus and other nutrients later in the season (Grosshans 2014). Therefore, phosphorus content of the harvested biomass was lower than with an earlier harvest.

The multiple environmental and economic values from utilizing *Typha* as an ecological biomass source not only create incentives for restoration of wetlands. *Typha* also offers an integrated approach to solve the problem of nutrient overloading.

11 The way out of the desert – What needs to be done

Christian Schröder, Hans Joosten &
Wendelin Wichtmann

The origin of agriculture lies in the steppes and semi-deserts of our planet, which also were the cradle of many cultivated plants. For mainstream agriculture, mires have to be transformed into steppe-like landscapes. Soils must be continuously moved, must periodically dry out, and both soil texture and moisture content must be adapted to the demands of plants that are mostly annuals. The disastrous outcomes of such practices are similar everywhere in the world, namely peat loss and subsidence, gigantic greenhouse gas emissions, abandonment and poverty. A sustainable utilisation of peatlands appears to be only possible under wet conditions. Otherwise, the peatland soil and area will disappear as a basis of production. A change of paradigm is needed to preserve peatland productivity for the future.

Now, the concept of paludiculture demonstrates that this is possible. Paludiculture aims to preserve the peat body as a sustainable base of production while generating marketable products (Box 1.1). Simultaneously, the provision of other ecosystem services, such as climate change mitigation, provision of water, regulation of biochemical cycles, and conservation of biodiversity is guaranteed (Chapter 5). A cost-effective utilisation of wet and rewetted peatlands is feasible (Chapter 6.1, Chapter 10.7). This raises the question why – in spite of all these advantages compared to conventional land use, paludiculture has not yet been implemented on a large scale. We must analyse and understand why wet peatland utilisation has hitherto not established and which obstacles are blocking its implementation. On the basis of such analysis, opportunities have to be recognized, avenues indicated and a change in peatland use realised.

The large-scale implementation of paludiculture requires answers on many practical questions with respect to the management of wet areas (Chapter 11.2). A critical consideration of these practical challenges makes quickly clear that the implementation of paludiculture is impossible without adaptation of the present boundary conditions. It is necessary to raise awareness among stakeholders at many levels and to show them alternatives (Chapter 11.3). Only then the sustainable management of the peatland resource will be addressed by politics and requested by society (Chapter 11.4). The role of science in this process it to provide facts and develop practicable alternatives (Chapter 11.5).

11.1 Problems of peatland management and the necessity of paludiculture

Drainage is the root cause of many problems associated with peatland utilisation (Chapter 2.2). Drainage is justified with the misconception that wet peatlands are useless wasteland. Furthermore, usually, there is a natural aversion towards wet peatlands (Chapter 8.1). The ingrained belief that peatlands must be dry, has legitimised public investments in drainage over the centuries, as illustrated respectively by the peatland colonisation under Frederic the Great in Prussia, the complex melioration in the GDR (Chapter 2.1) or the Mega Rice Project in Indonesia (Chapter 10.6). As the most modern technical possibilities were – and are applied, the negative consequences increase proportionally. The natural functions of mires in the landscape (Chapter 2.4) are only marginally considered (or not at all) by these 'modern' land use practices. The use is focused only on the exploitation of the production function (Chapter 6.4). The general public, including farmers, consider the problems of drainage-based peatland cultivation to be rather academic. There is a distorted perception and a substantial lack of knowledge about the matter (Chapter 8.2). Greenhouse gas emissions are as imperceptible as the loss of nutrients with the discharged drainage water. The change in species composition is taken for granted, or seen as part of the triumph of man over nature. Although subsidence and mineralisation in Central Europe lead to soil losses of 1–2 cm per year (in the tropics even of more than 5 cm per year, Chapter 2.2), this is barely noticed. The industrial nations point alarmingly to the carbon losses from tropical deforestation and to the expansion of deserts, while in their backyard the carbon-rich, fertile peat soils dissolve into air at a tearing pace. Problems that arise from continuous subsidence – such as the successive water logging of sites, are technically addressed by deepening drainage ditches, poldering, and establishing pumping stations. Even problems that are directly caused by peatland drainage, such as desiccation of crops, peat fires and more frequent floods, are attributed to climate change. Thus, the real root causes are blocked out. When peatlands have to be abandoned after a destructive use, this situation is indeed deplored but accepted, as these areas were anyhow considered 'wasteland' before they were drained.

The benefits of wet peatlands are inadequately assessed (or not at all). Currently, no instruments are available to remunerate these ecosystem services, and the gradual destruction of the production base is tolerated or even indirectly supported (Chapter 6.4, Chapter 7.3). Peatlands have been ruined as production facilities in less than a generation (Chapter 2.3). In many places, they are already lost for further utilisation. The formerly extensive, almost area-wide utilisation of wet peatlands shows, that these destructive practices are wrong.

Wet peatlands continue to be taken into cultivation, traditional uses are blocked out and their potentials completely ignored. The limited perception of the 'peatland issue' prevents an active reversal of land use policy. Instead, the degrading management of peatlands still is favoured by the prevailing conditions (Chapter 7.2). In the European Union, direct payments are granted to drainage-based peatland cultivation, despite the highest soil loss rates; peatland reclamation projects in developing countries are instigated, and in Germany, the use of biomass from drained peatlands (e. g. maize or grass silage for biogas plants) is supported by the Renewable Energy Act (Erneuerbare Energien Gesetz). In contrast, regulatory options (e. g. penalties for violation of the cross-compliance rules, or the Soil Protection Act, are not applied (Chapter 7.3).

11.2 Challenges for practice

The management of wet areas places high demands on the managers. The trafficability of the area is restricted and cannot be addressed with conventional agricultural techniques (Chapter 4.1). The requirements for harvest technology are higher (Chapter 4.2) and the logistics of biomass removal more complex (Chapters 4.3 and 4.4). The usable plant species (Chapter 3.1) require alternative, possibly new concepts for utilisation (Chapters 3.4, 3.5 and 3.6). Furthermore, the conversion to paludiculture and the necessary changes in infrastructure are associated with extensive planning and technical measures (Chapters 9.4 and 9.5). For many new cultures there is still a lack of clarity about optimal water and crop management, which in the end leads to uncertainty about the quantity and quality of the yields that can be attained. Whereas the cultivation of dry sites can rely on crop varieties that over millennia have been purposely selected and bred, paludiculture so far largely uses autochthonous, wild plant material (Chapter 3.1). This is not necessarily a handicap because, similarly to forestry, regional populations might be adapted better to regionally prevailing environmental conditions. However, it does show that major breeding progress can still be achieved, especially in terms of productivity.

A completely different sort of practical problems is the lack of planning security for the cultivation of new crops (Chapter 8.3). Regarding this matter, conflicts can arise with legal habitat protection (Chapters 7.1) besides the fact that agricultural administrations may deny the classification of paludiculture land as agricultural land (Chapters 7.2). These examples of existing gaps and contradictions illustrate the urgent need of a review. The existing legal framework needs to be adjusted, like it has been done in Germany for the cultivation of short-rotation plantations (Chapter 7.3).

Another barrier is the uncertain marketability of the goods produced. Until a complete market is developed, there will be a strong interdependence between purchasers and producers (Chapter 8.3). However, this may be a major opportunity to generate regional added value because producers and consumers have to cooperate closely, which can favour the regional processing of paludiculture biomass (Chapter 6.3).

Furthermore, it still is unclear who should bear the costs of rewetting plus the required infrastructure (Chapter 9.4). The current promotion of drainage-based peatland utilisation rewards a production system that leads to high economic costs, which must be borne by society and future generations (Chapter 6.4). This policy must be gradually phased out and the saved expenses should be used for rewetting measures and the implementation of paludiculture. Supplementary a system of payment for ecosystem services that are associated with wet cultivation has to be established (Chapter 2.4, Chapter 7.3).

The most important steps to develop the implementation of paludiculture from a practical perspective are:
- The establishment of pilot farms and demonstration plots for the further development of crop management;
- The creation of planning security by adjusting the legal framework;
- The adaptation and development of harvesting and transport technology;
- The development of market incentives for the utilisation of paludiculture biomass;
- The covering of the costs of rewetting and infrastructure by the public sector;
- The rewarding of provided ecosystem services.

11.3 Awareness raising and communication

Land is getting scarce. The problem of land scarcity attracts worldwide public attention with the discussion about land grabbing. The question how to deal with land, particularly with regard to soil preservation for the production of food and renewable raw materials, has not yet been satisfactorily solved. In the case of peatlands, the soil is not treated as a basis of production but instead is consumed as a product (Chapter 2.3). Land users are not aware of the particularity of time-limited usability, and its causes and consequences are insufficiently conveyed in trainings and consultancies (Chapter 8.4). The problems resulting from the use of peat soils, especially subsidence and the emission of greenhouse gases, not only have to be demonstrated to land users, but also be communicated on a broad level (Chapter 8.2). The importance of protecting peat soils is even less visible than the general interests of soil conservation. It is not recognised enough that peatlands function differently from other ecosystems. If a forest is cleared, the major loss of carbon stock (biomass) and the associated greenhouse gas emissions will take place simultaneously with the activity of clearing. In contrast, drained peatlands keep losing carbon and emit greenhouse gases continuously (Chapter 5.1). Everyone involved must be informed about these peculiarities of wet peat soils as carbon stocks. If this information is not shared, then it will be difficult to convey awareness about the opportunities for land use change. Paludiculture is not only able to maintain the function of peatlands as carbon stock, but can also provide important ecosystem services such as nutrient retention (Chapter 5.4) and regulation of the regional water balance (Chapter 5.3), as well as contribute to the conservation of biodiversity (Chapter 5.2). Thus, paludiculture does justice to the multisectorial demands of sustainable land use (e.g. Climate Smart Agriculture) (Chapter 9.1, Chapter 10.1). Many examples from all over the world show that the need for wet peatland cultivation is widely acknowledged (Chapter 10).

In order to promote awareness about the benefits of paludiculture among land users, the general public, and decision-makers, the following steps need to be taken:
- Development of a lobby for Sustainable Land Management, in particular for the protection of peat soils;
- Networking and sensitisation of stakeholders;
- Consulting and training of practitioners and multipliers;
- Development of strategies for knowledge transfer;
- Provision of illustrative examples of best practice.

11.4 Politics and society

The continuously growing world population poses a major challenge to land use. More people have to satisfy their needs from the available land. In other words, the production factor soil must be used more efficiently, more sensibly and above all, more sustainably. The gradual destruction of usable land by peatland drainage can no longer be tolerated.

So far, the land use sector enjoys the privilege of impunity with respect to greenhouse gas emissions. The European Union, a pioneer in climate protection, is the world's second largest hotspot when it comes to emissions from peatlands. 300 million tonnes of CO_2 e are emitted every year by the 12 countries of the European Union that are most rich in peatlands (Chapter 10.2). The greenhouse gas emissions from peatlands account for 99 % of all emissions caused by the agricultural use of land.

The first steps go into the right direction: In Mecklenburg-West Pomerania 30,000 ha have been rewetted already and, in the next few years, a further 70,000 ha will follow while retaining their production function (Chapter 10.3). Rewettings of similar size are planned in the other northern federal states of Germany.

In the future land must not just provide arable soil for the production of food and renewables, but has to meet a wide range of needs of humanity. Therefore, a unidimensional view on land use can no longer be justified. Land has to be handled in such a way that multifunctionality is guaranteed. Paludiculture fully satisfies this demand (Chapter 2.4 and 9.1). In many cases, cultivation of renewable raw materials in rewetted peatlands may also decrease the competition between food production and the cultivation of renewable resources, as the latter is carried out at locations that are not suitable for food production.

One of the key challenges that society poses to politics is to satisfy the increasing demands for land while the pressure of utilisation is growing. With respect to peatland use, all possible levers for large-scale implementation of alternative land use must be used. For decades, the focus has solely been on the production function of peatland soils. To date, significant sums are made available for social transfer payments to enable uses that require drainage.

This leads to high social damage. Only the cost of greenhouse gas emissions from peatlands for Germany is estimated to amount to approximately 3.6 billion EUR per year (Chapter 6.4). The current agricultural policy framework of the European Union supports the above mentioned damage with more than 300 million EUR of direct payments per year (Chapter 6.4). Thus the public not only bears the costs, but also pays for creating the damage. The 'polluter pays' principle is turned upside down and public funds are counterproductively and irresponsibly used (Chapter 7.3). As if this were not enough, alternative farming concepts are hampered by the existing legal framework. If land users want to implement paludiculture as a sustainable land use alternative, they face a serious setback, namely the loss of recognition as agricultural land and thus exclusion of agricultural support (Chapter 7.2). Moreover, their constructed paludiculture fields may become subject to legal habitat protection (Chapter 7.1), which not only restricts their future use but also leads to a massive depreciation of the land. As a consequence, they stick to drainage-based peatland management.

A way out of this contradictory situation requires adjustment of the current framework conditions, including:
- Recognition of paludiculture as a form of agriculture;
- Exempting established paludiculture fields from legal habitat protection;
- Discontinuation of counterproductive incentives;
- Consistent application of the polluter pays principle;
- Rewarding the land use associated to ecosystem services;
- Accounting of greenhouse gas emissions from peatland use in the reporting for the Convention on Climate Change (UNFCCC).

11.5 Research questions

The change in land use on peatlands to implement paludiculture implies a paradigm shift. It is not about just changing crop rotation or optimising the management. The use of peat soils must completely be rethought and partly redeveloped. The associated questions cannot be solved mono-disciplinarily, but require the cooperation of many scientific and technical disciplines in close feedback with practice (Chapter 8.4). The issues to be addressed go far beyond the management of land. New crops that can strive under permanently wet conditions have to be developed for, or adapted to, new production concepts (Chapter 3.4–3.6). Furthermore, the complex interactions and effects of management under wet and drained conditions have to be captured, and the positive and negative effects on various ecosystem services associated with land use have to be evaluated (Chapter 2.4).

The reorientation of land use on peatlands will involve considerable public means. The effects of the use of public funds must be recorded and evaluated, and the sustainability of different usage alternatives has to be assessed (Chapter 9.1). Science is called upon to provide facts and to lay the foundations for crucial decisions and their justification.

It has been shown through several research projects that wet peatlands can be used. In some parts of the world, wet utilisation systems are already successfully established (Chapter 10). In order to answer the many unresolved questions, the following research-related measures are needed:
- Establishment of demonstration and experimental sites;
- Promotion of practice and knowledge transfer;
- Development of payment for ecosystem services approaches;
- Instigation of interdisciplinary research programmes;
- Institutionalisation of paludiculture research.

11.6 Outlook

Paludiculture is a new, future-orientated concept for the sustainable use of peatlands. The examples presented in this book provide a first glance of the multiple possibilities. Ways to arrive at a large-scale implementation of paludiculture are indicated and solutions for open questions are suggested.

The large-scale implementation of paludiculture requires a consequent and consensually pursued paradigm change. The solution of problems that originate from drainage-based peatland utilisation will depend decisively on political will and successful best practice examples. Politics must create effective incentives for a change in peatland use and stop counterproductive subsidies. Such policy can only be based on awareness and sensitisation of a wide range of stakeholders. The importance of peatlands and the consequences of drainage have to be communicated, as well as the possibilities of a sustainable land use to be demonstrated in pilot projects.

Millennia-old natural peatlands are still being drained for ephemeral use and abandoned after they have been destroyed. Further land development has to be redirected urgently to degraded peatlands. Paludiculture can revive devastated peatlands and may recover the area of productive land that has been lost. Where population pressure is too high, natural peatlands may also have to be claimed for production. Also there, utilisation with continuing high water levels will be the only way to protect the peatland and to guarantee their continuous sustainable productive use.

Worldwide, wet cultivation of peatlands must become the norm whereas drainage-based cultivation must be an exception subject to licensing.

Paludiculture, the way out of the desert!

References

Abdoellah, O. (1996): Social and environmental impacts of transmigration: A case study in Barambai, South Kalimantan. In: Padoch, C. & Peluso, N.L. (eds.): Borneo in transition. People, forests, conservation, and development. South-East Asian social science monographs. Oxford University Press, Kuala Lumpur, New York: 266–279.

Abel, S., Haberl, A. & Joosten, H. (2011): A decision support system for degraded abondoned peatlands illustrated by reference to peatlands of the Russian Federation. Michael-Succow-Stiftung zum Schutz der Natur. 52 p.

Abel, S., Couwenberg, J., Dahms, T. & Joosten, H. (2013): The Database of Potential Paludiculture Plants (DPPP) and results for Western Pomerania. Plant Diversity and Evolution 130: 219–228.

Adapa, P.K. (2011): Densification of selected agricultural crop residues as feedstock for the biofuel industry. PhD thesis. University of Saskatchewa, Saskatchewa, Canada. Deparment of Chemical and Biological Engineering. 329 p.

Afreen, F., Zobayed, S., Armstrong, J. & Armstrong, W. (2007): Pressure gradients along whole culms and leaf sheaths, and other aspects of humidity-induced gas transport in *Phragmites australis*. Journal of Experimental Botany 58: 1651–1662.

Aldrian, E. & Susanto, R.D. (2003): Identification of three dominant rainfall regions within Indonesia and their relationship to sea surface temperature. International Journal of Climatology 23: 1435–1452.

Al-Hakkak, J.S. & Barbooti, M.M. (1989): Thermogravimetric study on typha (*Typha angustifolia* L.). Journal of Thermal Analysis 35: 815–821.

Alvira, P., Tomás-Pejó, E., Ballesteros, M. & Negro, M.J. (2010): Pretreatment technologies for an efficient bioethanol production process based on enzymatic hydrolysis: A review. Bioresource Technology 101: 4851–4861.

Ambak, K. & Melling, L. (2000): Management practices for sustainable cultivation of crop plants on tropical peatland. In: Iwakuma, T., Inoue, T., Kohyama, T., Osaki, M., Simbolon, H., Tachibana, H., Takahashi, H., Tanaka, N. & Yabe, K. (eds.): Proceedings of the International Symposium on Tropical Peatlands, Bogor, Indonesia, 22–23 November 1999. Hokkaido University, Indonesian Institute of Sciences, Bogor, Indonesia: 119–134.

Ammermann, K. & Mengel, A. (2011): Energetischer Biomasseanbau im Kontext von Naturschutz, Biodiversität, Kulturlandschaft. Informationen zur Raumentwicklung 5/6: 323–337.

Andrianandrasana, H.T., Randriamahefasoa, J., Durbin, J., Lewis, R.E. & Ratsimbazafy, J.H. (2005): Participatory ecological monitoring of the Alaotra wetlands in Madagascar. Biodiversity and Conservation 14: 2757–2774.

Andriesse, J.P. (1988): Nature and management of tropical peat soils. Soils Bulletin 59. FAO, Rome. 165 p.

Anonymus (2005): Rote Liste der gefährdeten Höheren Pflanzen Mecklenburg-Vorpommerns. Ministerium für Landwirtschaft, Umwelt und Verbraucherschutz Mecklenburg-Vorpommern, Schwerin. 59 p.

Arny, A., Hodgson, R. & Nesom, G. (1929): Reed Canary Grass for meadows and pastures. Minnesota Bulletin 263: 1–27.

Asaeda, T., Rajapakse, L., Manatunge, J. & Sahara, N. (2006): The effect of summer harvesting of *Phragmites australis* on growth characteristics and rhizome resource storage. Hydrobiologia 553: 327–335.

Asian Development Bank (ADB) (ed.) (2012): Basic Statistics. Asian Development Bank (ADB), Manila. 6 p.

Askaer, L., Elberling, B., Friborg, T., Jørgensen, C.J. & Hansen, B.U. (2011): Plant-mediated CH_4 transport and C gas dynamics quantified in-situ in a *Phalaris arundinacea*-dominant wetland. Plant and Soil 343: 287–301.

Augustin, J. & Chojnicki, B. (2008): Austausch von klimarelevanten Spurengasen, Klimawirkung und Kohlenstoffdynamik in den ersten Jahren nach der Wiedervernässung von degradiertem Niedermoorgrünland. In: Gelbrecht, J., Zak, D. & Augustin, J. (eds.): Phosphor- und Kohlenstoff-Dynamik und Vegetationsentwicklung in wiedervernässten Mooren des Peenetals in Mecklenburg-Vorpommern. Status, Steuergrößen und Handlungsmöglichkeiten. Leibniz-Institut für Gewässerökologie und Binnenfischerei (IGB) im Forschungsverbund Berlin e.V. Berichte des IGB 26, Berlin: 50–67.

Autorengemeinschaft (2013): „Zukunft des Thurbruchs – Ein Leben mit dem Moor". Citizens' jury assessment. 19 pp.; Authors team of participants of the citizens jury „Paludikultur – Eine Alternative für Mensch und Moor?" during February and March 2013, hand-in date 17.03.2013, Client: Ernst-Moritz-Arndt-University Greifswald / VIP-Project. Available online: http://www.botanik.uni-greifswald.de/3505.html (last access: 4.10.2015)

Ba, A.O., Diongue, N., Fall, A., Sow, M.A., Courteau, B. & Diao, D. (2009): Transforming invarding plants into fuel pellets in Ross-Bethio (Senegal). JADE project Report, S3IC-2009-001, Senegal.

Babula, P., Adam, V., Havel, L. & Kizek, R. (2009): Noteworthy secondary metabolites naphthoquinones – their occurrence, pharmacological properties and analysis. Current Pharmaceutical Analysis 5: 47–68.

Bakoariniaina, L.N., Kusky, T. & Raharimahefa, T. (2006): Disappearing Lake Alaotra: Monitoring catastrophic erosion, waterway silting, and land degradation hazards in Madagascar using Landsat imagery. Journal of African Earth Sciences 44: 241–252.

Balat, M., Balat, H. & Öz, C. (2008): Progress in bioethanol processing. Progress in Energy and Combustion Science 34: 551–573.

Bambalov, N.N., Tanovitskaya, N., Kozulin, A.V. & Rakovich, V.A. (2016): Belarus. In: Joosten, H., Tanneberger, F. & Moen, A. (eds.): Mires and peatlands of Europe. Schweizerbart, Stuttgart.

Baranyai, B. (2013): Literaturrecherche und Pilotversuch zur Projektidee: Untersuchung der Machbarkeit der Kultivierung von Drosera spp. – Biomasse für medizinale Zwecke auf wiedervernässten Hochmooren in Deutschland. Final report for DBU.

Bärisch, S. & Tanneberger, F. (2011): Recommended research activities. In: Tanneberger, F. & Wichtmann, W. (eds.): Carbon credits from peatland rewetting. Climate – biodiversity – land use. Schweizerbart, Stuttgart: 189–193.

Barr, C. (2004): Risk analysis and impact assessment for pulp and plantation investments: The case of indonesia. International forum on finance and investment in China's forestry sector, Beijing.

Barthelmes, A. (2010): Vegetation dynamics and carbon sequestration of Holocene alder (Alnus glutinosa) carrs of NE Germany. PhD thesis. Universität Greifswald, Greifswald. Institut für Botanik und Landschaftsökologie & Botanischer Garten. 240 p.

Barthelmes, A., Dommain, R. & Joosten, H. (2012): Global potential of paludiculture as land use alternative for rewetted peatlands. In: Proceedings of the 14th International Peat Congress, Stockholm. International Peat Society (IPS), Stockholm: 387–394.

Barthelmes, A., Couwenberg, J., Risager, M., Tegetmeyer, C. & Joosten, H. (2015): Peatlands and climate in a Ramsar context. A Nordic-Baltic perspective. Nordic Council of Ministers. 245 p.

Barz, M., Tanneberger, F. & Wichtmann, W. (2012): Sustainable production of Common Reed as an energy source from wet peatlands. In: Proceedings of the 4th International Conference on Sustainable Energy and Environment (SEE 2011): A Paradigm Shift to Low Carbon Society, Bangkok, Thailand: 780–787.

Bauriegel, A. (2013): Referenzierte Moorkarte Brandenburg.

Bayerisches Landesamt für Umwelt (BLU) (ed.) (2005): Leitfaden der Niedermoorrenaturierung in Bayern. Bayerisches Landesamt für Umwelt, Augsburg. 140 p.

Beetz, S., Liebersbach, H., Glatzel, S., Jurasinski, G., Buczko, U. & Höper, H. (2013): Effects of land use intensity on the full greenhouse gas balance in an Atlantic peat bog. Biogeosciences 10: 1067–1082.

Behrendt, D., Kleinhückelkotten, S., Kloten, M. & Neitzke, H.-P. (2010): Kriterien für die Nachhaltigkeit der Nutzung und die Vermarktbarkeit städtischer Brachflächen. In: Frerichs, S., Lieber, M. & Preuß, T. (eds.): Flächen- und Standortbewertung für ein nachhaltiges Flächenmanagement. Methoden und Konzepte. Beiträge aus der REFINA-Forschung 5. Deutsches Institut für Urbanistik, Berlin: 149–163.

Behrendt, D. & Neitzke, H.-P. (2013): Indikatoren für ein nachhaltiges Landmanagement. VIP project report. ECOLOG-Institut, Hannover.

Behrendt, D., Kleinhückelkotten, S. & Neitzke, H.-P. (2012): Soziale Anschlussfähigkeit von Entwicklungsstrategien für degradierte Moore. Überregionale und regionale Akteursfeldanalyse. Project report. ECOLOG-Institut, Hannover. 18 p.

Behrens, H. & Rösler, M. (1999): Egon auf dem Weg zum 9. Weltwunder? Zu den Auseinandersetzungen um die Zukunft der Friedländer Große Wiese mit dem Galenbecker See. In: IUGR e.V. (ed.): Studienarchiv Umweltgeschichte. Gemeinsame Mitteilungen vom Bund für Natur und Umwelt e.V. und Institut für Umweltgeschichte und Regionalentwicklung e.V. Bund für Natur und Umwelt e.V. (BNU), Institut für Umweltgeschichte und Regionalentwicklung e.V. (IUGR). Studienarchiv Umweltgeschichte 5 12, Berlin: 52–60.

Bellebaum, J. (pers. communication): Hierarchical models reveal positive effects of recurring habitat management on aquatic warbler Acrocephalus paludicola breeding numbers.

Beltz, R. (1916a): Moorkunde und Heimatschutz. In: Heimatbund Mecklenburg (ed.): Mecklenburg. Zeitschrift des Heimatbundes Mecklenburg 1, Schwerin: 8–12.

Beltz, R. (1916b): Über die Notwendigkeit der Schaffung von Moorschutzgebieten. In: Heimatbund Mecklenburg (ed.): Mecklenburg. Zeitschrift des Heimatbundes Mecklenburg 1, Schwerin: 12.

Berg, Å.K.E. (1992): Factors affecting nest-site choice and reproductive success of Curlews Numenius arquata on farmland. Ibis 134: 44–51.

Berg, C., Dengler, J., Abdank, A. & Isermann, M. (eds.) (2004): Die Pflanzengesellschaften Mecklenburg-Vorpommerns und ihre Gefährdung. Textband. Weissdorn, Jena. 568 p.

Berg, E. (1999): Moornutzung und -erschließung. In: Fansa, M. (ed.): Weder See noch Land. Moor – eine verlorene Landschaft. Schriftenreihe des Staatlichen Museums für Naturkunde und Vorgeschichte 2. Isensee, Oldenburg: 80–131.

Berg, H., Burger, A. & Thiele, K. (2008): Umweltschädliche Subventionen in Deutschland. Umweltbundesamt (UBA), Dessau-Rosslau.

Betha, R., Pradanib, M., Lestarib, P., Joshic, U., Reidd, J. & Balasubramania, R. (2013): Chemical speciation of trace metals emitted from Indonesian peat fires for health risk assessment. Atmospheric Research 122: 571–578.

Biancalani, R. & Avagyan, A. (eds.) (2014): Toward climate-responsible peatlands management. Food and Agriculture Organization of the United Nations (FAO). Mitigation of climate change in agriculture series, 9, Rome. 100 p.

Bibby, C.J. & Lunn, J. (1982): Conservation of reed-beds and their avifauna in England and Wales. Biological Conservation 23: 167–186.

Bic Room Inc. (ed.) (2013): Ecolabel Index. All Ecolabels, Vancouver, Canada. http://www.ecolabelindex.com/ecolabels/, last access 06 September 2015.

BioGrace (2013): Excel Greenhouse gas emission calculation tool. BioGrace. http://www.biograce.net/content/ghgcalculationtools/calculationtool, last access 30 June 2014.

Bioland e.V. (ed.) (2011): Bioland Richtlinien. Bioland e.V., Mainz. 52 p.

Bischoff, A., Selle, K. & Sinning, H. (2005): Informieren, Beteiligen, Kooperieren – Kommunikation in Planungsprozessen. Eine Übersicht zu Formen, Verfahren und Methoden. KiP – Kommunikation im Planungsprozess 1. Dortmunder Vertrieb für Bau- und Planungsliteratur, Dortmund. 334 p.

Bissels, S. & Oppermann, R. (2011): Analyse und Bewertung von Reformvorschlägen zur gemeinsamen Agrarpolitik (GAP) im Hinblick auf Ressourcenschutz und Nachhaltigkeit. In Rentenbank (ed.): Die Gemeinsame Agrarpolitik (GAP) der Europäischen Union nach 2013. Schriftenreihe der Rentenbank 27. Rentenbank, Frankfurt am Main: 141–177.

Björk, S. & Granéli, W. (1978): Energy, reeds and the environment. Ambio 7: 150–156.

Blab, J. (ed.) (1994): Grundlagen des Biotopschutzes für Tiere. 4th edition. Schriftenreihe Landschaftspflege und Naturschutz 24. Kilda, Greven. 479 p.

Blain, D., Murdiyarso, D., Couwenberg, J., Nagata, O., Renou-Wilson, F., Sirin, A.A., Strack, M., Tuittila, E.-S. & Wilson, D. (2014): Rewetted organic soils. In: Hirashi, T., Krug, T., Tanabe, K., Srivastava, N., Baasansuren, J., Fukuda, M. & Troxler, T.G. (eds.): 2013 Supplement to the 2006 IPCC Guidelines for National Greenhouse Gas Inventories: Wetlands. Intergovernmental Panel on Climate Change (IPCC), Geneva, Switzerland: 3.1–3.42.

Blicher-Mathiesen, G. & Hoffmann, C.C. (1999): Denitrification as a sink for dissolved nitrous oxide in a freshwater riparian fen. Journal of Environment Quality 28: 257–262.

Blievernicht, A., Irrgang, S., Zander, M. & Ulrichs, C. (2011): Produktion von Torfmoosen (*Sphagnum sp.*) als Torfersatz im Erwerbsgartenbau. Gesunde Pflanzen 62: 125–131.

Böcher, M. (2009): Faktoren für den Erfolg einer nachhaltigen und integrierten ländlichen Regionalentwicklung. In: Friedel, R. & Spindler, E.A. (eds.): Nachhaltige Entwicklung ländlicher Räume. Chancenverbesserung durch Innovation und Traditionspflege. VS research. Verlag für Sozialwissenschaften, Wiesbaden: 127–138.

Bockholt, R. (2001): Futterwert und Siliereignung der häufigsten autochthonen Pflanzenarten des Niedermoorgrünlandes. Archives of Agronomy and Soil Science 47: 183–199.

Bockholt, R. & Buske, F. (1997): Variationsbreite des Futterwertes von Niedermoorgrünland unter Berücksichtigung der häufigsten autochthonen Pflanzen. Das wirtschaftseigene Futter 43: 5–20.

Boeckx, P., van Cleemput, O. & Villaralvo, I. (1997): Methane oxidation in soils with different textures and land use. Nutrient Cycling in Agroecosystems 49: 91–95.

Böhm, H. & Siegert, F. (2001): Ecological impact of the one million hectare rice project in Central Kalimantan, Indonesia, using remote sensing and GIS. In: Centre for Remote Imaging, Sensing and Processing (CRISP) (ed.): Proceedings of the 22nd Asian Conference on Remote Sensing, Singapore, 05–09 November 2001. Centre for Remote Imaging, Sensing and Processing (CRISP). National University of Singapore, Singapore: 1–6.

Bonneville, M.-C., Strachan, I.B., Humphreys, E.R. & Roulet, N.T. (2008): Net ecosystem CO_2 exchange in a temperate cattail marsh in relation to biophysical properties. Agricultural and Forest Management 148: 69–81.

Bork (2013): Personal communication.

Bortfeldt, A., Homberger, J., Kopfer, H., Pankratz, G. & Strangmeier, R. (2008): Intelligent decision support. Current challenges and approaches. Intelligente Entscheidungsunterstützung. Aktuelle Herausforderungen und Lösungsansätze. Gabler Edition Wissenschaft. Gabler, Wiesbaden. 521 p.

Boström, M. & Klintman, M. (2008): Eco-standards, product labelling and green consumerism. Palgrave Macmillan, New York. 247 p.

Bowen, M., Bompard, J., Anderson, I., Guizol P. & Gouyon A. (2001): Anthropogenic fires in Indonesia: A view from Sumatra. In: Eaton, P. & Radojevic, M. (eds.): Forest fires and regional haze in Southeast Asia. Nova Science Publishers, Huntington, N.Y.

Boyd, J. & Banzhaf, S. (2007): What are ecosystem services? The need for standardized environmental accounting units. Ecological Economics 63: 616–626.

Brandhuber, R., Demmel, M., Koch, H.-J. & Brunotte, J. (2008): DLG Merkblatt 344. Bodenschonender Einsatz von Landmaschinen: Empfehlungen für die Praxis. Deutsche Landwirtschaftsgesellschaft (DLG) & Bayrische Landesanstalt für Landwirtschaft (LfL), Frankfurt am Main. 18 p. http://www.lfl.bayern.de/mam/cms07/publikationen/daten/informationen/p_32450.pdf, last access 24 June 2014.

Brandstaka, T., Helenius, J., Hovi, J., Kivelä, J., Koppelmäki, K., Simojoki, A., Soinne, H. & Tammeorg, P. (2010): Biochar filter: use of biochar in agriculture as soil conditioner. Report for BSAS Comittment 2010. Baltic Sea Action Summit, Helsinki. 22 p.

Briemle, G. & Ellenberg, H. (1994): Zur Mahdverträglichkeit von Grünlandpflanzen. Möglichkeiten der praktischen Anwendung von Zeigerwerten. Natur und Landschaft 69: 139–147.

Brix, H., Ye, S., Laws, E.A., Sun, D., Li, G., Ding, X., Yuan, H., Zhao, G., Wang, J. & Pei, S. (2014): Large-scale management of Common Reed, *Phragmites australis*, for paper production. A case study from the Liaohe Delta, China. Ecological Engineering 73: 760–769.

Broads Authority (2004): Purpose and Use of a supplement to the Fen Management Strategy. Broads Authority. Broads Authority, Norwich (GB). 41 p. http://www.broads-authority.gov.uk/__data/assets/pdf_file/0011/424838/Supplement_to_Fen_Management_Strategy.pdf, zuletzt geprüft am 06.09.2015.

Brooks, S. & Stoneman, R.E. (1997): Conserving bogs. The management handbook. Stationery Office, Edinburgh. 286 p.

Brünig, E. (1990): Oligotrophic forested wetlands in Borneo. In: Lugo, A.E., Brown, S. & Brinson, M.M. (eds.): Forested wetlands. Ecosystems of the world 15. Elsevier, Amsterdam, New York: 299–334.

Brünig, E. (1996): Conservation and management of tropical rainforest. An integrated approach to sustainability. CAB International, Wallingford. 339p.

Bubier, J.L. (1995): The relationship of vegetation to methane emission and hydrochemical gradients in northern peatlands. The Journal of Ecology 83: 403–420.

Bund Ökologische Lebensmittelwirtschaft e.V. (BÖLW) (2012) Zahlen, Daten, Fakten – Die Bio-Branche 2012. BÖLW, Berlin. 19 p.

Bundesinstitut für Bau-, Stadt- und Raumforschung (BBSR) (ed.) (2012): Raumordnungsbericht 2011, Bonn. 254 p.

Bundesministerium für Ernährung, Landwirtschaft und Verbraucherschutz (BMELV) (ed.) (2009a): Aktionsplan der Bundesregierung zur stofflichen Nutzung nachwachsender Rohstoffe. BMELV, Bonn. 40 p.

Bundesministerium für Ernährung, Landwirtschaft und Verbraucherschutz (BMELV) (ed.) (2009b: Handlungskonzept der Bundesregierung zur Weiterentwicklung der ländlichen Räume. BMELV, Bonn. 29 p.

Bundesministerium für Ernährung, Landwirtschaft und Verbraucherschutz (BMELV) (ed.) (2009c): Statistisches Jahrbuch über Ernährung, Landwirtschaft und Forsten. Wirtschaftsverlag NW GmbH, Bremerhaven. 645 p.

Bundesministerium für Ernährung, Landwirtschaft und Verbraucherschutz (BMELV) (ed.) (2011): Die wirtschaftliche Lage der landwirtschaftlichen Betriebe. Buchführungsergebnisse der Testbetriebe. BMELV, Bonn. 21 p.

Bundesministerium für Ernährung, Landwirtschaft und Verbraucherschutz (BMELV) (ed.) (2012): Fortschrittsbericht der Bundesregierung zur Entwicklung ländlicher Räume. BMELV, Bonn. 24 p.

Bundesministerium für Umwelt, Naturschutz und Reaktorsicherheit (BMU) (ed.) (2012): Erneuerbare Energien in Zahlen. Nationale und Internationale Entwicklung. BMU, Berlin. 80 p.

Burgstaler, J. & Wiedow, D. (2012): Neue Verfahren der P-Rückgewinnung aus Nebenprodukten der Landwirtschaft, Rest- und Abfallstoffen. Vortrag. Phosphor-Leibnitz-Campus. Universität Rostock, Rostock, 2012.

Burvall, J. (1997): Influence of harvest time and soil type on fuel quality in reed canary grass (*Phalaris arundinacea* L.). Biomass and Bioenergy 12: 149–154.

Burvall, J. & Hedman, B. (1998): Perennial rhizomatous grass. The delayed harvest system improves fuel characteristics for reed canary grass. In: El Bassam, N., Behl, R.K. & Prochnow, B. (eds.): Sustainable agriculture for food, energy and industry. Strategies towards achievement. James & James Science Publishers, London: 916–918.

Cattin, M.-F., Blandenier, G., Banašek-Richter, C. & Bersier, L.-F. (2003): The impact of mowing as a management strategy for wet meadows on spider (Araneae) communities. Biological Conservation 113: 179–188.

Central Kalimantan Peatland Project (CKPP) (ed.) (2008) Provisional Report of the Central Kalimantan Peatland Project. CKPP Consortium, Palangka Raya. 70 p.

Charman, D.J., Beilman, D.W., Blaauw, M., Booth, R.K., Brewer, S., Chambers, F.M., Christen, J.A., Gallego-Sala, A., Harrison, S.P., Hughes, P., Jackson, S.T., Korhola, A., Mauquoy, D., Mitchell, F., Prentice, I., van der Linden, M., Vleeschouwer, F. de, Yu, Z.C., Alm, J.P., Bauer, I.E., Corish, Y., Garneau, M., Hohl, V., Huang, Y., Karofeld, E., Le Roux, G., Loisel, J., Moschen, R., Nichols, J.E., Nieminen, T.M., MacDonald, G.M., Phadtare, N.R., Rausch, N., Sillasoo, Ü., Swindles, G.T., Tuittila, E.-S., Ukonmaanaho, L., Väliranta, M., van Bellen, S., van Geel, B., Vitt, D.H. & Zhao, Y. (2013): Climate-related changes in peatland carbon accumulation during the last millenium. BioSciences 10: 929–944.

Chistotin, M.V., Sirin, A.A. & Dulov, L.E. (2006): Seasonal dynamics of carbon dioxide and methane emission from peatland of Moscow region drained for peat extraction and agricultural use. Agrochemistry 6: 54–62.

Chivu, I. (1963): The use of reeds as raw material for pulp and paper. In: UNESCO Middle East Science Cooperation Office (ed.): Proceedings of the UNESCO/FAO Regional Symposium on Pulp and Paper Research and Technology in the Middle East and North Africa, Cairo: 227–256.

Chivu, I. (1968a): Practical experiment in the cropping of reeds for the manufacture of pulp and paper – economic results. In: Food and Agriculture Organization of the United Nations (FAO) (ed.): Pulp and paper development in Africa and the Near East. FAO, Rome: 877–899.

Chivu, I. (1968b): The use of reeds as raw material for pulp and paper production. In: Practical experiment in the cropping of reeds for the manufacture of pulp and paper – Economic results. In: Food and Agriculture Organization of the United Nations (FAO) (ed.): Pulp and paper development in Africa and the Near East. FAO, Rome: 227–289.

Christian, D.G., Yates, N.E. & Riche, A.B. (2006): The effect of harvest date on the yield and mineral content of *Phalaris arundinacea* L. (Reed Canary Grass) genotypes screened for their potential as energy crops in southern England. Journal of the Science of Food and Agriculture 86: 1181–1188.

Christoph, K.-H., Gläß, K., Kachelmaier, R., Lebek, G., Oehler, E. & Supranowitz, S. (1973): Kommentar zum Gesetz über die planmäßige Gestaltung der sozialistischen Landeskultur in der Deutschen Demokratischen Republik vom 14. Mai 1970. Landeskulturgesetz. Staatsverlag der DDR, Berlin.

Cicek, N., Lambert, S., Venema, H.D., Snelgrove, K.R., Bibeau, E.L. & Grosshans, R. (2006): Nutrient removal and bio-energy production from Netley-Libau Marsh at Lake Winnipeg through annual biomass harvesting. Biomass and Bioenergy 30: 529–536.

Clymo, R.S. (1983): Peat. In: Gore, A. (ed.): Mires: Swamp, bog, fen and moor. Regional studies. Ecosystems of the world 4A. Elsevier Scientific, Amsterdam, New York: 159–224.

Coase, R.H. (1960): The problem of social cost. Journal of Law and Economics 3: 1–44.

Cochrane, M.A. (2009): Fire in the Tropics. In: Cochrane, M.A. (ed.): Tropical fire ecology. Climate change, land use, and ecosystem dynamics. Springer-Praxis books in environmental sciences. Springer, Berlin, Heidelberg: 1–23.

Cong, N., Pfriem, A. & Wagenfuhr, A. (2006): Alternatives Verfahren zur Zerfaserung von Einjahrespflanzen für klein- und mittelständische Unternehmen, Teil 1: Vorstellung des Zerfaserungsprozesses. Holztechnologie 4: 11–16.

Conwentz, H. (ed.) (1916): Bericht über die siebente Konferenz für Naturdenkmalpflege in Preußen, Berlin, am 3. und 4. Dezember 1915. Beiträge zur Naturdenkmalpflege 5. Bornträger, Berlin. 356 p.

Coordinating Ministry for Economic Affairs (2011): Masterplan for acceleration and expansion of Indonesia economic development 2011–2025. Coordinating Ministry for Economic Affairs, Jakarta. 211 p.

Couwenberg, J. (2007): Biomass energy crops on peatlands: On emissions and perversions. IMCG Newsletter 2007/3: 12–14.

Couwenberg, J. & Fritz, C. (2012): Towards developing IPCC methane 'emission factors' for peatlands (organic soils). Mires & Peat 10, Art. 1: 1–17.

Couwenberg, J. & Hooijer, A. (2013): Towards robust subsidence-based soil carbon emission factors for peat soils in south-east Asia, with special reference to oil palm plantations. Mires & Peat 12: 1–13.

Couwenberg, J., Augustin, J., Michaelis, D., Wichtmann, W. & Joosten, H. (2008): Entwicklung von Grundsätzen für eine Bewertung von Niedermooren hinsichtlich ihrer Klimarelevanz. Studie im Auftrag des Ministerium für Landwirtschaft und Naturschutz Mecklenburg-Vorpommern. DUENE e.V., Greifswald. 33 p.

Couwenberg, J., Dommain, R. & Joosten, H. (2010): Greenhouse gas fluxes from tropical peatlands in south-east Asia. Global Change Biology 16: 1715–1732.

Couwenberg, J., Thiele, A., Tanneberger, F., Augustin, J., Bärisch, S., Dubovik, D., Liashchynskaya, N., Michaelis, D., Minke, M., Skuratovich, A. & Joosten, H. (2011): Assessing greenhouse gas emissions from peatlands using vegetation as a proxy. Hydrobiologia 674: 67–89.

Cowie, N.R., Sutherland, W.J., Ditlhogo, K.M. & James R. (1992): The effect of conservation management of reed bed. II. The flora and litter disappearance. Journal of Applied Ecology 29: 277–284.

Cox, P. (1996): Some issues in the design of agricultural decision support systems. Agricultural Systems 52: 355–381.

Cradle to Cradle Products Innovation Institute (C2C) (2011): Cradle to Cradle Products Program. Cradle to Cradle Products Innovation Institute (C2C). http://www.c2ccertified.org/, last access 26 April 2013.

Cris, R., Buckmaster, S., Bain, C. & Reed, M. (2014): Global Peatland Restoration demonstrating SUCCESS. IUCN UK National Committee Peatland Programme, Edinburgh. 35 p.

Daatselaar, C., Hoogendam, K. & Poppe, K.J. (2009): De economie van het veenrietweide bedrijf – Een quickscan voor West-Nederland. The economics of reed cultivation on peat soils – an overview for the Western Netherlands, No.09.2.218. Innovatie Netwerk, Utrecht.

Dahms, T. (2009): Bestandesetablierung. In: Wichmann, S. & Wichtmann, W. (eds.): Bericht zum Forschungs- und Entwicklungsprojekt Energiebiomasse aus Niedermooren (ENIM). Universität Greifswald, Greifswald: 117–122.

Davidson, E.A., Keller, M., Erickson, H.E., Verchot, L.V. & Veldkamp, E. (2000): Testing a conceptual model of soil emissions of nitrous and nitric oxides. BioScience 50: 667–680.

Davidsson, T.E., Trepel, M. & Schrautzer, J. (2002): Denitrification in drained and rewetted minerotrophic peat soils in Northern Germany (Pohnsdorfer Stauung). Journal of Plant Nutrition and Soil Science 165: 199–204.

Decleer, K. (1990): Experimental cutting of reedmarsh vegetation and its influence on the spider (Araneae) fauna in the blankaart nature reserve, Belgium. Biological Conservation 52: 161–185.

Demsetz, H. (1967): Toward a theory of property rights. The American Economic Review 57: 347–359.

Den Haan, E. J., Hooijer, A. & Erkens, G. (2012): Consolidation settlements of tropical peat domes by plantation development. Deltares, 1202415-041-Geo-0001, Delft. 44 p.

Deutsche Landwirtschaftsgesellschaft (DLG) (2000): Richtlinie für die Prüfung von Siliermitteln auf DLG-Gütezeichen-Fähigkeit. DLG Merkblatt. DLG, Frankfurt. 16 p.

Deutsche Landwirtschaftsgesellschaft (DLG) (2012): Biogas aus Gras. Wie Grünlandaufwüchse zur Energieerzeugung beitragen können. DLG Merkblatt 386. DLG, Frankfurt. 20 p.

Deutscher Bundestag (1996): Gesetzentwurf der Bundesregierung zur Neuregelung des Rechts des Naturschutzes und der Landespflege. Drucksache 13/6441: 51.

Deutscher Bundestag (1998b): Abschlussbericht der Enquete-Kommission ‚Schutz des Menschen und der Umwelt – Ziele und Rahmenbedingungen einer nachhaltig zukunftsverträglichen Entwicklung'. Drucksache 13/11200. Bundesanzeiger Verlagsgesellschaft mbH, Bonn. 252 p.

Deutscher Bundestag (1998a): Gesetzentwurf zur Neuregelung des Rechts des Naturschutzes und der Landschaftspflege. Drucksache 13/10186: 8.

Deutscher Bundestag (2001): Gesetzentwurf zur Neuregelung des Rechts des Naturschutzes und der Landschaftspflege. Drucksache 14/6378: 66.

Deutscher Bundestag (2009): Gesetzentwurf zur Neuregelung des Rechts des Naturschutzes und der Landschaftspflege. Drucksache 16/12274: 63;67.

Deutscher Verband für Wasserwirtschaft und Kulturbau (DVWK) (1996): Ermittlung der Verdunstung von Land- und Wasserflächen. Merkblatt/Arbeitsblatt. DWA Regelwerk, 238. Deutscher Verband für Wasserwirtschaft und Kulturbau (DVWK), Bonn. 240 p.

de Ventering, H., Pieterse, N.M., Belgers, J., Wassen, M.J. & Ruiter, P.C. de (2002): N, P, and K budgets along nutrient availability and productivity gradients in wetlands. Ecological Applications 12: 1010–1026.

devriescornjum (2015): http://www.devriesconjum.nl, last access 03 May 2015.

DIN 51624 (2008): Automotive fuels – Natural Gas – Requirements and test procedures, 01 February 2008.

Dias, A.T.C., Hoorens, B., Logtestijn, R.S.P., Vermaat, J.E. & Aerts, R. (2010): Plant species composition can be used as a proxy to predict methane emissions in peatland ecosystems after land-use changes. Ecosystems 13: 526–538.

Diefenbacher, H., Teichert, V. & Wilhelmy, S. (2005): Leitfaden Indikatoren im Rahmen einer Lokalen Agenda 21. FEST, Heidelberg. 80 p.

Dierks, A. & Fischer, K. (2009): Habitat requirements and niche selection of *Maculinea nausithous* and *M. teleius* (Lepidoptera: Lycaenidae) within a large sympatric metapopulation. Biodiversity and Conservation 18: 3663–3676.

Dierschke, H. & Briemle, G. (2002): Kulturgrasland. Wiesen, Weiden und verwandte Staudenfluren. Ökosysteme Mitteleuropas aus geobotanischer Sicht. Ulmer, Stuttgart. 239 p.

Dietze, M. & Heilmann, H.H. (2010): Die Wirtschaftlichkeit der Rindfleischproduktion in Mecklenburg-Vorpommern. Landesforschungsanstalt für Landwirtschaft und Fischerei Gülzow. 30 p.

Ding, W., Cai, Z. & Tsuruta, H. (2004): Diel variation in methane emissions from the stands of *Carex lasiocarpa* and *Deyeuxia angustifolia* in a cool temperate freshwater marsh. Atmospheric Environment 38: 181–188.

Ding, W., Cai, Z. & Tsuruta, H. (2005): Plant species effects on methane emissions from freshwater marshes. Atmospheric Environment 39: 3199–3207.

Dirlich, S. (2011): Integration der Bestandsqualität in die Zertifizierung von Gebäuden. Entwicklung eines ökonomisch-ökologischen Bewertungssystems für nachhaltiges Bauen unter besonderer Berücksichtigung von Bestandsbauten und traditionellen Bauweisen. IÖR-Schriften 55. Leibniz Institute of Ecological Urban and Regional Development, Dresden. 230 p.

Dommain, R., Couwenberg, J. & Joosten, H. (2010): Hydrological self-regulation of domed peat swamps in south-east Asia and consequences for conservation and restoration. Mires & Peat 6, Art. 5: 1–17.

Dommain, R., Couwenberg, J. & Joosten, H. (2011): Development and carbon sequestration of tropical peat domes in south-east Asia: links to post-glacial sea-level changes and Holocene climate variability. Quaternary Science Reviews 30: 999–1010.

Dommain, R., Barthelmes, A., Tanneberger, F., Bonn, A., Bain, C. & Joosten, H. (2012): Country-wise opportunities. In: Joosten, H., Tapio-Biström, M.-L. & Tol, S. (eds.): Peatlands – guidance for climate change mitigation through conservation, rehabilitation and sustainable use. FAO. Mitigation of climate change in agriculture 5, Rome: 45–84.

Dommain, R., Couwenberg, J., Glaser, P.H., Joosten, H. & Suryadiputra, I.N.N. (2014): Carbon storage and release in Indonesian peatlands since the last deglaciation. Quaternary Science Reviews 97: 1–32.

Dovel, R.L. (1996): Cutting height effects on wetland meadow forage yield and quality. Journal of Range Management 49: 151–156.

Drösler, M., Verchot, L.V., Freibauer, A., Pan, G., Evans, C.D., Bourbonniere, R.A., Alm, J.P., Page, S.E., Agus, F., Hergoualc'h, K., Couwenberg, J., Jauhiainen, J., Sabiham, S. & Wang, C. (2014): Drained inland organic soils. In: Hirashi, T., Krug, T., Tanabe, K., Srivastava, N., Baasansuren, J., Fukuda, M. & Troxler, T.G. (eds.): 2013 Supplement to the 2006 IPCC Guidelines for National Greenhouse Gas Inventories: Wetlands. Intergovernmental Panel on Climate Change (IPCC), Geneva, Switzerland: 2.1–2.79.

Duan, X., Wang, X. & Ouyang, Z. (2006): Plant-mediated CH_4 Emission from a Phragmites-dominated wetland in an arid region, China. Journal of Freshwater Ecology 21: 139–145.

Dubbe, D.R., Garver, E.G. & Pratt, D.C. (1988): Production of cattail (*Typha* spp.) biomass in Minnesota, USA. Biomass 17: 79–104.

Duncan, P. (1983): Determinants of the use of habitat by horses in a Mediterranean wetland. Journal of Animal Ecology 52: 93–109.

Dunker, M. (2011): Grundfutter 2011. Energie, mikrobielle Belastung und mehr. Futterbautagung. Landwirtschaftliche Untersuchungs- und Forschungsanstalt Rostock (LUFA), Retzin, 11 October 2011. http://www.landwirtschaft-mv.de/cms2/LFA_prod/LFA/content/de/Fachinformationen/Gruenland_und_Futterwirtschaft/Veranstaltungen_und_Jahresberichte/12Raminer_FBT_2011/Dunker_Grundfutterqualitt.pdf, last access 13 June 2014.

Dürr, H.-J., Petelkau, H. & Sommer, C. (1995): Literaturstudie Bodenverdichtung. Umweltbundesamt. UBA-Texte 55/95. 203 p.

Edom, F. (2001): Moorlandschaften aus hydrologischer Sicht. In: Succow, M. & Joosten, H. (eds.): Landschaftsökologische Moorkunde. 2nd edition. Schweizerbart, Stuttgart: 185–288.

Edom, F., Münch, A., Dittrich, I., Keßler, K. & Peters, R. (2010): Hydromorphological analysis and water balance modelling of ombro- and mesotrophic peatlands. Advances in Geosciences 27: 131–137.

EG (2000): Richtlinie 2000/60/EG des Europäischen Parlaments und des Rates vom 23.Oktober 2000 zur Schaffung eines Ordnungsrahmens für Maßnahmen der Gemeinschaft im Bereich der Wasserpolitik (ABl. L 327 vom 22.12.2000, S. 1). 83 p. http://www.bmu.de/fileadmin/bmu-import/files/pdfs/allgemein/application/pdf/wasserrichtlinie.pdf, last access 06 September 2015.

EG (2007): Richtlinie 2007/60/EG des Europäischen Parlaments und des Rates vom 23. Oktober 2007 über die Bewertung und das Management von Hochwasserrisiken. Amtsblatt der Europäischen Union L 288/27. 8 p. http://eur-lex.europa.eu/LexUriServ/LexUriServ.do?uri=OJ:L:2007:288:0027:0034:de:PDF, last access 06 September 2015.

EG (2008): Richtlinie 2008/56/EG des Europäischen Parlaments und des Rates vom 17. Juni 2008 zur Schaffung eines Ordnungsrahmens für Maßnahmen der Gemeinschaft im Bereich der Meeresumwelt (Meeresstrategie-Rahmenrichtlinie). http://eur-lex.europa.eu/LexUriServ/LexUriServ.do?uri=OJ:L:2008:164:0019:0040:de:PDF, last access 06 September 2015.

Egan, P.A. & van der Kooy, F. (2013): Phytochemistry of the carnivorous sundew genus Drosera (Droseraceae) – future perspectives and ethnopharmacological relevance. Chemistry and Biodiversity 10: 1774–1790.

Eggelsmann, R. (1981): Dränanleitung für Landbau, Ingenieurbau und Landschaftsbau. 2nd edition. Paul Parey, Hamburg, Berlin. 265 p.

Eggelsmann, R. (1990a): Moor und Wasser. In: Göttlich, K. (ed.): Moor- und Torfkunde. 3rd edition. Schweizerbart, Stuttgart: 288–320.

Eggelsmann, R. (1990b): Mikroklima der Moore. In: Göttlich, K. (ed.): Moor- und Torfkunde. 3rd edition. Schweizerbart, Stuttgart: 374–384.

El Bassam, N. (2010): Handbook of bioenergy crops. A complete reference to species, development and applications. Earthscan, London, Washington, D.C. 516 p.

Ellenberg, H. (1992): Zeigerwerte von Pflanzen in Mitteleuropa. 2nd edition. Scripta geobotanica 18. Goltze, Göttingen. 258 p.

Elsgaard, L., Görres, C.-M., Hoffmann, C.C., Blicher-Mathiesen, G., Schelde, K. & Petersen, S.O. (2012): Net ecosystem exchange of CO_2 and carbon balance for eight temperate organic soils under agricultural management. Agriculture, Ecosystems & Environment 162: 52–67.

Emmel, M. (2008): Growing ornamental plants in Sphagnum biomass. Acta Horticulturae 779: 173–178.

Endres, A. & Holm-Müller, K. (1997): Die Bewertung von Umweltschäden. Theorie und Praxis sozioökonomischer Verfahren. Kohlhammer, Stuttgart. 209 p.

Endres, E. (2011): §30 Rn. 8. In: Frenz, W. & Müggenborg, H.-J. (eds.): Kommentar zum Bundesnaturschutzgesetz. Erich Schmidt, Berlin.

Enge, D. (2009): Landschaftspflege mit Wasserbüffeln. Naturschutz und Landschaftsplanung 41: 277–285.

Engel, J. (2007): 100 Jahre Naturschutzgebiet Plagefenn. Landesforstanstalt Eberswalde. Eberswalde Forstliche Schriftreihe, XXXI. Ministerium für Ländliche Entwicklung, Umwelt und Verbraucherschutz des Landes Brandenburg (MLUV), Eberswalde. 134 p.

Euroconsult Mott MacDonald/Deltares/Delft Hydraulics (2008): Master Plan for the Rehabilitation and Revitalisation of the Ex-Mega Rice Project Area in Central Kalimantan. Main synthesis report. Government of Indonesia & Royal Netherlands Embassy. Euroconsult Mott MacDonald / Deltares / Delft Hydraulics, DHV, Wageningen UR, Witteveen+Bos, PT MLD & PT INDEC, Jakarta.

European Commission (EC) (2008): Commission Recommendation of 10 April 2008 on the management of intellectual property in knowledge transfer activities and Code of Practice for universities and other public research organisations. Official Journal of the European Union, Brussels..

European Commission (EC) (2011a): Our life insurance, our natural capital: an EU biodiversity strategy to 2020. COM (2011) 244 final. Brussels.

European Commission (EC) (2011b): Commission Regulation (EU) No 1006/2011 of 27 September 2011 amending Annex I to Council Regulation (EEC) No 2658/87 on the tariff and statistical nomenclature and on the Common Customs Tariff. No 1006/2011, of 27 September 2011.

European Court of Auditors (ECA) (ed.) (2009): Is cross compliance an effective policy? Special report No 8/2008, Luxembourg.

European Court of Auditors (ECA) (ed.) (2011a): Single Payment Scheme (SPS): issues to be addressed to improve its sound financial management. Special report No 5/2011, Luxembourg.

European Court of Auditors (ECA) (ed.) (2011b): On whether agri-environment support policy is well designed and managed. Special report No 7/2011, Luxembourg.

European Parliament (ed.) (2013): Erschließung des EP vom 12. März 2013 über Anrechnungsvorschriften und Aktionspläne für Emission und den Abbau von Treibhausgasen infolge von Tätigkeiten im Sektor Landnutzung, Büssel.

European Union (EU) (2008): Council Regulation (EEC) N° 834/2007 of 28 June 2007 on organic production and labelling of organic products and repealing Regulation (EEC) No 2092/91.

European Union (EU) (2009): Directive 2009/28/EC of the European Parliament and of the Council of 23 April 2009 on the promotion of the use of energy from renewable sources and amending and subsequently repealing Directives 2001/77/EC and 2003/30/EC. Official Journal of the European Union: 16–62.

European Union (EU) (2013): Regulation (EU) No 1305/2013 of the European Parliament and of the Council of 17 December 2013 on support for rural development by the European Agricultural Fund for Rural Development (EAFRD) and repealing Council Regulation (EC) No 1698/2005.

Ewers, H.-J. & Schulz, W. (1982): Die monetären Nutzen gewässergüteverbessernder Maßnahmen. Dargestellt am Beispiel des Tegeler Sees in Berlin. Pilotstudie zur Bewertung des Nutzens umweltverbessernder Massnahmen. Umweltforschungsplan des Bundesministeriums des Innern, Umweltplanung. Forschungsbericht 101 03 037/2 UBA-FB 82-041. Berichte / Umweltbundesamt 3/82. Schmidt, Berlin. 358 p.

EWG (1992): Richtlinie E92/43/EWG des Rates vom 21. Mai 1992 zur Erhaltung der natürlichen Lebensräume sowie der wildlebenden Tiere und Pflanzen (ABl.L206 vom 22.7.1992, S.7). 60 p. http://eur-lex.europa.eu/LexUriServ/LexUriServ.do?uri=CONSLEG:1992L0043:20070101:DE:PDF, last access 06 September 2015.

Fachagentur Nachwachsende Rohstoffe e.V. (FNR) (ed.) (2006): Marktanalyse Nachwachsende Rohstoffe. Studie für das Bundesministerium für Ernährung, Landwirtschaft und Verbraucherschutz. Fachagentur Nachwachsende Rohstoffe e. V. (FNR), Gülzow. 577 p.

Fachagentur Nachwachsende Rohstoffe e.V. (FNR) (ed.) (2007): Leitfaden Bioenergie. Planung und Wirtschaftlichkeit von Bioenergieanlagen. Fachagentur Nachwachsende Rohstoffe e. V. (FNR), Gülzow. 353 p.

Fährmann, B., Grajewski, R., Koch, B. & Peter, H. (2008): Die Politik zur ländlichen Entwicklung im Rahmen der Gemeinsamen Agrarpolitik. Schriftenreihe Europäisches Verwaltungsmanagement 7. Fachhochschule für Verwaltung und Rechtspflege Berlin, Berlin. 86 p.

Fansa, M. (ed.) (1999): Weder See noch Land. Moor – eine verlorene Landschaft. Schriftenreihe des Staatlichen Museums für Naturkunde und Vorgeschichte 2. Isensee, Oldenburg. 184 p.

Finnish FSC Association (ed.) (2010): Forest Stewardship Council (FSC) Standart for Finland. Finnish FSC Association. 66 p. http://ic.fsc.org/download.fsc-std-fin-01-1-2010-finland-natural-forests.136.htm, last access 12 August 2014.

Fischer, K., Busch, R., Fahl, G., Kunz, M. & Knopf, M. (2013): Habitat preferences and breeding success of Whinchats (*Saxicola rubetra*) in the Westerwald mountain range. Journal of Ornithology 154: 339–349.

Fischer, U. (1999): Zur Vegetationsentwicklung naturnaher Flußtalmoore am Beispiel des NSG "Peenewiesen bei Gützkow" (Mecklenburg-Vorpommern). Feddes Repertorium 110: 287–324.

Fischer-Hüftle, P. & Czybulka, D. (2010): §14 Rn. 57,58,65. In: Fischer-Hüftle, P. & Schumacher, J.: Kommentar zum Bundesnaturschutzgesetz. 2nd edition. W. Kohlhammer, Stuttgart.

Flade, M. (1994): Die Brutvogelgemeinschaften Mittel- und Norddeutschlands. IHW Verlag, Eching.

Flade, M. & Lachmann, L. (2008): Internatonal Species Action Plan for the aquatic Warbler Acrocephalus paludicola. Birdlife International. European Commission (EC), Cambridge, Brüssel. http://ec.europa.eu/environment/nature/conservation/wildbirds/action_plans/docs/acrocephalus_paludicola.pdf.

Flessa, H., Wild, U., Klemisch, M. & Pfadenhauer, J. (1998): Nitrous oxide and methane fluxes from organic soils under agriculture. European Journal of Soil Science 49: 327–335.

Foley, J., DeFries, R., Asner, G., Barford, C., Bonan, G., Carpenter, S., Chapin, S., Coe, M., Daily, G., Gibbs, H., Helkowski, J., Holloway, T., Howard, E., Kucharik, C., Monfreda, C., Patz, J., Prentice, I., Ramankutty, N. & Snyder, P. (2005): Global consequences of land use. Science 309: 570–574.

Food and Agriculture Organization of the United Nations (FAO) (2000): FAO/WFP Mission to assess the impact of cyclones and drought on the food supply situation in Madagascar. Special Report, FAO Global information and early warning system on food and agriculture, World Food Programme. http://www.fao.org/docrep/004/x7379e/x7379e00.htm#P297_23766.

Food and Agriculture Organization of the United Nations (FAO) (ed.) (2013): FAOSTAT (Food and Acriculture Organization of the United Nations – Statistics). FAO. http://faostat.fao.org., last access 13 June 2014.

Foreign Agricultural Service (FAS)/US Department of Agriculture (USDA) (ed.) (2011): Oilseeds. FAS/USDA. 38 p. http://www.fas.usda.gov/psdonline/circulars/oilseeds.pdf, last access 06 September 2015.

Foresight (2011):The Future of Food and Farming: Challenges and choices for global sustainability. Final Project Report. The Government Office for Science, London.

211 p. https://www.gov.uk/government/uploads/system/uploads/attachment_data/file/288329/11-546-future-of-food-and-farming-report.pdf, zuletzt geprüft am 06.09.2015.

Forster, P., Ramaswamy, V., Artaxo, P., Berntsen, T., Betts, R., Fahey, D.W., Haywood, J., Lean, J., Lowe, D.C., Myhre, G., Nganga, J., Prinn, R.G., Raga, G., Schulz, M. & van Dorland, R. (2007): Changes in atmospheric constituts and in radiative forcing. Chapter 2. In: Solomon, S., Qin, D., Manning, M., Chen, Z., Marquis, M., Averyt, K.B., Tignor, M. & Miller, H.L. (eds.): Climate change 2007. The physical science basis. Contribution of Working Group I. Cambridge University Press, Cambridge.

Förster, A. (1998): Paul Holz. Zeichner. Hinstorff, Rostock. 326 p.

Forstner, B., Deblitz, C., Kleinhanß, W., Nieberg, H., Offermann, F., Röder, N., Salomon, P., Sanders, J. & Weingarten, P. (2012): Analyse der Vorschläge der EU Komission vom 12.Oktober 2011 zur künftigen Gestaltung der Direktzahlungen i Rahmen der GAP nach 2013. Arbeitsberichte aus der vTI-Agrarökonomie 04/2012, Braunschweig.

Förtsch, G. & Meinholz, H. (2011): Handbuch betriebliches Umweltmanagement. Vieweg+Teubner, Wiesbaden. 587 p.

Freeman, A.M. III (2007): The measurement of environmental and resource values. Theory and methods. 2nd edition. Resources for the Future, Washington, D.C. 491 p.

Frenz, W. (2004): Landwirtschaftlicher Bodenschutz und Agrarreform. Natur und Recht 26: 642–649.

Frick, A., Steffenhagen, P., Zerbe, S., Timmermann, T. & Schulz, K. (2011): Monitoring of the vegetation composition in rewetted peatland with iterative decision tree classification of satellite imagery. Photogrammetrie – Fernerkundung – Geoinformation 2011: 109–122.

Fricke, K., Bahr, T., Thiel, T. & Kugelstadt, O. (2009): Stoffliche oder energetische Verwertung. Ressourceneffizientes Handeln in der Abfallwirtschaft. Seminare. Graßner, Groth, Siederer & Coll, Berlin. 22 p. http://www.ggsc-seminare.de/pdf/Fricke-Ressourceneffizientes-Handeln-in-der-Abfallwirtschaft.pdf, last access 13 June 2014.

Friedel, R. & Spindler, E.A. (eds.) (2009): Nachhaltige Entwicklung ländlicher Räume. Chancenverbesserung durch Innovation und Traditionspflege. VS research. Verlag für Sozialwissenschaften, Wiesbaden. 503 p.

Fritsch, M., Wein, T. & Ewers, H.-J. (2007): Marktversagen und Wirtschaftspolitik. Mikroöko-nomische Grundlagen staatlichen Handelns. 7th edition. Vahlen, München. 451 p.

Fritsch, M., Henning, T., Slavtchev, V. & Steigenberger, N. (2008): Hochschulen als regionaler Entwicklungsmotor? Arbeitspapier 158. Hans Böckler Stiftung, Düsseldorf. 44 p.

Fritsch, M., Wein, T. & Ewers, H.-J. (2010): Marktversagen und Wirtschaftspolitik. Mikroökonomische Grundlagen staatlichen Handelns. 8th edition. Vahlen, München. 449 p.

Frohn, H.-W. (2012): Von der "Urnatur" zum Ökosystemdienstleister, Moorschutz am Beispiel der Esterweger Dose von 1900–2005. Natur und Landschaft 87: 24–28.

Frolking, S. & Roulet, N.T. (2007): Holocene radiative forcing impact of northern peatland carbon accumulation and methane emissions. Global Change Biology Bioenergy 13: 1079–1088.

Frolking, S., Roulet, N.T. & Fuglestvedt, J. (2006): How northern peatlands influence the Earth's radiative budget. Sustained methane emission versus sustained carbon sequestration. Journal of Geophysical Research 111: 1–10.

Frolking, S., Talbot, J., Jones, M.C., Treat, C.C., Kauffman, J.B., Tuittila, E.-S. & Roulet, N.T. (2011): Peatlands in the Earth's 21st century climate system. Environmental Reviews 18: 371–396.

FSC Working Group Germany (2012): German FSC-Standard and Small Forest Standard, Freiburg. 73 p. https://ic.fsc.org/download.fsc-std-deu-02-2012-german-natural-and-plantations.a-1742.pdf, last access 06 September 2015

Fukarek, F. (1969): Ein Beitrag zur potentiellen natürlichen Vegetation in Mecklenburg. Mitteilungen der floristisch-sozialistischen Arbeitsgemeinschaft N.F.: 231–237.

Furubotn, E.G. & Pejovich, S. (eds.) (1974): The economics of property rights. Ballinger Pub. Co, Cambridge, Massachusetts. 367 p.

Gade, R., Fischer, M., Kaiser, R., Rabe, O., Grage, A., Knaack, J., Petri, G., Trepel, M., Brockmann, U., Beusekom, J. von & Kuhn, U. (2011): Konzept zur Ableitung von Nährstoffenreduzierungszielen in den Flussgebieten Ems, Weser, Elbe, Eider aufgrund von Anforderungen an den ökologischen Zustand der Küstengewässer gemäß Wasserrahmrichtlinie. Bund Länder Messprogramm (BLMP). Ad-hoc-AG Nährstoffreduzierung des BLMP. 50 p.

Gaisler, J., Hejcman, M. & Pavlů, V. (2004): Effect of different mulching and cutting regimes on the vegetation of upland meadow. Plant, Soil and Environment 50: 324–331.

Galambosi, B. & Jokela, K. (2002): Uhanalaisten lääkekasvien markkinat ja viljely: Kirjallisuusselvitys. Maa- ja elintarviketalous, 17. 88 p. http://urn.fi/URN:ISBN:951-729-712-2.

Gańko, E., Kopczyński, P., Wnęk, K. & Wróbel, A. (2008): Studium alternatywnego i efektywnego kosztowo zagospodarowania biomasy z późnego koszenia roślinności uzyskanej w wyniku zarządzania podmokłymi łąkami dla celów ochrony przyrody w Polsce. Project report, Warsaw. 68 p.

Gaudig, G. & Wichmann, S. (2011): Sphagnum as a renewable resource. In: Tanneberger, F. & Wichtmann, W. (eds.): Carbon credits from peatland rewetting. Climate – biodiversity – land use. Schweizerbart, Stuttgart: 109.

Gaudig, G., Fengler, F., Krebs, M., Prager, A., Schulz, J., Wichmann, S. & Joosten, H. (2014): Sphagnum farming in Germany. A review of progress. Mires & Peat 13, Art. 8: 1–11.

Geber, U. (2002): Cutting frequency and stubble height of reed canary grass (Phalaris arundinacea L.): influence on quality and quantity of biomass for biogas production. Grass and Forage Science 57: 389–394.

Geiger, B., Brückl, O., Tzscheutschler, T., Hardi, M. & Roth, H. (2004): CO_2-Vermeidungskosten im Kraftwerksbereich, bei den erneuerbaren Energien sowie bei nachfrageseitigen Energieeffizienzmassnahmen. IfE-Schriftenreihe / hrsg. von: Lehrstuhl für Energiewirtschaft und Anwendungstechnik, Technische Universität München 47. Energie & Management, Herrsching. 157 p.

Geinitz, E. (1916): Gegen übertriebene Melioration. In: Heimatbund Mecklenburg (ed.): Mecklenburg. Zeitschrift des Heimatbundes Mecklenburg 1, Schwerin: 5–8.

Gellermann, M. (2007): Die "kleine Novelle" des Bundesnaturschutzgesetzes. Natur und Recht 29: 783–789.

Gellermann, M. (2012): §44 Rn. 15. In: Landmann, R. & Rohmer, G.: Umweltrecht. Umweltrecht 4. C.H.Beck, München.

George, W., Bonow, M., Hoppenbrock, C. & Moser, P. (2009): Regionale Energieversorgung. Chance für eine zukunftsfähige Ziel- und Ressourcensteuerung in der Energiewirtschaft. Zeitschrift für angewandte Geographie 33: 13–21.

Georgoudis, A.G., Papanastasis, V.P. & Boyazoglu, J. (1999): Use of Water Buffalo for environmental conservation of waterland. Asian-Australian Journal of Animal Science 12: 1324–1331.

Gerstmeier, R. & Lang, C. (1996): Beitrag zu Auswirkungen der Mahd auf Arthropoden. Zeitschrift für Ökologie und Naturschutz 5: 1–14.

Gesellschaft zur Qualitätssicherung Reet (QSR) (2008): Reet als Dacheindeckungsmaterial. Qualitätssicherung und -erhaltung eines Baustoffs aus nachwachsenden Rohstoffen. Bericht zum Forschungsvorhaben der Deutschen Bundesstiftung Umwelt, Aktenzeichen: 25018-25. Gesellschaft zur Qualitätssicherung Reet (QSR). 182 p.

Giesen, W. (2013): Paludiculture: sustainable alternatives on degraded peat land in Indonesia. Quick Assessment and Nationwide Screening of Peat and Lowland Resources and Action Planning for the Implementation of a National Lowland Strategy. Euroconsult Mott MacDonald and Deltares, Indonesia.

Giesen, W. & van der Meer, P. (2009): Guidelines for the Rehabilitation of Degraded Peat Swamp Forest in Central Kalimantan. Technical Guideline Number 5. Master Plan for the Rehabilitation and Revitalisation of the Ex-Mega Rice Project Area in Central Kalimantan. Delft Hydraulics in association with DHV, Wageningen University & Research, Witteven+Bos Indonesia, PT.MLD & PT.Indec. Government of Indonesia & Royal Netherlands Embassy. Euroconsult Mott MacDonald and Deltares, Jakarta.

Gilbert, G., Tyler, G.A., Dunn, C.J. & Smith, K.W. (2005): Nesting habitat selection by bitterns *Botaurus stellaris* in Britain and the implications for wetland management. Biological Conservation 124: 547–553.

Gilsenbach, R. (1971): Wasser. Probleme, Projekte, Perspektiven. Urania, Leipzig, Jena, Berlin. 399 p.

Glatzel, S., Forbrich, I., Krüger, C., Lemke, S. & Gerold, G. (2008): Small scale controls of greenhouse gas release under elevated N deposition rates in a restoring peat bog in NW Germany. Biogeosciences 5: 925–935.

Glatzel, S., Koebsch, F., Beetz, S., Hahn, J., Richter, P. & Jurasinski, G. (2011): Maßnahmen zur Minderung der Treibhausgasfreisetzung aus Mooren im Mittleren Mecklenburg. Telma Beiheft: 85–106.

Göbel, F. (2000): Die Rimpau'sche Moordammkultur – Untersuchungen ausgewählter Standorte aus landeskultureller Sicht. Diploma thesis. Humboldt-Universität, Berlin. Geographisches Institut. 135 p.

Goldammer, J.G. (2010): Preliminary assessment of the fire situation in Western Russia. Analysis of 15 August 2010, presented at the State Duma, Moscow, 23 September 2010. International Forest Fire News 40: 20–42.

Golze, M. (2004): Ergebnisse der Büffelhaltung in Deutschland (Teil 1). 4. Deutscher Büffeltag, Chursdorf. 21 p.

Gore, A. (ed.) (1983): Mires: Swamp, bog, fen and moor. Regional studies. Ecosystems of the world 4A. Elsevier Scientific, Amsterdam, New York. 440 p.

Gorham, E. (1991): Northern peatlands: Role in the carbon cycle and probable responses to climatic warming. Ecological Applications 1: 182–195.

Görn, S., Dobner, B., Suchanek, A. & Fischer, K. (2014): Assessing human impact on fen biodiversity. Effects of different management regimes on butterfly, grasshopper, and carabid beetle assemblages. Biodiversity and Conservation 23: 309–326.

Grace, J.B. & Wetzel, R.G. (1981): Habitat partitioning and competitive displacement in cattails (*Typha*): Experimental field studies. The American Naturalist 118: 463–474.

Graebner, P., Medlewska, E. & Zinz, A. (1919): Typha als Nutzpflanze. Angewandte Botanik, Zeitschrift für Erforschung der Nutzpflanzen: 30–48, 98–103.

Grajewski, R., Bathke, M., Bergschmidt, A., Bormann, K., Eberhardt, W., Ebers, H., Fährmann, B., Fengler, B., Fitschen-Lischewski, A., Forstner, B., Kleinhanß, W., Nitsch, H., Osterburg, B., Plankl, R., Raue, P., Reiter, K., Röder, N., Sander, A., Schmidt, T., Tietz, A. & Weingarten, P. (2011): Ländliche Entwicklungspolitik ab 2014. Eine Bewertung der Verordnungsvorschläge der Europäischen Komission vom Oktober 2011. Arbeitsberichte aus der vTI-Agrarökonomie 08/2011, Braunschweig.

Granéli, W. (1984): Reed *Phragmites australis* (Cav.) Trin. ex Steudel as an energy source in Sweden. Biomass 4: 183–208.

Grantzau, E. & Gaudig, G. (2005): Torfmoos als Alternative. TASPO Magazin: 8–10.

Graveland, J. (1999): Effects of reed cutting on density and breeding success of reed warbler *Acrocephalus scirpaceus* and sedge warbler *Acrocephalus schoenobaenus*. Journal of Avian Biology 30: 469–482.

Gray, K.A., Zhao, L. & Emptage, M. (2006): Bioethanol. Current Opinion in Chemical Biology 10: 141–146.

Grootjans, A.P., Schipper, P.C. & van der Windt, H.J. (1985): Influence of drainage on N-mineralization and vegetation response in wet meadows. I: *Calthion palustris* stands. Oecol. Plant. 6: 403–417.

Grosser, D. (2004): Das Holz der Schwarzerle – Eigenschaften und Verwendung. LWF-Wissen 42: 51–55.

Grosshans, R. (2013): Ertrag Typha in Kanada. Email, Greifswald, 15 April 2013.

Grosshans, R. (2014): Cattail (*Typha* spp.) biomass harvesting for nutrient capture and sustainable bioenergy for integrated watershed management. PhD thesis. University of Manitoba, Winnipeg, Manitoba, Canada.

Grosshans, R., Venema, H.D., Cicek, N. & Goldsborough, G. (2011a): Cattail farming for water quality: Harvesting cattails for nutrient removal and phosphorous recovery in the watershed. In: WEF-IWA (ed.): Proceedings of the WEF-IWA Nutrient Recovery and Management 2011 Conference: inside and outside the fence, Miama, Florida, USA, 09.–12. Januar 2011. Water Environment Federation-International Water Association (WEF-IWA).

Grosshans, R., Zubrycki, K., Hope, A., Roy, D. & Venema, H.D. (2011b): Netley-Libau Nutrient-Bioenergy Project Winnipeg. International Institute for Sustainable

Development (IISD). International Institute for Sustainable Development (IISD), Winnipeg. 6 p. www.iisd.org/pdf/2011/brochure_iisd_wic_netley_libau_2011.pdf.

Grosvernier, P. & Staubli, P. (2009): Regeneration von Hochmooren. Grundlagen und technische Massnahmen. Bundesamt für Umwelt (BAFU). Vollzug Umwelt VU, UV-0918-D. Bundesamt für Umwelt (BAFU), Bern.

Grüner Strom Label e.V. (ed.) (2013): Grünes Gas Label. Grüner Strom Label e.V., Bonn. http://www.gruenerstromlabel.de/gruenes-gas-label/, last access 12 August 2014.

Grünfeld, S. & Brix, H. (1999): Methanogenesis and methane emissions: effects of water table, substrate type and presence of *Phragmites australis*. Aquatic Botany 64: 63–75.

Gryseels M. (1989): Nature management experiments in a derelict reedmarsh. II: effects of summer mowing. Biological Conservation: 85–99.

Grzelak, M., Waliszewska, B., Sieradzka, A. & Speak-Dz´wigała, A. (2011): Ecological meadow communities with participation of species from sedge (*Carex*) family. Journal of Research and Applications in Agricultural Engineering 56: 122–126.

Günther, A.B., Huth, V., Jurasinski, G. & Glatzel, S. (2014): Scale-dependent temporal variation in determining the methane balance of a temperate fen. Greenhouse Gas Measurement and Management 4: 41–48.

Günther, A.B., Huth, V., Jurasinski, G. & Glatzel, S. (2015): The effect of biomass harvesting on greenhouse gas emissions from a rewetted temperate fen. GCB Bioenergy 7: 1092–1106.

Güsewell, S., Le Nédic, C., Buttler, A. (2000): Dynamics of common reed (*Phragmites australis* Trin.) in Swiss fens with different management. Wetlands Ecology and Management 8: 375–389.

Hadders, G. & Olsson, R. (1997): Harvest of grass for combustion in late summer and in spring. Biomass quality for power production. Biomass and Bioenergy 12: 171–175.

Hagelberg, E. & Lyytinen, S. (2007): Turning reed into bioenergy. The long and winding road from beach to boiler. In: Ikonen, I. & Hagelberg, E. (eds.): Read up on reed! Vammalan Kirjapaino Oy, Turku: 94–101.

Haines-Young, R. & Potschin, M. (2011): Common International Classification of Ecosystem Services (CICES). 2011 Update. Paper prepared for discussion at the. Centre for Environmental Management & University of Nottingham, EEA/BSS/07/007, Nottingham. 17 p. http://unstats.un.org/unsd/envaccounting/seeaLES/egm/Issue8a.pdf.

Hammer, T. & Leng, M. (2008): Moorlandschaften im Bedeutungswandel. Zur gesamtgesellschaftlichen Aufwertung naturnaher Kulturlandschaften. Allgemeine Ökologie zur Diskussion gestellt 10. Universität Bern, Interfakultäre Koordinationsstelle für Allgemeine Ökologie (IKAÖ), Bern. 199 p.

Hampicke, U. (2009): Kosten der Renaturierung. In: Zerbe, S. & Wiegleb, G. (eds.): Renaturierung von Ökosystemen in Mitteleuropa. Spektrum Akademischer Verlag, Heidelberg: 441–457.

Hampicke, U. (2013): Kulturlandschaft und Naturschutz. Probleme – Konzepte – Ökonomie. Springer; Spektrum, Wiesbaden. 356 p.

Hamsun, K. (1928): Der Landstreicher. Albert Langen, München. 494 p.

Han, K., Albrecht, K.A., Muck, R. & Kim, D. (2000): Moisture effect on fermentation characteristics of cup-plant silage. Asian-Australasian Journal of Animal Sciences 13: 636–640.

Hanganu, J., Gridin, M. & Coops, H. (1999): Responses of ecotypes of *Phragmites australis* to increased sea water influence: A field study in the Danube Delta, Romania. Aquatic Botany 64: 351–358.

Hanley, N. & Barbier, E.B. (2009): Pricing nature. Cost-benefit analysis and environmental policy. Edward Elgar Publishing, Cheltenham. 360 p.

hanzewetlands (2015): http://hanzewetlands.com, last access 03 May 2015.

Hardman, C.J., Harris, D.B., Sears, J. & Droy, N. (2012): Habitat associations of invertebrates in reedbeds, with implications for management. Aquatic Conservation: Marine and Freshwater Ecosystems 22: 813–826.

Hartmann, H. (2009a): Brennstoffzusammensetzung und -eigenschaften. In: Kaltschmitt, M., Hartmann, H. & Hofbauer, H. (eds.): Energie aus Biomasse. Grundlagen, Techniken und Verfahren. 2nd edition. Springer, Dordrecht, Heidelberg, London, New York: 333–374.

Hartmann, H. (2009b): Grundlagen der thermo-chemischen Umwandlung biogener Festbrennstoff. In: Kaltschmitt, M., Hartmann, H. & Hofbauer, H. (eds.): Energie aus Biomasse. Grundlagen, Techniken und Verfahren. 2nd edition. Springer, Dordrecht, Heidelberg, London, New York: 333–373.

Hartmann, H. (2009c): Transport, Lagerung, Konservierung und Trocknung. In: Kaltschmitt, M., Hartmann, H. & Hofbauer, H. (eds.): Energie aus Biomasse. Grundlagen, Techniken und Verfahren. 2nd edition. Springer, Dordrecht, Heidelberg, London, New York: 277–332.

Hartmann, H. & Witt J. (2009): Mechanische Aufbereitung. In: Kaltschmitt, M., Hartmann, H. & Hofbauer, H. (eds.): Energie aus Biomasse. Grundlagen, Techniken und Verfahren. 2nd edition. Springer, Dordrecht, Heidelberg, London, New York: 264–267.

Hartmann, H., Nussbaumer, T., Hofbauer, H. & Good, J. (1996): Automatisch beschickte Feuerungen. In: Kaltschmitt, M. (ed.): Biomasse als Festbrennstoff. Anforderungen – Einflussmöglichkeiten – Normung. Schriftenreihe Nachwachsende Rohstoffe 6. Landwirtschaftsverlag, Münster: 492–532.

Hasch, B., Schütze, S Haberl, A., Schuilz, J., Wichtmann, W. & Schwill, S. (2012): Paludikultur und angepasste Moornutzung Uckertal/Prenzlau. Machbarkeitsstudie Paludikulturen Brandenburg. ARGE Machbarkeitsstudie Paludikultur Brandenburg p2mberlin & Michael Succow Stiftung zum Schutz der Natur, Berlin.

Hasch, B., Zeitz, J., Lotsch, H., Luthardt, V. & Meier, R. (2008): A decision support system for management of mires in the forest. In: International Peat Society (IPS) (ed.): After Wise Use – The Future of Peatlands, Proceedings of the 13th International Peat Congress, Tullamore, Ireland. International Peat Society (IPS).

Haslam, S.M. (1969): The development and emergence of buds in *Phragmites communis* Trin. Annals of Botany 33: 289–301.

Haslam, S.M. (2003): Understanding wetlands. Fen, bog and marsh. Taylor & Francis, London, New York. 312 p.

Haslam, S.M. (2010): A Book of reed. (*Phragmites australis* (Cav.) Trin. ex Steudel, *Phragmites communis* Trin.). Forrest Text, Cardigan. 261 p.

Hawke, C. & José, P. (1996): Reedbed management for commercial and wildlife interests. Royal Society for the Protection of Birds, Sandy. 212 p.

Hebauer, C., Gandorfer, M., Hoffmann, H. & Heißenhuber, A. (eds.) (2011): Gemeinsame Agrarpolitik (GAP) der europäischen Union nach 2013. Schriftenreihe der Rentenbank 27. Rentenbank, Frankfurt am Main. 180 p.

Hegi, G. (1923): Illustrierte Flora von Mitteleuropa – Mit besonderer Berücksichtigung von Deutschland, Oesterreich und der Schweiz IV/2. J.F. Lehmanns, München. 1112 p.

Heindorf, V. (2010): Der Einsatz moderner Informationstechnologien in der Automobilproduktentwicklung. Produktivitätspotenziale und Systemkomplementaritäten. Gabler Research. Markt- und Unternehmensentwicklung, Markets and Organisations. Gabler, Wiesbaden. 184 p.

Heinsoo, K., Hein, K., Melts, I., Holm, B. & Ivask, M. (2011): Reed canary grass yield and fuel quality in Estonian farmers' fields. Biomass and Bioenergy 35: 617–625.

Heinz, S. (2011): Population biology of *Typha latifolia* L. and *Typha angustifolia* L.: establishment, growth and reproduction in a constructed wetland. PhD thesis. TU München, München. 110 p.

Held, M. & Nutzinger, H.G. (1998): Eigentumsrechte verpflichten – Zum inneren Zusammenhang von Rechten und Pflichten. In: Held, M. & Nutzinger, H.G. (eds.): Eigentumsrechte verpflichten. Individuum, Gesellschaft und die Institution Eigentum. Campus, Frankfurt: 7–35.

Henckel, D., Kuczkowski, K. & Lau, P. (2010): Identität. In: Henckel, D., Kuczkowski, K., Lau, P., Pahl-Weber, E. & Stellmacher, F. (eds.): Planen – Bauen – Umwelt. Ein Handbuch. Verlag für Sozialwissenschaften / Springer Fachmedien Wiesbaden GmbH, Wiesbaden: 216–253.

Hendriks, D.M.D., van Huissteden, J., Dolman, A.J. & van der Molen, M.K. (2007): The full greenhouse gas balance of an abandoned peat meadow. Biogeosciences 4: 411–424.

Hendriks, D.M.D., van Huissteden, J. & Dolman, A.J. (2010): Multi-technique assessment of spatial and temporal variability of methane fluxes in a peat meadow. Agricultural and Forest Meteorology 150: 757–774.

Hennings, H. (1995): Einfluss der Vernässung und freier Vegetationsentwicklung auf den Wärmehaushalt von Niedermooren. Zeitschrift Kulturtechnik und Landentwicklung 36: 141–144.

Herbst, M., Friborg, T., Ringgaard, R. & Soegaard, H. (2011): Interpreting the variations in atmospheric methane fluxes observed above a restored wetland. Agricultural and Forest Meteorology 151: 841–853.

Hering, T. (2012): Energetische Halmgutnutzung in Deutschland. In: Fachagentur Nachwachsende Rohstoffe e.V. (FNR) (ed.): Gülzower Fachgespräche. 2. Internationale Fachtagung Strohenergie, Berlin, 29–30 March 2012. Gülzower Fachgespräche 38. Fachagentur Nachwachsende Rohstoffe e. V. (FNR), Gülzow-Prüzen: 17–19.

Hernández Allica, J., Mitre, A.J., González Bustamante, J.A., Itoiz, C., Blanco, F., Alkorta, I. & Garbisu, C. (2001): Straw quality for its combustion in a straw-fired power plant. Biomass and Bioenergy 21: 249–258.

Herold, B. (2012): Neues Leben in alten Mooren. Brutvögel wiedervernässter Flusstalmoore. Bristol-Schriftenreihe 34. Haupt, Bern. 200 p.

Hirschl, B., Aretz, A. & Böther, T. (2011a): Regionalökonomische Effekte Erneuerbarer Energien. Kommunale Wertschöpfung und Beschäftigung durch dezentrale Energieerzeugung. Solarzeitalter 3: 45–51.

Hirschl, B., Aretz, A. & Böther, T. (2011b): Wertschöpfung und Beschäftigung durch Erneuerbare Energien in Mecklenburg-Vorpommern 2010 und 2030. Kurzstudie im Auftrag der SPD-Landtagsfraktion Mecklenburg-Vorpommern. Institut für ökologische Wirtschaftsforschung (IÖW), Schwerin. 60 p. http://www.ioew.de/uploads/tx_ukioewdb/Studie-Wertschoepfung_EE-MV.pdf, last access 06 September 2015.

Hiss, T. (2013): Schriftliche Mitteilung, 12 June 2013.

Hock, W. & Vorbrüggen, W. (1997): Röhrichte und Großseggenrieder. In: Landesanstalt für Ökologie, Bodenordnung und Forsten (ed.): Praxishandbuch Schmetterlingsschutz. LÖBF-Reihe Artenschutz 1. LÖBF, Recklinghausen: 60–65.

Hock, W. & Weidner, A. (1997): Sumpf-, Feucht-, und Naßwiesen, uferbegleitende Staudenfluren, Sümpfe. In: Landesanstalt für Ökologie, Bodenordnung und Forsten (ed.): Praxishandbuch Schmetterlingsschutz. LÖBF-Reihe Artenschutz 1. LÖBF, Recklinghausen: 46–59.

Hoffmann, J., Krawczynski, R. & Wagner, H.-G. (2010): Wasserbüffel in der Landschaftspflege. Lexxion, Berlin. 196 p.

Holmes, S., Speirs, S., Berney, P. & Rose, H. (2009): Guidelines for grazing in the Gwydir Wetlands & Macquarie Marshes. New South Wales (NSW) Department of Primary Industries, Orange, N.S.W. 5 booklets, 2 sheets.

Holst, H. (2003): Zur möglichen Bedeutung von Niederungsstandorten für die regionale Entwicklung in Mecklenburg-Vorpommern. In: Timmermann, T. & Wichtmann, W. (eds.): Alternative Nutzungsformen für Moorstandorte in Mecklenburg-Vorpommern. Beiträge zur Tagung, Greifswald, 23 November 2002. Universität Greifswald, Geographisches Institut. Greifswalder Geographische Arbeiten.-Greifswald: Univ, 1980- 31, Greifswald: 115–123.

Holsten, B., Ochsner, S., Schäfer, A. & Trepel, M. (2012): Praxisleitfaden für Maßnahmen zur Reduzierung von Nährstoffausträgen aus dränierten landwirtschaftlichen Flächen. CAU, Kiel. 99 p.

Holsten, K. (1940): Wie sieht der Pommer die heimische Landschaft? In: Gesellschaft für pommersche Landesgeschichte und Altertumskunde (ed.): Baltische Studien. Neue Folge 42. Ludwig, Stettin: 36–61.

Holsten, T. (2012): Geographische Untersuchung zur Anwendbarkeit von Paludikultur in Schleswig-Holstein. Diploma thesis. Universität Greifswald

Hölzel, N. (2009): Ökologische Grundlagen und limitierende Faktoren der Renaturierung. In: Zerbe, S. & Wiegleb, G. (eds.): Renaturierung von Ökosystemen in Mitteleuropa. Spektrum Akademischer Verlag, Heidelberg: 24–29.

Holzmann, G. & Wangelin, M. (2009): Natürliche und pflanzliche Baustoffe. Rohstoff – Bauphysik – Konstruktion. Vieweg + Teubner, Wiesbaden. 225 p.

Hooijer, A., Silvius, M., Wösten, H. & Page, S.E. (2006): PEAT-CO_2, assessment of CO_2 emissions from drained peatlands in SE Asia. Delft Hydraulics report Q3943. Delft Hydraulics, Delft. 66 p.

Hooijer, A., Page, S.E., Jauhiainen, J., Lee, W.A., Lu, X.X., Idris, A. & Anshari, G. (2012): Subsidence and carbon

loss in drained tropical peatlands. Biogeosciences 9: 1053–1071.
Hope, G., Chokkalingam, U. & Anwar, S. (2005): The stratigraphy and fire history of the Kutai peatlands, Kalimantan, Indonesia. Quaternary Research 64: 407–417.
Höper, H. (2009): Die Rolle von Organischen Böden als Kohlenstoffspeicher. In: Alfred Toepfer Akademie für Naturschutz (ed.): Bodenschutz im Spannungsfeld von Umwelt- und Naturschutz. Heft 1. NNA. NNN-Berichte 22: 91–97.
Höper, H., Augustin, J., Cagampan, J.P., Drösler, M., Lundin, L., Moors, E.J., Vasander, H., Waddington, J.M. & Wilson, D. (2008): Restoration of peatlands and greenhouse gas balances. In: Strack, M. (ed.): Peatlands and climate change. International Peat Society, Jyväskylä: 182–210.
Horn, R. (1984): Die Vorhersage des Eindringwiderstandes von Böden anhand von multiplen Regressionsanalysen. Zeitschrift für Kulturtechnik und Flurbereinigung 25: 377–380.
Houghton, J., Jenkins, G. & Ephraums, J. (eds.) (1990): Climate change. The IPCC scientific assessment. Cambridge University Press, Cambridge. 365 p.
Howard-Williams, C. (1985): Cycling and retention of nitrogen and phosphorus in wetlands: a theoretical and applied perspective. Freshwater Biology 15: 391–431.
Huemer, P. (1996): Frühzeitige Mahd, ein bedeutender Gefährdungsfaktor für Schmetterlinge der Streuwiesen (NSG Rheindelta, Vorarlberg, Österreich). Vorarlberger Naturschau 1: 265–300.
Humbert, J.-Y., Ghazoul, J. & Walter, T. (2009): Meadow harvesting techniques and their impacts on field fauna. Agriculture, Ecosystems & Environment 130: 1–8.
Husemann, C. (1947): Die landwirtschaftliche Bewertung der Moorböden und ihre natürlichen Grundlagen: ein Beitrag zur Kultivierung und Besiedlung der deutschen Moore. Kinau Verlag, Lüneburg. 105 p.
Huth, V., Günther, A.B., Jurasinski, G. & Glatzel, S. (2013): The impact of an extraordinarily wet summer on methane emissions from a 15-year re-wetted fen in northeast Germany. Mires & Peat 13, Art. 2: 1–7.
Ilnicki, P. & Zeitz, J. (2003): Irreversible loss of organic soil functions after reclamation. In: Parent, L.-E. & Ilnicki, P. (eds.): Organic soils and peat materials for sustainable agriculture. CRC Press, Boca Raton, Florida: 15–33.
Immirzi, C.P., Maltby, E. & Clymo, R.S. (1992): The global status of peatlands and their role in carbon cycling. A report for Friends of the Earth by the Wetland Ecosystems Research Group. Friends of the Earth, London. 145 p.
Intergovernmental Panel on Climate Change (IPCC) (2013): The physical science basis. Contribution of Working Group I to the Fifth Assessment Report of the Intergovernmental Panel on Climate Change. Intergovernmental Panel on Climate Change (IPCC). T.F. Stocker, D. Qin, G.-K. Plattner, M. Tignor, S.K. Allard, Boschung, J. Nauels, A., Y. Xia, V. Bex & P.M. Midgley, Cambridge, New York. 1535 p. http://www.ipcc.ch/report/ar5/wg1/, last access 06 May 2015.
Intergovernmental Panel on Climate Change (IPCC) (2014a): 2013 Supplement to the 2006 IPCC Guidelines for National Greenhouse Gas Inventories: Wetlands. ed. by Hiraishi, T., Krug, T., Tanabe, K., Srivastava, N., Baasansuren, J., Fukuda, M. & Troxler, T.G. IPCC, Switzerland. 354 p.
Intergovernmental Panel on Climate Change (IPCC) (2014b): 2013 Revised Supplementary Methods and Good Practice Guidance arising from the Kyoto Protocol. ed. by Hiraishi, T., Krug, T., Tanabe, K., Srivastava, N., Baasansuren, J., Fukuda, M. & Troxler, T.G. IPCC, Switzerland. 268 p.
International Institute for Sustainable Development (IISD) (ed.) (2013): The Netley-Libau Nutrient-Bioenergy Project. The Water Innovation Centre. http://www.iisd.org/wic/research/wetlands/netleylibau.asp.
International Organization for Standardization (ISO) (ed.) (2001): Umweltkennzeichnung und -deklarationen. Allgemeine Grundsätze, EN ISO 14020:2001, Genf. 5 p.
International Sustainability & Carbon Certification (ISCC) (ed.) (2013): International Sustainability & Carbon Certification. International Sustainability & Carbon Certification (ISCC). http://www.iscc-system.org/, last access 26 April 2013.
Isermeyer, F. (2012): Erst die Mittel, dann das Ziel? Wie sich die EU-Agrarpolitik in eine Sackgasse manövriert und wie sie dort wieder herauskommen kann. In: Lange, J. (ed.): Die Begrün(d)ung der gemeinsamen Agrarpolitik? Die kommende Reform der GAP. Loccumer Protokolle 5/12, Rehburg-Loccum: 19–62.
Isermeyer, F. & Weingarten, P. (2012): GAP Reform – Stellungnahme im Rahmen einer öffentlichen Anhörung des Ausschusses für Ernährung, Landwirtschaft und Verbraucherschutz des Deutschen Bundestages am 22. oktober 2012 Braunschweig. Deutscher Bundestag, Berlin. 25 p. http://www.bundestag.de/bundestag/ausschuesse17/a10/anhoerungen/2012_10_22_GAP-Reform/Stellungnahmen/A-Drs__983-D-E_v_Th__nen_Institut.pdf, last access 15 October 2012.
Isermeyer, F., Otte, A., Christen, O., Dabbert, S., Frohberg, K., Grabski-Kieron, U., Hartung, J., Heißenhuber, A., Hess, J., Kirschke, D., Schmitz, P., Spiller, A., Sundrum, A. & Thoroe, C. (2008): Nutzung von Biomasse zur Energiegewinnung – Empfehlungen an die Politik: Gutachten, Münster-Hiltrup. Landwirtschaftsverlag (Berichte über Landwirtschaft, Sonderheft 216).
Jacob, M. (2012): Informationsorientiertes Management. Ein Überblick für Studierende und Praktiker. Gabler Verlag, Wiesbaden. 388 p.
Jaekel, O. (1922): Die Gefahren der Entwässerung unseres Landes. Mitteilungen aus dem Geologisch-Palaeontologischen Institut der Ernst-Moritz-Arndt-Universität Greifswald 4. Bamberg, Greifswald. 29 p.
Jaenicke, J., Rieley, J.O., Mott, C., Kimman, P. & Siegert, F. (2008): Determination of the amount of carbon stored in Indonesian peatlands. Geoderma 147: 151–158.
Janakiraman, V.S. & Sarukesi, K. (2004): Decision support systems. Prentice-Hall of India, New Delhi. 236 p.
Janinhoff, A. (2008): Wann sind die Äcker zu weit weg? DLG-Mitteilungen 5: 34–40.
Janke, W. (2002): Zur Genese der Flußtäler zwischen Uecker und Warnow (Mecklenburg-Vorpommern). Greifswalder Geographische Arbeiten 26: 39–44.
Jassal, R.S., Black, T.A., Roy, R. & Ethier, G. (2011): Effect of nitrogen fertilization on soil CH_4 and N_2O fluxes, and soil and bole respiration. Geoderma 162: 182–186.
Jauhiainen, J., Hooijer, A. & Page, S.E. (2012): Carbon dioxide emissions from an *Acacia* plantation on peatland in Sumatra, Indonesia. Biogeosciences 9: 617–630.

Jensen, R., Landgraf, L., Lenschow, U., Paterak, B., Permien, T., Schiefelbein, U., Sorg, U., Thormann, J., Trepel, M., Wälter, T., Wreesmann, H. & Ziebarth, M. (2012): Potentiale und Ziele zum Moor- und Klimaschutz – Positionspapier der Länderfachbehörden von Brandenburg, Bayern, Mecklenburg-Vorpommern, Niedersachsen und Schleswig-Holstein. Natur und Landschaft 87: 87–88.

Jeschke, L. (1987): Vegetationsdynamik des Salzgraslandes im Bereich der Ostseeküste der DDR unter Einfluß des Menschen. Hercynia 24: 321–328.

Ji, Y.H., Zhou, G.S., Lv, G., Zhao, X.L. & Jia, Q.Y. (2009): Expansion of *Phragmites australis* in the Liaohe Delta, north-east China. Weed Research 49: 613–620.

Jong, J.J. de, Schaafsma, A.H., Aertsen, E. & Hoksbergen, F. (2003): Machines voor het beheer van natte graslanden. Een studie naar de kosten van beheer van natte en vochtige graslanden met aangepaste machines. Alterra, Research Instituut voor de Groene Ruimte. Alterra-rapport, 747. Alterra, Research Instituut voor de Groene Ruimte, Wageningen. 45 p. http://edepot.wur.nl/31476, last access 24 June 2014.

Joosten, H. (2009): The global peatland CO_2 picture. Peatland status and emissions in all countries of the world. Wetlands International, Ede. 10 S.

Joosten, H. (2012): Weighed and found wanting. Peer review of the report: PeatImpact. Greenhouse gas calculation methodologies for fuels based on peat and peat grown biomass. Service contract to improve understanding of greenhouse gas impacts of using peat or peat grown biomass for transport fuels or other types of energy. Contract no 070307/2009/546431/SER/C3. 12 S.

Joosten, H. & Clarke, D. (2002): Wise use of mires and peatlands. Background and principles including a framework for decision-making. International Peat Society; International Mire Conservation Group. 304 S.

Joosten, H. & Couwenberg, J. (2008): Peatlands and carbon. In: Parish, F., Sirin, A., Charman, D., Joosten, H., Minaeva, T. & Silvius, M. (eds): Assessment on peatlands, biodiversity and climate change. Global Environment Centre, Kuala Lumpur and Wetlands International Wageningen, pp. 99-117.

Joosten, H. & Couwenberg, J. (2009): Are emission reductions from peatlands MRV-able? Wetlands International, Ede. 14 S.

Joosten, H. & Couwenberg, J. (2001): Bilanzen zum Moorverlust – das Beispiel Europa. In: Succow, M. & Joosten, H. (eds.): Landschaftsökologische Moorkunde. 2. Auflage. Schweizerbart, Stuttgart: 406–409.

Joosten, H., Haberl, A. & Schumann, M. (2008): Degradation and restoration of peatlands on the Tibetan Plateau. Peatlands International: 31–35.

Joosten, H., Couwenberg, J., Schäfer, A., Wichmann, S. & Wichtmann, W. (2012a): Perspektiven der Regeneration und Nutzbarmachung von Mooren. Mitteilungen der Gesellschaft für Pflanzenbauwissenschaften 24: 13–16.

Joosten, H., Tapio-Biström, M.-L. & Tol, S. (eds.) (2012b): Peatlands – guidance for climate change mitigation through conservation, rehabilitation and sustainable use. 2. Auflage. FAO. Mitigation of climate change in agriculture, 5, Rome. 114 S.

Joosten, H., Berghöfer, A., Couwenberg, J., Doetrich, K., Holsten, B., Permien, T., Schäfer, A., Tanneberger, F., Trepel, M. & Wahren, A. (2015): Die neuen MoorFutures® – Kohlenstoffzertifikate mit ökologischen Zusatzleistungen. Natur und Landschaft 90: 170–175.

Jungkunst, H.F. & Fiedler, S. (2007): Latitudinal differentiated water table control of carbon dioxide, methane and nitrous oxide fluxes from hydromorphic soils: feedbacks to climate change. Global Change Biology 13: 2668–2683.

Jungkunst, H.F., Freibauer, A., Neufeldt, H. & Bareth, G. (2006): Nitrous oxide emissions from agricultural land use in Germany – a synthesis of available annual field data. Journal of Plant Nutrition and Soil Science 169: 341–351.

Juutinen, S., Alm, J., Larmola, T., Saarnio, S., Martikainen, P.J., Silvola, J. (2004): Stand-specific diurnal dynamics of CH_4 fluxes in boreal lakes: Patterns and controls. Journal of Geophysical Research 109.

Kadlec, R.H. & Knight, R.L. (1996): Treatment wetlands. Lewis Publishers, Boca Raton. 893 p.

KAE – Komplexlabor Alternative Energien Fachhochschule Stralsund (2009): Charakterisierung und Analyse der Brennstoffe Schilfrohr und Rohrglanzgras. Zwischenbericht zum Forschungs- und Entwicklungsprojekt Energiebiomasse aus Niedermooren (ENIM). Deutsche Bundesstiftung Umwelt (DBU), Greifswald.

Kaeker, H. (1919): Vom Land am Meer. Ein Büchlein von Holden und Unholden. Norddeutscher Verlag für Literatur und Kunst, Stettin. 64 p.

Kaimer, M. & Schade, D. (2000): Abfallentsorgung zu Lasten der Bürger? Probleme der Kreislaufwirtschaft und Lösungsansätze für eine Entlastung der Haushalte. Arbeitsbericht / Akademie für Technikfolgenabschätzung in Baden-Württemberg. Akademie für Technikfolgenabschätzung in Baden-Württemberg, Stuttgart. 21 p.

Kalisz, B., Lachacz, A. & Glaziewski, R. (2010): Transformation of some organic matter components in organic soils exposed to drainage. Turkish Journal of Agriculture and Forestry 34: 245–256.

Kaltschmitt, M. (2012): Biomass as renewable source of energy, possible conversion routes. In Meyers, R.A. (ed.): Encyclopedia of sustainability science and technology. Springer, New York, Dordrecht, Heidelberg, London.

Kaltschmitt, M., Hartmann, H. & Hofbauer, H. (eds.) (2009): Energie aus Biomasse. Grundlagen, Techniken und Verfahren. 2nd edition. Springer, Dordrecht, Heidelberg, London, New York. 1030 p.

Kalzendorf, C. (2011): Blühmischungen, Schilf und Seggen. Erste Bewertung des Biomassepotentiales, der Silierbarkeit und des Gasbildungsvermögens. Interreg IV b: enerCOAST, Torgau, 28 June 2011.

Kampichler, C., Misslinger B. & Waitzbauer, W. (1994): Der Einfluß des Schnittes auf die endophage Fauna des Schilfes. Zeitschrift für Ökologie und Naturschutz 3: 1–9.

Karl, H. & Orwat, C. (1999): Economic aspects of environmental labelling. In: Folmer, H. & Tietenberg, T. (eds.): The international yearbook of environmental and resource economics. Edward Elgar Publishing, Cheltenham: 107–170.

Kask, Ü., Kask, L. & Paist, A. (2007): Reed as an energy source in Estonia. In: Ikonen, I. & Hagelberg, E. (eds.): Read up on reed! Vammalan Kirjapaino Oy, Turku: 102–114.

Kasten, H. (1906): Pommersche Dichtung der Gegenwart. Festgabe zur 33. Pommerschen Provinzial-Lehrerversammlung; mit Bild, Buchschmuck und Musikbeigaben. Verlag des Kösliner Lehrervereins, Köslin. 55 p.

Kerkmann, J. (ed.) (2010): Naturschutzrecht in der Praxis. 2nd edition. Lexxion, Berlin. 833 p.

Kern, J., Dicke, C., Libra, J. & Mumme, J. (2011): Biokohle: ein Kohlenstoffspeicher, der die Böden verbessert. Forschungsreport. ForschungsReport: 28–30.

Kersten, U., Lindner, H., Melzer, R., Rehberg, U., Staak, R. & Werner, W. (1999): Ergebnisse des Projektes "Regeneration und alternative Nutzung von Niedermoorflächen im Landkreis Ostvorpommern". Kurzfassung. Stiftung Odermündung, Anklam. 57 p.

Kieckbusch, J., Schrautzer, J. & Trepel, M. (2006): Spatial heterogeneity of water pathways in degenerated riverine peatlands. Basic and Applied Ecology 7: 388–397.

Kim, J., Verma, S.B., Billesbach, D.P. & Clement, R.J. (1998): Diel variation in methane emission from a midlatitude prairie wetland. Significance of convective throughflow in *Phragmites australis*. Journal of Geophysical Research 103: 28029.

Kirschke, S., Bousquet, P., Ciais, P., Saunois, M., Canadell, J.G., Dlugokencky, E.J., Bergamaschi, P., Bergmann, D., Blake, D.R., Bruhwiler, L., Cameron-Smith, P., Castaldi, S., Chevallier, F., Feng, L., Fraser, A., Heimann, M., Hodson, E.L., Houweling, S., Josse, B., Fraser, P.J., Krummel, P.B., Lamarque, J.-F., Langenfelds, R.L., Le Quéré, C., Naik, V., O'Doherty, S., Palmer, P.I., Pison, I., Plummer, D., Poulter, B., Prinn, R.G., Rigby, M., Ringeval, B., Santini, M., Schmidt, M., Shindell, D.T., Simpson, I.J., Spahni, R., Steele, L.P., Strode, S.A., Sudo, K., Szopa, S., van der Werf, Guido R., Voulgarakis, A., van Weele, M., Weiss, R.F., Williams, J.E. & Zeng, G. (2013): Three decades of global methane sources and sinks. Nature Geoscience 6: 813–823.

Kitzler, H., Pfeifer, C. & Hofbauer, H. (2012): Combustion of reeds in a 3 MW district heating plant. International Journal of Environmental Science and Development 3: 407–411.

Kivinen, E. & Pakarinen, P. (1980): Peatland areas and the proportion of virgin peatlands in different countries. In: Proceedings of the 6th International Peatland Congress. Peatland restoration and reclamation, Duluth, Minnesota, Jyska, Finnland: 52–54.

Kivinen, E. & Pakarinen, P. (1981): Geographical distribution of peat resources and major peatland complexes in the world. Annales Academiæ Scientiarum Fennicæ Series A, III Geologica, Geographica 132: 1–28.

Klapp, E., Boeker, P., König, F., Stählin, A. (1953): Wertzahlen der Grünlandpflanzen. Das Grünland 2: 38-40.

Klapp, E. & von Boberfeld, W.O. (1995): Gräserbestimmungsschlüssel für die häufigsten Grünland- und Rasengräser. 4th edition. Blackwell Wissenschaftsverlag, Berlin. 82 p.

Kleinhückelkotten, S. (2005): Suffizienz und Lebensstile. Ansätze für eine milieuorientierte Nachhaltigkeitskommunikation. Umweltkommunikation 2. BWV, Berliner Wissenschafts-Verlag, Berlin. 208 p.

Kleinhückelkotten, S. & Neitzke, H.-P. (2005): Lokale Nachhaltigkeitskommunikation – Soziale Milieus als Zielgruppen in lokalen Agenda-Prozessen. In: Michelsen, G. & Godemann, J. (eds.): Handbuch Nachhaltigkeitskommunikation. Grundlagen und Praxis. Oekom, München: 689–697.

Kleinhückelkotten, S. & Wegner, E. (2008): Nachhaltigkeit kommunizieren. Zielgruppen, Zugänge, Methoden. 2nd edition. ECOLOG-Institut, Hannover. 134 p.

Kleinhückelkotten, S. & Neitzke, H.-P. (2013): Landnutzungskonkurrenzen und -konflikte. Nutzung von Niedermoorflächen. Project report. ECOLOG-Institut, Hannover.

Kleinhückelkotten, S. & Neitzke, H.-P. (2014): Nachhaltige Nutzung von Niedermoorstandorten: Szenarien für das Thurbruch/Insel Usedom. VIP – Project report. ECOLOG-Institut, Hannover.

Kleinhückelkotten, S., Neitzke, H.-P. & Wippermann, C. (2009): Einstellungen der Deutschen zu Wald und Forstwirtschaft. Ergebnisse einer bevölkerungsrepräsentativen Befragung differenziert nach sozialen Milieus. Forst und Holz 64: 12–19.

Klepper, G. (2002): Nachhaltigkeit und technischer Fortschritt – Die Perspektive der neoklassischen Ökonomie. In: Grunwald, A. (ed.): Technikgestaltung für eine nachhaltige Entwicklung. Von der Konzeption zur Umsetzung. Global zukunftsfähige Entwicklung – Perspektiven für Deutschland 4. Edition Sigma, Berlin: 21–36.

Klinck, S. (2012): Agrarumweltrecht im Wandel. Schriften zum Umweltrecht 174. Duncker & Humblot, Berlin. 166 p.

Kloskowski, J. & Krogulec, J. (1999): Habitat selection of Aquatic Warbler *Acrocephalus paludicola* in Poland: consequences for conservation of the breeding areas. Vogelwelt 120: 113–120.

Klotz, S., Kühn, I. & Durka, W. (eds.) (2002): BIOLFLOR – Eine Datenbank zu biologisch-ökologischen Merkmalen der Gefäßpflanzen in Deutschland. Schriftenreihe für Vegetationskunde 38. Bundesamt für Naturschutz, Bonn. 334 p.

Knieper, M. (1999): Tragfähigkeit der Niedermoore in der Nuthe-Nieplitz-Niederung. In: Prochnow, A. (ed.): Angepasstes Befahren von Niedermoorgrünland. Landschaftspflege in der Nuthe-Nieplitz-Niederung 3. Landschafts-Fördervereins Nuthe-Nieplitz-Niederung e.V., Stücken: 20–34.

Knieß, A. (2007): Development and application of a semi-quantitative decision support system to predict long-term changes of peatland functions. PhD thesis. Christian-Albrechts-Universität, Kiel. Mathematisch-Naturwissenschaftliche Fakultät. 179 p.

Knobelsdorff-Brenkenhoff, B.v. (1988): Pommern im 18. Jahrhundert. Eine Provinz im Frieden erobert. In: Rothe, H. (ed.): Ostdeutsche Geschichts- und Kulturlandschaften. Teil III, Pommern. Studien zum Deutschtum im Osten 3. Böhlau Verlag, Köln: 131–152.

Knoll, T. (1986): Der Schilfschnitt am Neusiedler See. Analyse einer Landschaftsnutzung für Landschaftsplanung. In: Vereinigung Burgenländischer Geographen (ed.): Geographisches Jahrbuch Burgenland. Geographisches Jahrbuch Burgenland. Universität für Bodenkultur, Wien: 34–67.

Köbbing, J.F. (2010): Schilfverarbeitung, Firma Hiss Reet. Interview with O. Jedack, Bad Oldesloe (25 November 2010).

Köbbing, J.F., Beckmann, V., Thevs, N., Peng, H. & Zerbe, S. (2014): Investigation of a reed economy (*Phragmites australis*) under threat: pulp and paper market, values and netchain at Wuliangsuhai Lake, Inner Mongolia, China. Journal of Environmental Management, re-submitted June 2015.

Koehler, A.-K., Sottocornola, M. & Kiely, G. (2011): How strong is the current carbon sequestration of an Atlantic blanket bog? Global Change Biology 17: 309–319.

Koh, L.P. & Ghazoul, J. (2010): Spatially explicit scenario analysis for reconciling agricultural expansion, forest protection, and carbon conservation in Indonesia. Proceedings of the National Academy of Sciences 107: 11140–11144.

Komulainen, V.-M., Simi, P., Hagelberg, E., Ikonen, I., Lyytinen, S. & Salmela, P. (2008): Reed energy. Possibilities of using the Common Reed for energy generation in Southern Finland. Reports 67. Turku University of Applied Sciences, Turku. 81 p.

Könker, H. (2007): Komplexe Standortmelioration. In: Behrens, H. & Hoffmann, J. (eds.): Umweltschutz in der DDR. Analysen und Zeitzeugenberichte. Oekom-Verlag, München: 45–58.

Koppisch, D., Roth, S. & Hartmann, M. (2001): Vom Saatgrasland zum wieder torfspeichernden Niedermoor – Die Experimentalanlage Am Fleetholz/Friedländer Große Wiese. In: Succow, M. & Joosten, H. (eds.): Landschaftsökologische Moorkunde. 2nd edition. Schweizerbart, Stuttgart: 497–504.

Korthals, E. (1928): Zur Kenntnis des Futterwertes der Süßgräser Wiesenfuchsschwanz (*Alopecurus pratensis*), Fruchtbare Rispe (*Poa serotina*), Beckmannia (*Beckmannia eruciformis* Host.), Rohrglanzgras (*Phalaris arundinacea*) und der Sauergräser unter besonderer Berücksichtigung der Milchleistung. PhD thesis. Albertus-Universität, Königsberg. 43 p.

Kosfeld, R. & Gückelhorn, F. (2012): Ökonomische Effekte erneuerbarer Energien auf regionaler Ebene. Raumforschung und Raumordnung 70: 437–449.

Koska, I. (2001): Ökohydrologische Kennzeichnung. In: Succow, M. & Joosten, H. (eds.): Landschaftsökologische Moorkunde. 2nd edition. Schweizerbart, Stuttgart: 92–111.

Kösling, S. (2000): Baustoff Reet – Geerntet in Europa, vermarktet in Bad Oldesloe. Lübecker Nachrichten (LN), 20 May 2000.

Kotowski, W., Dembek, W. & Pawlikowski, P. (2016): Poland. In: Joosten, H., Tanneberger, F. & Moen, A. (eds.): Mires and peatlands of Europe. Schweizerbart, Stuttgart.

Kotowski, W. & Piórkowski, H. (2003): Poland. In: Bragg, O., Lindsay, R. & Risager, M. (eds.): Strategy and action plan for mire and peatland conservation in Central Europe. Wetlands International global series 18. Wetlands International, Wageningen: 49–53.

Kotowski, W., van Andel, J., van Diggelen, R. & Hogendorf, J. (2001): Responses of fen plant species to groundwater level and light intensity. Plant Ecology 155: 147–156.

Kotowski, W., Jabłońska E. & Bartoszuk, H. (2013): Conservation management in fens: do large tracked mowers impact functional plant diversity? Biological Conservation 167: 292–297.

Kowalewsky, H.H. (2009): Landwirtschaftliche Transporte mit Schlepper oder Lkw. Landwirtschaftskammer Niedersachsen. Landwirtschaftskammer Niedersachsen. http://www.lwk-niedersachsen.de/index.cfm/portal/6/nav/1082/article/13273.html.

Kowalewsky, H.H. (2011): Schlepper, Unimog und Lkw im Vergleichstest. Biogas Journal 4: 44–50.

Kowatsch, A. (2007): Moorschutzkonzepte und -programme in Deutschland. Ein historischer und aktueller Überblick. Naturschutz und Landschaftsplanung 39: 197–204.

Kowatsch, A., Schäfer, A. & Wichtmann, W. (2008): Nutzungsmöglichkeiten auf Niedermoorstandorten. Umweltwirkungen, Klimarelevanz und Wirtschaftlichkeit sowie Anwendbarkeit und Potenziale in Mecklenburg-Vorpommern. Final report. Im Auftrag des Landes Mecklenburg-Vorpommern, Ministerium für Landwirtschaft. DUENE e.V. & Institut für Landschaftsökologie und Botanik, Universität Greifswald, Schwerin. 57 p. http://duene-greifswald.de/doc/moornutzung_endbericht.pdf. last access 06 September 2015.

Kozulin, A.V. (2011): Rewetting of peatland. In: Tanneberger, F. & Wichtmann, W. (eds.): Carbon credits from peatland rewetting. Climate – biodiversity – land use. Schweizerbart, Stuttgart: 9–12.

Kozulin, A.V., Tanovitskaya, N. & Vershitskaya, I.N. (2010): Methodical Recommendations for ecological rehabilitation of damaged mires and prevention of disturbances to the hydrological regime of mire ecosystems in the process of drainage. Belarus Project Guidebook on Peatland Rehabilitation. United Nation Development Program (UNDP). 29 p. http://www.undp.org/content/undp/en/home/librarypage/environment-energy/ecosystems_and_biodiversity/belarus-project-guidebook-on-peatland-rehabilitation.html, last access 02 June 2015.

Krägenow, P. & Wiesehöfer, G. (1999): Vögel der Binnengewässer und Feuchtgebiete. Eugen Ulmer, Stuttgart. 257 p.

Krahmer, U. (1997): Penetrometermessungen auf rekultivierten Böden. Ergebnisse eines Pilotprojektes. In: Geologisches Landesamt Nordrhein-Westfalen (ed.): Fünf Beiträge zur Geologie und Bodenkunde. Schriftenreihe Nachwachsende Rohstoffe 2. Geologisches Landesamt Nordrhein-Westfalen, Krefeld: 40–50.

Kraschinski, S., Prochnow, A., Tölle, R. & Hahn, J. (1999): Verfahrenstechnische Arbeiten zur Befahrbarkeit von Niedermoorgrünland. In: Prochnow, A. (ed.): Angepasstes Befahren von Niedermoorgrünland. Landschaftspflege in der Nuthe-Nieplitz-Niederung 3. Landschafts-Fördervereins Nuthe-Nieplitz-Niederung e.V., Stücken: 35–57.

Krasuska, E. & Rosenqvist, H. (2012): Economics of energy crops in Poland today and in the future. Biomass and Bioenergy 38: 23–33.

Kratsch, D. & Czybulka, D. (2010): §30 Rn. 21. In: Fischer-Hüftle, P. & Schumacher, J.: Kommentar zum Bundesnaturschutzgesetz. 2nd edition. W. Kohlhammer, Stuttgart.

Kratz, R. & Pfadenhauer, J. (eds.) (2001): Ökosystemmanagement für Niedermoore. Strategien und Verfahren zur Renaturierung. Ulmer, Stuttgart (Hohenheim). 317 p.

Krawczynski, R., Biel, P. & Zeigert, H. (2008): Wasserbüffel als Landschaftspfleger - Erfahrungen zum Einsatz in Feuchtgebieten. Naturschutz und Landschaftsplanung 40: 133–139.

Krebs, M., Gaudig, G. & Joosten, H. (2012): Sphagnum farming on bog grassland in Germany - first results. In: Proceedings of the 14th International Peat Congress. Peatlands in balance, Stockholm.

Kreil, W., Simon, W. & Wojahn, E. (1982): Futterpflanzenanbau 1. Empfehlungen, Richtwerte, Normative 1. VEB Deutscher Landwirtschaftsverlag, Berlin.

Kreisner, P. (1919): Die deutsche Textilindustrie in und nach dem Kriege. Dresden: Mitteilung aus dem deutschen Forschungsinstitut für Textilindustrie. Zeitschrift für angewandte Chemie 1: 1–8.

Kropf, P. (1985): Die Erle und die Verwendung ihres Holzes. Holz-Zentralblatt 111: 114–125.

Krüger, E. (1916): Die Meliorirung der Moore in Preußen, ihre Technik und ihr Einfluß auf die Wasserverhältnisse. In: Conwentz, H. (ed.): Bericht über die siebente Konferenz für Naturdenkmalpflege in Preußen, Berlin, am 3. und 4. Dezember 1915. Beiträge zur Naturdenkmalpflege 5. Bornträger, Berlin: 120–129.

Krüger, F., Meissner, R., Gröngröft, A. & Grunewald, K. (2005): Flood induced heavy metal and arsenic contamination of Elbe River floodplain soils. Acta hydrochimica et hydrobiologica 33: 455–465.

Kubacka, J., Oppel, S., Dyrcz, A., Lachmann, L., Barros da Costa, J.P.D., Kail, U. & Zdunek, W. (2014): Effect of habitat management on productivity of the endangered aquatic warbler *Acrocephalus paludicola*. Bird Conservation International 24: 45–58.

Kühl, H. & Kohl, J.-G. (1992): Nitrogen accumulation, productivity and stability of reed stands (*Phragmites australis* (Cav.) Trin. ex Steudel) at different lakes and sites of lake districts of Uckermark and Mark Brandenburg (Germany). Hydrobiologia - Hydrogeographica 77: 85–107.

Kuhlman, T., Diogo, V. & Koomen, E. (2013): Exploring the potential of reed as a bioenergy crop in the Netherlands. Biomass and Bioenergy 55: 41–52.

Kuhnert, H., Behrens, G., Hamm, U., Müller, H., Nieberg, H., Sandders, J. & Strohm, R. (2013): Ausstiege aus dem ökologischen Landbau. Umfang, Gründe, Handlungsoptionen. Thünen-Report 3. Johann Heinrich von Thünen-Institut, Braunschweig. 69 p.

Kujawski & G.R. (1972): Traditionelles Wohnhaus – Japan. Untersuchung der typischen Behausung einer naturvölkischen Kultur heutiger Zeit. Völkerkundliche Studie. Universität GH Essen, Essen. 43 p. http://due-publico.uni-duisburg-essen.de/servlets/DerivateServlet/Derivate-12819/Ethnolog.JAPAN.pdf.

Kullmann, A. (2012): Bio International!? – Marketing potentials of Organic Products from UNESCO Biosphere Reserves. Discussions and results of an international expert workshop 04–07 October 2011, Gülstorf/Amt Neuhaus. Biosphere Reserve Niedersächsische Elbtalaue. Institut für Ländliche Strukturforschung (IfLS) an der Johann Wolfgang Goethe-Universität, Frankfurt am Main. 15 p.

Kuntze, H. (1984): Bewirtschaftung und Düngung von Moorböden. Niedersächsisches Landesamt für Bodenforschung – Bodentechnologisches Institut Bremen, Bremen. 80 p.

Kuntze, H. (1988): Nährstoffdynamik der Niedermoore und Gewässereutrophierung. Telma 18: 61–72.

Kuratorium für Technik und Bauwesen in der Landwirtschaft (KTBL) (ed.) (2004): Direktvermarktung 2004. Daten zur Kalkulation der Kosten und des Arbeitszeitbedarfs. KTBL-Datensammlung. 3rd edition. Kuratorium für Technik und Bauwesen in der Landwirtschaft, Münster. 110 p.

Kuratorium für Technik und Bauwesen in der Landwirtschaft (KTBL) (ed.) (2009): Faustzahlen in der Landwirtschaft. 14th edition. Kuratorium für Technik und Bauwesen in der Landwirtschaft, Darmstadt.

Kuratorium für Technik und Bauwesen in der Landwirtschaft (KTBL) (ed.) (2011): Die Leistung-Kostenrechnung in der landwirtschaftlichen Betriebsplanung. KTBL-Schrift 486. Kuratorium für Technik und Bauwesen in der Landwirtschaft, Münster. 96 p.

Kuratorium für Technik und Bauwesen in der Landwirtschaft (KTBL) (ed.) (2012): Anbau und thermische Nutzung von Miscanthus. KTBL-Heft 95. Kuratorium für Technik und Bauwesen in der Landwirtschaft, Darmstadt. 52 p.

Lachmann, L., Marczakiewicz, P. & Grzywaczewski, G. (2010): Protecting Aquatic Warblers (*Acrocephalus paludicola*) through a landscape-scale solution for the management of fen peat meadows in Poland. Grassland Science in Europe 15: 711–713.

Lai, D. (2009): Methane dynamics in northern peatlands: A review. Pedosphere 19: 409–421.

Lake Winnipeg Stewardship Board (2005): Our Collective Responsibility. Reducing Nutrient Loading to Lake Winnipeg. An Interim Report to the Minister of Manitoba Water Stewardship. 66 p. http://gov.mb.ca/waterstewardship/questionnaires/surface_water_management/pdf/connected_docs/LWSBInterimReportJan05.pdf.

Lakner, S., Brümmer, B., Cramon-Taubadel, S., Heß, J., Isselstein, J., Liebe, U., Marggraf, E., Mußhoff, O., Theuvsen, L., Tscharntke, T., Westphal, C. & Weise, G. (2012): Der Komissionsvorschlag zur GAP-Reform 2013 aus Sicht von Göttinger und Witzenhäuser Agrarwissenschaftler(inne)n. Diskussionsbeitrag 1208. Georg-August-Universität, Göttingen. Department für Rurale Entwicklung.

Lakshman, G. (1984): A study to evaluate the potential of cattail as an energy crop. SRC technical report 162. Saskatchewan Research Council, Saskatoon. 178 p.

Länderfachbehörden (2012): Potenziale und Ziele zum Moor- und Klimaschutz. Gemeinsame Erklärung der Naturschutzbehördern. Landesamt für Landwirtschaft, Umwelt und ländliche Räume Schleswig-Holstein, Landesamt für Umwelt, Gesundheit und Verbraucherschutz Brandenburg, Landesamt für Umwelt, Naturschutz und Geologie Mecklenburg-Vorpommern (LUNG MV), Ministerium für Landwirtschaft, Umwelt und Verbraucherschutz Mecklenburg-Vorpommern (MLUV), Niedersächsischer Landesbetrieb für Wasserwirtschaft, Küsten- und Naturschutz & Bayerisches Landesamt für Umwelt (BLU), Kiel. 18 p. http://www.umweltdaten.landsh.de/nuis/upool/gesamt/moore/moorresolution.pdf, zuletzt geprüft am 22.06.2015.

Landesforschungsanstalt Mecklenburg-Vorpommern (LFA M-V) (lfd.): Testbetriebsergebnisse des Landes Mecklenburg-Vorpommern. http://www.landwirtschaft-mv.de/cms2/LFA_prod/LFA/content/de/Fachinformationen/Betriebswirtschaft/index.jsp?&artikel=970, last access 17 May 2015.

Landesumweltamt (LUA) Brandenburg (ed.) (2004): Leitfaden zur Renaturierung von Feuchtgebieten in Brandenburg. LUA Ökologie 50. Landesumweltamt Brandenburg, Potsdam. 192 p.

Landgesellschaft Mecklenburg-Vorpommern (2011): Stark fürs Land. Geschäftsbericht 2011, Leezen. 60 p. http://www.imcg.net/media/download_gallery/books/gprm_01.pdf.

Landry, J., Pouliot, R., Gaudig, G., Wichmann, S. & Rochefort, L. (2011): Sphagnum farming workshop in the

Canadian Maritimes: international research efforts and challenges. IMCG Newsletter 2011/2–3: 42–44.

Landström, S., Lomakka, L. & Andersson, S. (1996): Harvest in spring improves yield and quality of reed canary grass as a bioenergy crop. Biomass and Bioenergy 11: 333–341.

Langeveld, C.A., Segers, R., Dirks, B.O., van den Pol-van Dasselaar, A., Velthof, G.L. & Hensen, A. (1997): Emissions of CO_2, CH_4 and N_2O from pasture on drained peat soils in the Netherlands. European Journal of Agronomy 7: 35–42.

Langhoff, W. (1935): Die Moorsoldaten. 13 Monate Konzentrationslager. Schweizer Spiegel Press, Zürich.

Lappalainen, E. (ed.) (1996): Global peat resources. International Peat Society, Jyväskylä. 359 p.

Lassey, K.R., Ulyatt, M.J., Martin, R.V., Walker, C.F. & Shelton, I.D. (1997): Methane emissions measured directly from grazing livestock in New Zealand. Atmospheric Environment 31: 2905–2914.

LAWA (Bund/Länder Arbeitsgemeinschaft Wasser) (2014): Empfehlung zur Übertragung flussbürtiger, meeresökologischer Reduzierungsziele ins Binnenland. Produktdatenblatt 2.4.7. 17 p.

LAWA-AO (Bund/Länder Arbeitsgemeinschaft Wasser, ständiger Ausschuss "Oberirdische Gewässer und Küstengewässer") (2015): Rahmenkonzeption Monitoring. Teil B Bewertungsgrundlagen und Methodenbeschreibungen. Arbeitspapier II Hintergrund- und Orientierungswerte für physikalisch-chemische Qualitätskomponenten zur unterstützenden Bewertung von Wasserkörpern entsprechend EG-WRRL. 32 p. http://www.wasserblick.net/servlet/is/142684/?highlight=rakon, last access 06 September 2015.

Lebedeva, E.A. (1998): Waders in agricultural habitats of European Russia. International Wader Studies 10: 315–324.

LeBlanc, C. (2003): Ecolabelling in the fisheries sector. In: Mann Borgese, E., Chircop, A. & McConnell, M.L. (eds.): Ocean yearbook 17. University of Chicago Press, Chicago, IL: 93–141.

Leffler, S. (2007): Gaswechsel, Kohlenstoffbilanz und Biomasseproduktion bei *Typha angustifolia* L. PhD thesis. Universität Ulm, Ulm. Fakultät für Naturwissenschaften. 177 p.

Lehrkamp, H. (1987): Die Auswirkungen der Melioration auf die Bodenentwicklung im Randow-Welse-Bruch. PhD thesis. Humboldt-Universität, Berlin. Sektion Pflanzenproduktion.

Lehrkamp, H. & Zeitz, J. (2014): Landnutzung der Moore in der Region bis Anfang der 1990er Jahre: Historsicher Rahmen. In: Luthardt, V. & Zeitz, J. (eds.): Moore in Brandenburg und Berlin. Natur+Text, Rangsdorf: 93–97.

Leifeld, J., Gubler, L. & Grünig, A. (2011): Organic matter losses from temperate ombrotrophic peatlands: an evaluation of the ash residue method. Plant and Soil 341: 349–361.

Lemm, R. (2005): Abschlussbericht zum Forschungsvorhaben Anbau von Schilf als nachwachsender Rohstoff für die Verwendung auf Reithdächern. Fakultät Mathematik- und Naturwissenschaften, Universität Oldenburg, Oldenburg. 36 p.

Lenschow, U. (1997): Landschaftsökologische Grundlagen und Ziele zum Moorschutz in M-V. Materialien zur Umwelt in Mecklenburg-Vorpommern. Landesamt für Umwelt, Naturschutz und Geologie Mecklenburg-Vorpommern (LUNG MV). 72 p.

Lenz, V. (2012): Emissionsminderungsmaßnahmen bei Halmgutfeuerungsanalagen. In: Fachagentur Nachwachsende Rohstoffe e.V. (FNR) (ed.): Gülzower Fachgespräche. 2. Internationale Fachtagung Strohenergie, Berlin, 29–30 March 2012. Gülzower Fachgespräche 38. Fachagentur Nachwachsende Rohstoffe e. V. (FNR), Gülzow-Prüzen: 35–38.

Leponiemi, A. (2011): Fibres and energy from wheat straw by simple practice. Espoo 2011. VTT Publications.

Lewandowski, I., Scurlock, J.M., Lindvall, E. & Christou, M. (2003): The development and current status of perennial rhizomatous grasses as energy crops in the US and Europe. Biomass and Bioenergy 25: 335–361.

Ley, A. & Weitz, L. (eds.) (2004): Praxis Bürgerbeteiligung. Ein Methodenhandbuch. Arbeitshilfen für Selbsthilfe- und Bürgerinitiativen 30. Stiftung Mitarbeit, Bonn. 312 p.

Limin, S.H., Jentha, Y. & Ermiasi, Y. (2007): History of the development of tropical peatland in Central Kalimantan, Indonesia. Tropics 16: 291–301.

Lindner, H. (1963): Über die Abhängigkeit der Scherfestigkeit von Böden verschiedener mechanischer Zusammensetzung von der Dichte, der Porengrößenverteilung sowie dem Wassergehalt. Archives of Agronomy and Soil Science 7: 11–20.

Liu, Z., Jin, Z., Li, Y., Li, T., Gu, J. & Gao, S. (2007): Sediment phosphorus fractions and profile distribution at different vegetation growth zones in a macrophyte dominated shallow Wuliangsuhai Lake, China. Environmental Geology 52: 997–1005.

Lockow, K.-W. (1996): Ertragstafel für die Roterle (*Alnus glutinosa* [L.] Gaertn.) in Mecklenburg-Vorpommern. Ministerium für Landwirtschaft und Naturschutz, Eberswalde. 67 p.

LogLogic (2013): LogLogic Softrak. http://www.loglogic.co.uk/softrak.php.

Luamkanchanaphan, T., Chotikaprakhan, S. & Jarusombati, S. (2012): A study of physical, mechanical and thermal properties for thermal insulation from Narrow-leaved Cattail fibers. APCBEE Procedia 1: 46–52.

Luick, R. (2002): Möglichkeiten und Grenzen extensiver Weidesysteme mit besonderer Berücksichtigung von Feuchtgebieten. Laufener Seminarbeiträge 1: 5–21.

Lumkes, M.L. (1969): De Rietbundelmachine: een nieuw werktuig als sluitstuk van een onderzoekprogramma van het PAW tot mechanisatie van de rietoogst. Mededeling 161. Proefstation voor de Akker-en Weidebouw, Wageningen. 42 p.

Luthardt, V. (1987): Ökologische Untersuchungen an landwirtschaftlich genutzten tiefgründigen Niedermoorstandorten unterschiedlicher Bodenentwicklung. PhD thesis. Akademie der Landwirtschaftswissenschaften der DDR. 140 p.

Luthardt, V. & Zeitz, J. (eds.) (2014): Moore in Brandenburg und Berlin. Natur+Text, Rangsdorf. 384 p.

Lutze, G., Assmann, R., Wieland, R., Voß, M. & Wenkel, K.-O. (2000): Elanus – Prototyp für ein Entscheidungsunterstützungssystem zur Landschaftsanalyse und zur integrativen Bewertung alternativer Landnutzungsstrategien. Zeitschrift für Agrarinformatik 8: 28–35.

Maibach, M., Sieber, N., Bertenrath, R., Ewringmann, D., Koch, L., Thöne, M. & Bickel, P. (2007): Praktische Anwendung der Methodenkonvention: Möglichkeiten der Berücksichtigung externer Umweltkosten bei Wirtschaftlichkeitsrechnungen von öffentlichen Investitionen. Endbericht zum UFOPLAN-Vorhaben 203 14 127. FIFO Köln und INFRAS Zürich, Köln/Zürich. 109 p. http://www.umweltbundesamt.de/sites/default/files/medien/publikation/long/3194.pdf, last access 17 May 2015.

Maljanen, M., Sigurdsson, B.D., Guðmundsson, J., Óskarsson, H., Huttunen, J.T. & Martikainen, P.J. (2010): Greenhouse gas balances of managed peatlands in the Nordic countries – present knowledge and gaps. Biogeosciences 7: 2711–2738.

Malkus, J. (1997): Habitatpräferenzen und Mobilität der Sumpfschrecke (*Stethophyma grossum* L. 1758) unter besonderer Berücksichtigung der Mahd. Articulata 12: 1–18.

Mälson, K., Backéus, I. & Rydin, H. (2008): Long-term effects of drainage and initial effects of hydrological restoration on rich fen vegetation. Applied Vegetation Science 11: 99–106.

Mander, Ü., Järveoja, J., Maddison, M., Soosaar, K., Aavola, R., Ostonen, I. & Salm, J.-O. (2012): Reed canary grass cultivation mitigates greenhouse gas emissions from abandeoned peat extraction areas. Global Change Biology Bioenergy 4: 462–474.

Martikainen, P.J., Nykänen, H., Crill, P. & Silvola, J. (1993): Effect of a lowered water table on nitrous oxide fluxes from northern peatlands. Nature 366: 51–53.

Martin, R.V., Ginzky, H., Henke-Jelit, S., Kasten, H., Gabriel, A., Dahlmann, I., Oechtering, E., Kock, D. & Faensen-Thiebes, A. (2011): Klimawandel - Betroffenheit und Handlungsempfehlungen des Bodenschutzes. Möglichkeiten der rechtlichen Verankerung des Klimaschutzes im Bodenschutzrecht. Arbeitsgemeinschaft Bodenschutz des Bundes und der Länder, Ständiger Ausschuss Recht (BORA).

Massenbach, G. von (1887): Praktische Anleitung zur Rimpau'schen Moordammkultur. Paul Parey, Berlin. 31 p.

McCormick, K. & Kåberger, T. (2007): Key barriers for bioenergy in Europe: Economic conditions, know-how and institutional capacity, and supply chain co-ordination. Biomass and Bioenergy 31: 443–452.

McGinn, S.M., Beauchemin, K.A., Flesch, T.K. & Coates, T. (2009): Performance of a dispersion model to estimate methane loss from cattle in pens. Journal of environmental quality 38: 1796–1802.

MEA (2005) Millennium Ecosystem Assessment. Ecosystems and human well-being. Synthesis. Island Press, Washington, D.C. 137 p.

Meyer, K. (2012): Sparsam transportieren – Wer fährt günstiger: Traktor, Unimog oder LKW? Bauernzeitung 35: 44–45.

Meyerhoff, J., Angeli, D. & Hartje, V. (2012): Valuing the benefits of implementing a national strategy on biological diversity – The case of Germany. Environmental Science and Policy 23: 109–119.

Miettinen, J. & Liew, S.C. (2010): Status of peatland degradation and development in Sumatra and Kalimantan. Ambio 39: 394–401.

Miettinen, J., Hooijer, A., Shi, C., Tollenaar, D., Vernimmen, R., Liew, S.C., Malins, C. & Page, S.E. (2012a): Extent of industrial plantations on Southeast Asian peatlands in 2010 with analysis of historical expansion and future projections. GCB Bioenergy 4: 908–918.

Miettinen, J., Hooijer, A., Tollenaar, D., Page, S.E., Malins, C., Vernimmen, R., Chenghua, S. & Liew, S.C. (2012b): Historical Analysis and Projection of Oil Palm Plantation Expansion on Peatland in Southeast Asia. International Council of Clean Transportation (ICCT). White Paper, 17, Indirect Effects of Biofuel Production, Washington, D.C. 54 p. http://www.theicct.org/sites/default/files/publications/ICCT_palm-expansion_Feb2012.pdf.

Miettinen, J., Shi, C. & Liew, S.C. (2012c): Two decades of destruction in Southeast Asia's peat swamp forests. Frontiers in Ecology and the Environment 10: 124–128.

Miller, R.L. (2011): Carbon gas fluxes in re-established wetlands on organic soils differ relative to plant community and hydrology. Wetlands 31: 1055–1066.

Minister for agriculture, environment and consumer protection Mecklenburg-Vorpommern (2013): Personal communication, 17 April 2013.

Ministerium für Landwirtschaft, Umwelt und Verbraucherschutz Mecklenburg-Vorpommern (MLUV) (ed.) (2009): Konzept zum Schutz und zur Nutzung der Moore. Fortschreibung des Konzeptes zur Bestandssicherung und zur Entwicklung der Moore (Moorschutzkonzept). Ministerium für Landwirtschaft, Umwelt und Verbraucherschutz Mecklenburg-Vorpommern, Schwerin. 109 p.

Ministry of Environmental Protection of the People's Republic of China (MEP) (2008): Report on the State of the Environment in China. Ministry of Environmental Protection of the People's Republic of China (MEP), China. 52 p. http://english.mep.gov.cn/down_load/Documents/201002/P020100225377359212834.pdf.

Ministry of Forestry, Jakarta (MoF) (ed.) (2010): Laporan Perkembangan Pemanfaatan dan Penggunaan Hutan Produksi Triwulan IV. MoF Jakarta: Direktorat Jenderal Bina Usaha Keutanan, Direktorat Bina Rencana Pemanfaatan dan Usha Kawasan, Departmen Kehutanan, Jakarta. http://www.dephut.go.id/index.php?q=id/node/6981.

Minkkinen, K. & Laine, J. (2006): Vegetation heterogenity and ditches create spatial variability in methane fluxes from peatlands drained for forestry. Plant and Soil 285: 289–304.

Mishan, E.J. & Quah, E. (2007): Cost-benefit analysis. 5th edition. Routledge, London, New York. 316 p.

Mohaupt, V., Richter, S., Völker, J. & Borchardt, D. (2012): Bewirtschaftungspläne zur Wasserrahmenrichtlinie in Deutschland: Resultate und Schlussfolgerungen. Natur und Landschaft 87: 168–176.

Mohr, H.-J. (2007): Komplexe Standortmelioration. Mediale und sektorale Aspekte. In: Behrens, H. & Hoffmann, J. (eds.): Umweltschutz in der DDR. Analysen und Zeitzeugenberichte. Oekom-Verlag, München: 59–79.

Möhring, T. (2012): Abteilungsleiter Gut Darß, 2012.

Moore, T.R., Roulet, N.T. & Waddington, J.M. (1998): Uncertainty in predicting the effect of climatic change on the carbon cycling of Canadian peatlands. Climatic Change 40: 229–245.

Moore, T.R., Young, A. de, Bubier, J.L., Humphreys, E.R., Lafleur, P.M. & Roulet, N.T. (2011): A multi-year record of methane flux at the Mer Bleue Bog, Southern Canada. Ecosystems 14: 646–657.

Morrissey, L.A. & Livingston, G.P. (1992): Methane emissions from Alaska Arctic tundra. An assessment of local spatial variability. Journal of Geophysical Research 97: 16661–16670.

Mucha, Ł. (2011): Short history of ratrak in swamps. Aquatic Warbler & Biomass LIFE+ Project Newsletter 1: 6.

Müller, J. (2009): Forestry and water budget of the lowlands in northeast Germany – consequences for the choice of tree species and for forest management. Journal of Water and Land Development 13a: 133–148.

Müller, J. & Bauer, R. (2006): Futterkonservierung. In: Munzert, M. & Frahm, J. (eds.): Pflanzliche Erzeugung. Landwirtschaft 1. BLV, München: 865–934.

Müller, J. & Heilmann, S. (2011): Stand und Entwicklung der agrarischen Nutzung von Niedermoorgrünland in Mecklenburg-Vorpommern. Telma: 235–248.

Müller, J., Jantzen, C. & Kayser, M. (2012): The biogas potential of *Juncus effusus* L. using solid phase fermentation technique. Grassland Science in Europe 17: 387–389.

Müller, L., Behrendt, A., Shepherd, T.G., Schindler, U. & Kaiser, T. (2007): Implications of soil substrate and land use for properties of fen soils in North-East Germany Part III: Soil quality for grassland use. Archives of Agronomy and Soil Science 53: 137–146.

Müller-Motzfeld, G. & Schmidt, J. (2008): Rote Liste der gefährdeten Laufkäfer. Mecklenburg-Vorpommern. Ministerium für Landwirtschaft, Umwelt und Verbraucherschutz Mecklenburg-Vorpommern, Schwerin. 32 p.

Mundel, G. (1969): Untersuchungen zur Entstehung des Havelländischen Lurches und seiner Veränderungen durch Meliorationsmaßnahmen mit besonderer Berücksichtigung der Torfmineralisation. PhD thesis. Deutsche Akademie der Landwirtschaftswissenschaften zu Berlin, Berlin. Institut für Grünland- und Moorforschung Paulinenaue.

Mundel, G. (1976): Untersuchungen zur Torfmineralisation in Niedermooren. Archiv für Acker- und Pflanzenbau und Bodenkunde: 669–679.

Mußhoff, O. & Hirschauer, N. (2011): Modernes Agrarmanagement. Betriebswirtschaftliche Analyse und Planungsverfahren. 2nd edition. Franz Vahlen, München. 571 p.

Muster, C., Gaudig, G., Krebs, M. & Joosten, H. (2015): Sphagnum farming: the promised land for peat bog species? Biodiversity and Conservation 24: 1–21.

Mutschler, T. (2004): Lac Alaotra. In: Goodman, S.M. & Benstead, J.P. (eds.): The natural history of Madagascar. University of Chicago Press, Chicago, London: 1530–1534.

Muuß, U. (1996): Anreicherungspflanzungen im tropischen Feuchtwald Sumatras – eine waldbauliche Herausforderung. Forstarchiv 67: 65–70.

Myhre, G., Shindell, D., Breon, F.-M., Collins, W., Fuglestvedt, J., Huang, J., Koch, D., Lamarque, J.-F., Lee, D., Mendoza, B., Nakajima, T., Robock, A., Rotstayn, L., Stephens, G. & Zhan, H. (2013): Anthropogenic and natural radiative forcing. In: Stocker, T.F., Qin, D., Plattner, G.-K., Tignor, M., Allard, S.K., Boschung, J. Nauels, A., Xia, Y., Bex, V. & Midgley, P.M. (eds.): The physical science basis. Contribution of Working Group I to the Fifth Assessment Report of the Intergovernmental Panel on Climate Change. Intergovernmental Panel on Climate Change (IPCC), Cambridge, New York: 659–740. https://www.ipcc.ch/pdf/assessment-report/ar5/wg1/WG1AR5_Chapter08_FINAL.pdf.

naporo (2015): Naporo Klima Dämmstoff. http://www.naporo.com, last access 30 April 2015.

natureplus e.V. (2013): natureplus. http://www.natureplus.org/de/aktuelles/home/.

Naturland (2012): Naturland Richtlinien Erzeugung, Gräfelfing. 50 p. http://www.naturland.de/fileadmin/MDB/documents/Richtlinien_deutsch/Naturland-Richtlinien_Erzeugung.pdf, last access 08 April 2013.

Nehring, K. (1972): Lehrbuch der Tierernährung und Futtermittelkunde. 9th edition. Neumann, Radebeul. 599 p.

Neumann, D. & Krüger, M. (1991): Schilfhalme im Winter – Überwinterungsquartier für Insekten und Spinnen sowie Nahrungsquelle für insektivore Singvögel. Natur und Landschaft 66: 166–168.

Niedersächsisches Oberverwaltungsgericht (2011): Beschluss vom 30.3.2011, 4 LA24/10. In: Beck-online (BeckRS; Juristische Datenbank):

Nielsen, O.-K., Lyck, E., Mikkelsen, M.H., Hoffmann, L., Gyldenkærne, S., Winther, M., Nielsen, M., Fauser, P., Thomsen, M., Plejdrup, M.S., Illerup, J.B., Sørensen, P.B. & Vesterdal, L. (2008): Denmark's national inventory report 2008. Emission inventories 1990–2006 – submitted under the United Nations Framework Convention on Climate Change. National Environmental Research Institute, University of Aarhus, Aarhus. 707 p.

Nikiforov, M.E., Bambalov, N.N., Kozulin, A.V., Tanovitskaya, N. & Rakovich, V.A. (2013): Аналитический доклад Реабилитация и устойчивое управление торфяниками для обеспечения экологической безопасности и сохранения биологического разнообразия. [Analytical presentation on rehabilitation and sustainable use of peatlands for securing ecological safety and protection of biological diversity].

Nussbaum, M. (2003): Frauen und Arbeit – Der Fähigkeitenansatz. Zeitschrift für Wirtschafts- und Unternehmensethik 4: 8–31.

Oates, L.G., Jackson, R.D. & Allen-Diaz, B. (2008): Grazing removal decreases the magnitude of methane and the variability of nitrous oxide emissions from springfed wetlands of a California oak savanna. Wetlands Ecology and Management 16: 395–404.

Obernberger, I., Brunner, T. & Barnthaler, G. (2006): Chemical properties of solid biofuels – significance and impact. Standarisation of solid biofuels in Europe. Biomass and Bioenergy 30: 973–982.

Oberpaur, C., Puebla, V., Vaccarezza, F. & Arévalo, M.E. (2010): Preliminary substrate mixtures including peat moss (*Sphagnum magellanicum*) for vegetable crop nurseries. Ciencia e Investigación Agraria 37: 123–132.

Oehmke, C. & Wichtmann, W. (2011): Festbrennstoffe aus Paludikultur – Produktivität und Verbrennungseignung von Halmgut aus nassen und wiedervernässten Mooren. In: Deutsches BiomasseForschungsZentrum (DBFZ) (ed.): Energetische Nutzung von Landschaftspflegematerial. Dokumentation. Energetische Nutzung von Landschaftspflegematerial, Berlin, 01–02 March 2011. Bundesministerium für Umwelt, Naturschutz und Reaktorsicherheit (BMU). Schriftenreihe des BMU-Förderprogramms "Energetische Biomassenutzung" 1, Leipzig.

Oleszczuk, R., Regina, K., Szajdak, L., Höper, H. & Maryganova, V. (2008): Impacts of agricultural utilization of peat soils on the greenhouse gas balance. In: Strack, M. (ed.): Peatlands and climate change. International Peat Society, Jyväskylä: 70–97.

Organisation for Economic Co-operation and Development (OECD) (ed.) (1997): Investing in Biological Diversity, Paris.

Organisation for Economic Co-operation and Development (OECD) (ed.) (2006): Das neue Paradigma für den ländlichen Raum. Politik und Governance. Organisation for Economic Co-operation and Development (OECD), Paris. 187 p. http://edok.ahb.niedersachsen.de/07/525946535.pdf, last access 06 September 2015.

Ostendorp, W. (1993): Schilf als Lebensraum. Beihefte zu den Veröffentlichungen für Naturschutz und Landschaftspflege in Baden-Württemberg 68: 173–280.

Osterburg, B., Rüter, S., Freibauer, A., de Witte, T., Elsasser, P., Kätsch, S., Leischner, B., Paulsen, H.M., Rock, J., Röder, N., Sanders, J., Schweinle, J., Steuk, J., Stichnothe, H., Stürmer, W., Welling, J. & Wolff, A. (2013): Handlungsoptionen für den Klimaschutz in der deutschen Agrar- und Forstwirtschaft. Thünen-Report 11. Johann Heinrich von Thünen-Institut, Braunschweig. 162 p.

Özesmi, U.A.R.A. (2003): The ecological economics of harvesting sharp-pointed rush (Juncus acutus) in the Kizilirmak Delta, Turkey. Human-Ecology 31: 645–655.

Paepke, A. (1992): Untersuchungen zu Kennwerten der Torfzersetzung in flachgründigen Niedermooren. PhD thesis. Humboldt-Universität, Berlin. Fakultät für Landwirtschaft und Gartenbau. 260 p.

Page, S.E., Rieley, J.O. & Banks, C. (2011): Global and regional importance of the tropical peatland carbon pool. Global Change Biology 17: 798–818.

Pahkala, K. & Pihala, M. (2000): Different plant parts as raw material for fuel and pulp production. Industrial Crops and Products 11: 119–128.

Päivänen, J. & Vasander, H. (1994): Carbon balance in mire ecosystems. World Resource Review 6: 102–111.

Pantenius, W.H. & Schönert, C. (1999): Zwischen Haff und Heringsdorf. Das Thurbruch auf Usedom. Neuendorf, Neubrandenburg. 80 p.

Paulrud, S. & Nilsson, C. (2001): Briquetting and combustion of spring-harvested reed canary-grass: effect of fuel composition. Biomass and Bioenergy 20: 25–35.

Paulrud, S., Nilsson, C. & Öhman, M. (2001): Reed canary-grass ash composition and its melting behaviour during combustion. Fuel 80: 1391–1398.

Pearce, D.W. (1993): Economic values and natural world. MIT Press, London. 129 p.

Peatland Ecology Research Group (ed.) (2009): Production of Berries in Peatlands. Guide produced under the supervision of Line Rochefort and Line Lapointe. University of Laval, Quebec. 134 p.

Peet, R.K., Wentworth, T.R. & White, P.S. (1998): A flexible, multipurpose method for recording vegetation composition and structure. Castanea 63: 262–274.

Perera, B. (2011): Reproductive cycles in buffalo. Animal Reproduction Science 124: 194–199.

Permien, T. & Ziebarth, M. (2011): MoorFutures – Innovative Finanzierung von Projekten zur Moorwiedervernässung in Mecklenburg-Vorpommern. Natur und Landschaft 87: 77–80.

Petermann, S., Orban, S., Salge, H.-J., Pohlenz, F., Ringena, I., Zech, K., Brügemann, M. & Maiworm, K. (2008): Heckrindhaltung in Naturschutzgebieten – aktuelle Erfahrungen. Mellumrat 7: 68–73.

Petersen, A. (1952): Die neue Rostocker Grünlandschätzung. Abhandlungen der Deutschen Akademie d. Wissenschaften zu Berlin 1. Akademie Verlag, Berlin. 20 p.

Petersohn, H. (2005): Data mining. Verfahren, Prozesse, Anwendungsarchitektur. Oldenbourg, München. 342 p.

Picardt, J. (1660): Annales Drenthiæ: ofte een provisioneel ontworp en beginsel van seeckere antiquiteten, en beschrijvinghe sommigher ghedenckwaerdige gheschiedenissen, die in de landtschap Drenth gepasseert zijn, van de geboorte Christi af, tot op desen tijdt / t'samen vergadert, en aen 't licht ghebracht, door Johan Picardt. In: Picardt, J. & van der Sanden, W.A.B.: Korte Beschryvinge Van Eenige Vergetene En Verborgene Antiquiteten. Sidestone Press, Leiden. 302 p.

Pigou, A.C. (1920): Economics of welfare. Macmillan, London.

Pollex, A., Ortwein, A. & Kaltschmitt, M. (2012): Thermo-chemical conversion of solid biofuels. Biomass Conversion and Biorefinery 2: 21–39.

Poulin, B., Lefebvre, G. & Mathevet, R. (2005): Habitat selection by booming bitterns Botaurus stellaris in French Mediterranean reed-beds. Oryx 39: 265–274.

Poulin, B., Lefebvre, G., Allard, S.K. & Mathevet, R. (2009): Reed harvest and summer drawdown enhance bittern habitat in the Camargue. Biological Conservation 142: 689–695.

Pöyry (ed.) (2006a): Technical assistance for the sustainable development of non-wood pulp and paper industry. Review of fibre and raw material availability by province. Technical Report Module 3, China. http://s3.amazonaws.com/zanran_storage/www.tem.fi/ContentPages/16078944.pdf, last access 06 September 2015.

Pöyry (ed.) (2006b): Technical assistance for the sustainable development of non-wood pulp and paper industry. Review of existing production capacity by province and product group. Technical Report Module 2, China.

Pratt, D.C., Dubbe, D.R., Garver, E.G. & Linton, P.J. (1984): Wetland Biomass Production: Emergent Aquatic Menagement options and Evaluations. A final Subcontract Report. University of Minnesota. U.S. Department of Energy, Minnesota.

Pratt, D.C., Dubbe, D.R., Garver, E.G. & Johnson, W.D. (1988): Cattail (Typha spp.) Biomass production – Stand Management and Sustainable Yields – subcontract report. report. University of Minnesota. National Technical Information Service, Department of Commerce, Minnesota.

Preuß, J. D. E. (1834): Friedrich der Große. Eine Lebensgeschichte. Vol. 4. Nauckshe Buchhandlung, Berlin. 500 p.

Pries, M. (2007): Grobfutterqualität. Bedeutung, Anforderung, Folgerung. Grünlandtag. Deutsche Landwirtschaftsgesellschaft (DLG). Deutsche Landwirtschaftsgesellschaft (DLG), Arnstadt, 2007. http://www.dlg.org/fileadmin/downloads/dates/gruenland/3_Pries.pdf, last access 06 September 2015.

Prochnow, A., Heiermann, M., Plöchl, M., Linke, B., Idler, C., Amon, T. & Hobbs, P.J. (2009): Bioenergy from permanent grassland. A review: 1. Biogas. Bioresource Technology 100: 4931–4944.

Quinty, F. & Rochefort, L. (2003): Peatland restoration guide. 2nd edition. Canadian Sphagnum Peat Moss Association, St. Albert, Alberta. 106 p.

RAL (2013a): Der Blaue Engel. http://www.blauer-engel.de/, last access 26 April 2013.
RAL (2013b): EU Ecolabel. http://www.eu-ecolabel.de/, last access 26 April 2013.
Rappold, A.G., Stone, S.L., Cascio, W.E., Neas, L.M., Kilaru, V.J., Carraway, M.S., Szykman, J.J., Ising, A., Cleve, W.E., Meredith, J.T., Vaughan-Batten, H., Deyneka, L. & Devlin, R.B. (2011): Peat bog wildfire smoke exposure in rural North Carolina is associated with cardiopulmonary emergency department visits assessed through syndromic surveillance. Environmental Health Perspectives 119: 1415–1420.
Ratzke, U. & Mohr, H.-J. (2005): Böden in Mecklenburg-Vorpommern: Abriss ihrer Entstehung, Verbreitung und Nutzung. LUNG MV, Güstrow. 84 p.
Raussen, T., Hackländer, G. & Siepenkothen, H.-J. (2010): Konditionierung von Grünabfällen zur regionalen energetischen Nutzung. Bio- und Sekundärrohstoffverwertung stofflich-energetisch. In: Witzenhausen-Institut für Abfall, Umwelt und Energie (ed.): Bio- und Sekundärrohstoffverwertung. Stofflich und Energetisch. Neues aus Forschung und Praxis 5. Witzenhausen-Institut für Abfall, Umwelt und Energie, Witzenhausen: 561–575.
Rechberger, C. (2003): Schilf (*Phragmites australis*), Analyse der Ernte- und Verwertungsmöglichkeiten unter besonderer Berücksichtigung des Neusiedler Sees. Diploma thesis. Fachhochschule Wiener Neustadt für Wirtschaft und Technik, Studiengang Produkt- und Projektmanagement, Wieselburg. 131 p.
Reeves, P.N. & Champion, P.D. (2004): Effect of livestock grazing on wetlands. Literature Review. NIWA Client Report HAM 2004-059. National Institut of Water and Atmospheric Research (NIWA), Hamilton. 33 p.
Regierungspräsidium Gießen (2011): Verfahrensbuch Wasserrechtliches Planfeststellungs-/Plangenehmigungsverfahren. Stand: Januar 2011.
Regina, K., Nykänen, H., Silvola, J. & Martikainen, P.J. (1996): Fluxes of nitrous oxide from boreal peatlands as affected by peatland type, water table level and nitrification capacity. Biogeochemistry 35: 401–418.
Regina, K., Syvasalo, E., Hannukkala, A. & Esala, M. (2004): Fluxes of N_2O from farmed peat soils in Finland. European Journal of Soil Science 55: 591–599.
Rex, E. & Baumann, H. (2007): Beyond ecolabels: what green marketing can learn from conventional marketing. Journal of Cleaner Production 15: 567–576.
Rieley, J.O. & Page, S.E. (2008): Master Plan for the Rehabilitation and Revitalisation of the Ex-Mega Rice Project Area in Central Kalimantan. The Science of Tropical Peatlands and the Central Kalimantan Peatland Development Area. Technical Review No. 1. Euroconsult Mott MacDonald and Deltares & Delft Hydraulics, Jakarta. 66 p.
Rijksen, H. & Persoon, G. (1991): Food from Indonesia's swamp forest: ideology or rationality? Landscape and Urban Planning 20: 95–102.
Ritterbusch, D. (2011): Nutzung von Rohr/Schilf – ein umweltverträgliches Entwicklungspotential für die Fischerei? Institut für Binnenfischerei e.V. Potsdam-Sacrow, 30. Institut für Binnenfischerei e.V. Potsdam-Sacrow, Potsdam. 79 p.
Röder, N. & Grützmacher, F. (2012): Emissionen aus landwirtschaftlich genutzten Mooren – Vermeidungskosten und Anpassungsbedarf. Natur und Landschaft 87: 56–61.
Rodewald-Rodescu, L. (1974): Das Schilfrohr. Die Binnengewässer 17. Schweizerbart, Stuttgart. 302 p.
Rönsch, S. & Kaltschmitt, M. (2012): Bio-SNG production – Concepts and their assessment. Biomass Conversion and Biorefinery 2: 285–296.
Rönsch, S., Müller-Langer, F. & Kaltschmitt, M. (2009): Produktion des Erdgassubstitutes Bio-SNG im Leistungsbereich um 30 MWBWL – Eine techno-ökonomische Analyse und Bewertung. Chemie Ingenieur Technik 81: 1417–1428.
Roos, A., Graham, R.L., Hektor, B. & Rakos, C. (1999): Critical factors to bioenergy implementation. Biomass and Bioenergy 17: 113–126.
Rosenthal, G. (1992): Erhaltung und Regeneration von Feuchtwiesen. Vegetationsökologische Untersuchungen auf Dauerflächen. Dissertationes botanicae 182. Cramer, Berlin/Stuttgart. 283 p.
Roßkopf, N., Fell, H., Zeitz, J. (2015). Organic soils in Germany, their distribution and carbon stocks. Catena, 133, 157–170.
Roth, S. (2000): Etablierung von Schilfröhrichten und Seggenriedern auf wiedervernässtem Niedermoor. Shaker, Aachen. 154 p.
Roth, S., Seeger, T., Poschlod, P., Pfadenhauer, J. & Succow, M. (2001): Etablierung von Röhrichten und Seggenrieden. In: Kratz, R. & Pfadenhauer, J. (eds.): Ökosystemmanagement für Niedermoore. Strategien und Verfahren zur Renaturierung. Ulmer, Stuttgart (Hohenheim): 125–133.
Rotherham (2005): The Trade and Environmental Effects of Ecolabels. Assesment and Response. International Institute for Sustainable Development (IISD), Winnipeg. 56 p.
Roulet, N.T., Lafleur, P.M., Richard, P.J.H., Moore, T.R., Humphreys, E.R. & Bubier, J.L. (2007): Contemporary carbon balance and late Holocene carbon accumulation in a northern peatland. Global Change Biology 13: 397–411.
Rückert-John, J., Bormann, I. & John, R. (2013): Repräsentativumfrage zu Umweltbewusstsein und Umweltverhalten im Jahr 2012. Bundesministerium für Umwelt, Naturschutz und Reaktorsicherheit (BMU). Bundesministerium für Umwelt, Naturschutz und Reaktorsicherheit (BMU), Berlin, Marburg. 84 p.
Rühs, M., Schlauderer, R. & Hampicke, U. (2005): Die Ökonomie der Offenhaltung tiergebundener Verfahren – Ergebnisse von Untersuchungen auf Grünland und Truppenübungsplätzen. Naturschutz und Landschaftsplanung 37: 325–335.
Sächsisches Landesamt für Umwelt und Geologie (LfUG) (ed.) (2007): Bodenschutz: Bodenatlas des Freistaates Sachsen. Teil 4: Auswertungskarten zum Bodenschutz. Erläuterungsheft. Dresden. 62 p.
Sachverständigenrat für Umweltfragen (SRU) (ed.) (2009): Für zeitgemäße gemeinsame Agrarpolitik (GAP). Stellungnahme Nr. 14. Sachverständigenrat für Umweltfragen (SRU), Berlin. 28 p. http://www.umweltrat.de/SharedDocs/Downloads/DE/04_Stellungnahmen/2009_11_Stellung_14_GAP.pdf?__blob=publicationFile, zuletzt geprüft am 06.09.2015.
Sachverständigenrat für Umweltfragen (SRU) (ed.) (2012): Umweltgutachten 2012: Verantwortung in einer begrenz-

ten Welt. Sachverständigenrat für Umweltfragen (SRU), Berlin. 420 p. http://www.umweltrat.de/SharedDocs/Downloads/DE/01_Umweltgutachten/2012_06_04_Umweltgutachten_HD.pdf?__blob=publicationFile, zuletzt geprüft am 06 September 2015.

Sainty, G. (1985): Weed control and utilization of aquatic plants of lake Edku and barsik fish farm – Egypt. Food and Agriculture Organization of the United Nations (FAO). Food and Agriculture Organization of the United Nations (FAO). http://www.fao.org/docrep/field/003/R7236E/R7236E00.htm.

Sassner, P., Galbe, M. & Zacchi, G. (2008): Techno-economic evaluation of bioethanol production from three different lignocellulosic materials. Biomass and Bioenergy 32: 422–430.

Sauerbrey, R. & Zeitz, J. (2003): Moore. In: Blume, H.-P., Felix-Hennigsen, P., Fischer, W.R., Frede, H.-G. Horn, R., Stahr, K. (ed.): Handbuch der Bodenkunde. Wiley-VCH, Weinheim: Chapter 3.3.7.

Savcor Indufor Oy (ed.) (2006): Technical assistance for the sustainable development of non-wood pulp and paper industry. Recommendation for long-term concept of non wood pulp production. Conclusion report, China.

Schäfer, A. (1999): Schilfrohrkultur auf Niedermoor – Rentabilität des Anbaus und der Ernte von *Phragmites australis*. Archiv für Naturschutz und Landschaftsforschung 38: 193–216.

Schäfer, A. (2004): Umwelt als knappes Gut. Ökonomische Aspekte der Niedermoorrenaturierung und Gewässerschutz. Archiv für Naturschutz und Landschaftsforschung 43: 87–105.

Schäfer, A. (2009): Moore und Euros – die vergessenen Millionen. Archiv für Forstwesen und Landschaftsökologie 43: 156–160.

Schäfer, A. (2012): Den Nutzen von Ökosystemleistungen indirekt sichtbar machen: Ersatz- Schadens- und Vermeidungskosten. In: Hansjürgens, B., Neßhöver, C. & Schniewind, I. (eds.): Den Nutzen von Ökonomie und Ökosystemleistungen für die Naturschutzpraxis. Bundesamt für Naturschutz (BFN). BfN-Skripten 318, Bonn-Bad Godesberg: 59–66.

Schäfer, A. & Joosten, H. (2005): Erlenaufforstung auf wiedervernässten Niedermooren. Institut für Dauerhaft Umweltgerechte Entwicklung von Naturräumen der Erde (DUENE) e.V., Greifswald. 68 p.

Schäfer, A., Couwenberg, J. & Joosten, H. (2012): Moor-Futures. CO_2-Zertifikate aus Moorwiedervernässung. In: Bundesamt für Naturschutz (BFN) (ed.): Der Nutzen von Ökonomie und Ökosystemleistungen für die Naturschutzpraxis Workshop II: Gewässer, Auen und Moore. BfN Skripten 319. Bundesamt für Naturschutz (BFN), Bonn-Bad Godesberg: 72–82.

Schätzl, R., Schmitt, F., Wild, U. & Hoffmann, U. (2006): Gewässerschutz und Landnutzung durch Rohrkolbenbestände. Wasserwirtschaft 96: 24–27.

Schiefelbein, U., Lenschow, U. & Otto, D. (2011): Moorrevitalisierung in Mecklenburg-Vorpommern – Eine Bilanz der letzten 20 Jahre. Telma Beiheft 4: 73–84.

Schillberg, K. (1996): Altbausanierung mit Naturbaustoffen. AT-Verlag, Aarau. 260 p.

Schleuß, U., Trepel, M., Wetzel, H., Schimming, C.G. & Kluge, W. (2002): Interactions between hydrologic parameters, soils, and vegetation at three minerotrophic peat ecosystems. In: Broll, G., Merbach, W. & Pfeifer, E.M. (eds.): Wetlands in Central Europe. Soil organisms, soil ecological processes, and trace gas emissions. Springer, Berlin, New York: 117–132.

Schmidt, M.H., Lefebvre, G., Poulin, B. & Tscharntke, T. (2005): Reed cutting affects arthropod communities, potentially reducing food for passerine birds. Biological Conservation 121: 157–166.

Schmidt, M.H., Rocker, S., Hanafi, J. & Gigon, A. (2008): Rotational fallows as overwintering habitat for grassland arthropods: the case of spiders in fen meadows. Biodiversity and Conservation 17: 3003–3012.

Schmidt, R., Waybrink, W. v. d., Mundel, G. & Scholz, A. (1981): Kennzeichnung und Beurteilung der Bodenentwicklung auf Niedermoor unter besonderer Berücksichtigung der Degradierung. Forschungsabschlussbericht. Akademie der Landwirtschaftswissenschaften der DDR. Futterproduktion Paulinenaue der Akademie der Landwirtschaftswissenschaften der DDR, Berlin. 124 p.

Schmidt, W. (1980): Zur Bestimmung der Scherfestigkeit von Torfen und Mudden. Wissenschaftlich-technische Informationen für das Meliorationswesen 61: 28–41.

Schmidt, W. (1995): Einfluss der Wiedervernässung auf physikalische Eigenschaften des Moorkörpers der Friedländer Großen Wiese. Zeitschrift für Kulturtechnik und Landentwicklung 36: 107–122.

Schoenichen, W. (ed.) (1926): Bericht über den ersten deutschen Naturschutztag, München, 26–28 July 1925. Beiträge zur Naturdenkmalpflege 10. Gebrüder Borntraeger, Berlin. 560 p.

Scholz, A. (1986): Anzustrebende Grundwasserstände auf tiefgründigem Niedermoor und Möglichkeiten ihrer Realisierung mittels der rohrlosen Dränung. Institut für Futterproduktion Paulinenaue der Akademie der Landwirtschaftswissenschaften der DDR, Berlin. 128 p.

Schönfeld-Bockholt, R. (2005): Das Salzgrünland Mecklenburg Vorpommerns mit seinen häufigsten Pflanzen. Landwirtschaftlicher Wert, Naturschutzwert, ökologischer Zeigerwert, Erkennungsmerkmale der Pflanzen. Universität Rostock. Universität Rostock & Agrar- und Umweltwissenschaftliche Fakultät, Rostock. 129 p.

Schothorst, C. (1977): Subsidence of low moor peat soils in the Western Netherlands. Geoderma 17: 265–291.

Schrader, A. & Kaltofen, H. (1987): Gräser. Deutscher Landwirtschaftsverlag, Berlin. 360 p.

Schrautzer, J. (2004): Niedermoore Schleswig-Holsteins: Charakterisierung und Beurteilung ihrer Funktion im Landschaftshaushalt. Mitteilungen der AG Geobotanik Schleswig-Holstein und Hamburg 63: 1–350.

Schreiner, B.-G. (1967): Technique for estimating the performance of tracked vehicles in muskeg. Journal of Terramechanics 4: 23–29.

Schrier-Uijl, A.P., Veraart, A.J., Leffelar, P.A., Berendse, F. & Veenendaal, E.M. (2011): Release of CO_2 and CH_4 from lakes and drainage ditches in temperate wetlands. Biogeochemistry 102: 265–279.

Schröder, C., Dahms, T., Wichmann, S., Wichtmann, W. & Joosten, H. (2012): Paludikultur. Ein regionales Bioenergiekonzept für Mecklenburg-Vorpommern. Univerität Greifswald, Greifswald.

Schröder, C., Luthardt, V. & Jeltsch, F. (2013): Das Konzept der Ökosystem-Dienstleistungen und ihre Bewertung. Defizitanalyse und Lösungsansätze mittels einer holistischen ökosystemspezifischen Methodik. In: Ibisch, P.L. (ed.): Regionale Anpassung des Naturschutzes an den Klimawandel. Strategien und methodische Ansätze zur Erhaltung der Biodiversität und Ökosystemdienstleistungen in Brandenburg. Hochschule für nachhaltige Entwicklung, Eberswalde: 134–143.

Schroeder, P. (2012): Natürliches Moor oder Landwirtschaftsbrache. Eine Studie über die rezente Entwicklung ungenutzter Moorstandorte als Beitrag zur realistischen Einschätzung von Baseline-Szenarios für Moorwiedervernässung in Mecklenburg-Vorpommern. Diploma thesis. Universität Greifswald. 94 p.

Schulz, K. (2005): Vegetations- und Standortentwicklung des wiedervernässten Grünlandes im Anklamer Stadtbruch (Mecklenburg-Vorpommern). Diploma thesis. Univerität Greifswald, Greifswald. 168 p.

Schulz, K., Timmermann, T., Steffenhagen, P., Zerbe, S. & Succow, M. (2011): The effect of flooding on carbon and nutrient standing stocks of helophyte biomass in rewetted fens. Hydrobiologia 674: 25–40.

Schulze-Hagen, K. (1991): *Acrocephalus paludicola* (Vieillot 1817) – Seggenrohrsänger. In: Glutz von Blotzheim, U.N. & K.M. Bauer (ed.): Handbuch der Vögel Mitteleuropas 12. AULA-Verlag, Wiesbaden: 252–291.

Schuster, J. (1985): Schilfverwertung – Erntestudie. Naturraumpotential Neusiedler See. Auswirkungen des Grünschnittes auf den Seegürtel. In: Bundesministerium für Wissenschaft und Forschung und Gesundheit und Umweltschutz (BMWFGU) & Land Burgenland (eds.): Arbeitsgemeinschaft Gesamtkonzept Neusiedler See. Forschungsbericht 1981–1984. Landesmuseum. Wissenschaftliche Arbeiten aus dem Burgenland Sonderband 72, Mattersburg: 589–618.

Schwaiger, M. & Meyer, A. (eds.) (2011): Theorien und Methoden der Betriebswirtschaft. Handbuch für Wissenschaftler und Studierende. Franz Vahlen, München. 600 p.

Schwalm, U. (1999): Großer Erfolg mit kleinen Halmen. Die Renaissance des Reets. Die traditionsreiche Firma Hiss ist führend im deutschen Reethandel. Und das Geschäft wächst weiter: Immer mehr Bauherren entdecken das Naturmaterial. Ahrensburger Zeitung, 16 October 1999 (42).

Schwalm, U. (2005): Bauen mit Schilf: Der neue Trend. Firmenchef Tom Hiss hat wärmedämmende Wandelemente aus türkischem Schilf entwickelt. Ahrensburger Zeitung, 19 October 2005. http://www.hiss-reet.de/unternehmen/presse/ahrendsburger-19-10-2005.pdf, last access 18 June 2014.

Schwemmer, R. (2010): Entwicklung der Fertigungstechnologie für Rohrkolben-Dämmstoffe. Berichte aus Energie und Umweltforschung. Dynamik mit Verantwortung. Bundesministerium für Verkehr, Innovation und Technologie (bmvit). NACHHALTIGwirtschaften konkret, 69, Wien. 38 p. http://www.fabrikderzukunft.at/fdz_pdf/endbericht_1069_rohrkolben_daemmstoff.pdf, last access 18 June 2014.

Seeholzer, C. (1993): Biosystematische Untersuchungen an schweizerischen Drosera-Arten. Botanica Helvetica 103: 39–53.

Segeberg, H. (1960): Moorsackung durch Grundwasserabsenkung und deren Vorausberechnung mit Hilfe empirischer Formel. Zeitschrift für Kulturtechnik und Flurbereinigung 3: 144–161.

Shepherd, P.A., Rieley, J.O. & Page, S.E. (1997): The relationship between vegetation and peat characteristics in the upper catchment of Sungai Sebangau, Central Kalimantan, Indonesia. In: Rieley, J.O. & Page, S.E. (eds.): Biodiversity and sutainibility of tropical peatlands. Samara Publishing, Cardigan: 191–210.

Shim, J., Warkentin, M., Courtney, J.F., Power, D.J., Sharda, R. & Carlsson, C. (2002): Past, present, and future of decision support technology. Decision Support Systems 33: 111–126.

Shurpali, N.J., Hyvönen, N., Huttunen, J.T., Clement, R.J., Reichestein, M., Nykänen, H., Biasi, C. & Martikainen, P.J. (2009): Cultivation of a perennial grass for bioenergy on a boreal organic soil. Carbon sink and source? Global Change Biology Bioenergy 1: 35–50.

Silva Rodriguez, J. M. (2011): Personal communication, 21 December 2011.

Silvius, M., Simons, H.W. & Verheugt, W. J. M. (1984): Soils, vegetation, fauna and nature conservation of the Berbak Game Reserve, Sumatra, Indonesia. RIN contributions to research on management of natural resouces. Research Institute for Nature Management, Arnhem. 146 p.

Silvius, M. & Suryadiputra, I.N.N. (2005): Review of policies and practices in tropical peat swamp forest management in Indonesia. Wetlands International, Wageningen. 12 p.

Simula, M. (1996): Economics of Certification. In: Certification of Forest Products. Issues and Perspectives, Washington, D.C: 123–136.

Sinning, H. (2005): Planungskommunikation und Nachhaltigkeit in der Stadt-, Regional- und Umweltplanung. In: Michelsen, G. & Godemann, J. (eds.): Handbuch Nachhaltigkeitskommunikation. Grundlagen und Praxis. Oekom, München: 274–286.

Sirin, A.A. & Laine, J. (2008): Peatlands and Greenhouse Gases. In: Parish, F., Charman, D., Joosten, H., Minayeva, T., Silvius, M. & Stringer, L. (eds.): Assessment on peatlands, biodiversity and climate change. Main report. Wetlands International, Wageningen: 118–138.

Skøtt, T. (2011): Straw to Energy. Status, Technologies and Innovation in Denmark 2011. Agro Business Park A/S. Innovation Network for Biomass (INBIOM), Tjele. http://www.inbiom.dk/download/viden_biomasse/halmpjeceuk_2011.pdf, last access 18 June 2014.

Sobottke, T. & Strunk, J. (2007): Hiss sorgt mit Schilf für Charme. Lübecker Nachrichten (LN), 06 April 2007: 10. http://www.hiss-reet.de/unternehmen/presse/ln_6_04_2007.pdf, last access 18 June 2014.

Sohi, S., Loez-Capel, E., Krull, E. & Bol, R. (2009): Biochar's roles in soil and climate change. A review research needs. CSIRO Land and Water Science Report, 5. University press, Newcarstel. 65 p.

SooNahe (ed.) (2011): Regionalmarke SooNahe. Energie von Nahe und Hunsrück. Regionalbündnis Soonwald-Nahe e.V., Kirn.

Spiekers, H., Richter, W., Zimmermann, N. & Rößl, G. (2007): Silagebewertung. 8. Jahrestagung der Wissenschaftlichen Gesellschaft der Milcherzeugerberater e.V. Wissenschaftliche Gesellschaft der Milcherzeugerberater e.V., 17./18 October 2007, 2007.

Spielmans, S., Rühs, M., Schäfer, A., Hartje, V., Heiland, S., Druckenbrod, C. & Wüstemann, H. (2012): Ein Szenario von Maßnahmen zum Schutz der Biodiversität in

Deutschland. Working Paper on Management in Environmental Planning 31. TU Berlin, Berlin.

Spindler, B. (2008): Erhebung von Grund- und Planungsdaten für die heimische Haltung von Bisons und Wasserbüffeln. Interner Abschlussbericht im Rahmen des KTBL- Arbeitsprogramms. Tierärztliche Hochschule Hannover & Institut für Tierhygiene, Tierschutz, und Nutztierethologie, Hannover. 122 p. http://www.bueffelhof-heerdes.de/downloads/Bison_und_Bueffel.pdf, last access 13 August 2014.

Sponagel, H., Grottenthaler, W. & Hartmann, K.-J. (2005): Bodenkundliche Kartieranleitung. 5th edition. Schweizerbart, Stuttgart. 438 p.

Stachow, U., Werner, A., Rehbinder, E., Schäfer, A., Couwenberg, J. & Wichtmann, W. (2011): Möglichkeiten und zukünftige Erfordernisse einer Einbeziehung von Landnutzung in den Emissionshandel unter Berücksichtigung der Beziehung zur Biodiversität. BfN Skript 291. Bundesamt für Naturschutz (BFN), Bonn-Bad Godesberg.

Statistics Finland (2014): Greenhouse gas emissions in Finland 1990–2012. National Inventory Report under the UNFCCC and the Kyoto Protocol. https://unfccc.int/national_reports/annex_i_ghg_inventories/national_inventories_submissions/items/8108.php, last access 06 May 2015.

Statistisches Amt MV (ed.) (2012): Kaufwerte für landwirtschaftliche Grundstücke in Mecklenburg-Vorpommern. Statistisches Amt MV, Schwerin.

Statistisches Bundesamt (2012): Nachhaltige Entwicklung in Deutschland. Indikatorenbericht 2012. Statistisches Bundesamt, Wiesbaden. 80 p.

Statistisches Bundesamt (2013): Bodenfläche nach Art der tatsächlichen Nutzung. Fachserie 3 Reihe 5.1 – 2011. Statistisches Bundesamt, Wiesbaden. 37 p.

Statistisches Jahrbuch (2009): Statistisches Jahrbuch über Ernährung, Landwirtschaft und Forsten. Wirtschaftsverlag NW GmbH, Bremerhaven. 645 p. http://www.bmelv-statistik.de/fileadmin/sites/010_Jahrbuch/Stat_Jahrbuch_2009.pdf, last access 06 September 2015.

Steffenhagen, P., Timmermann, T., Schulz, K. & Zerbe, S. (2008): Biomassenreproduktion sowie Kohlenstoff- und Nährstoffspeicherung durch Sumpfpflanzen (Helophyten) und Wasserpflanzen (Hydrophyten). In: Gelbrecht, J., Zak, D. & Augustin, J. (eds.): Phosphor- und Kohlenstoff-Dynamik und Vegetationsentwicklung in wiedervernässten Mooren des Peenetals in Mecklenburg-Vorpommern. Status, Steuergrößen und Handlungsmöglichkeiten. Leibniz-Institut für Gewässerökologie und Binnenfischerei (IGB) im Forschungsverbund Berlin e.V. Berichte des IGB 26, Berlin: 145–154.

Steffenhagen, P., Zerbe, S., Frick, A., Schulz, K. & Timmermann, T. (2010): Wiederherstellung von Ökosystemdienstleistungen der Flusstalmoore in Mecklenburg-Vorpommern. Naturschutz und Landschaftsplanung 42: 304–311.

Stegmann, H. & Zeitz, J. (2001): Bodenbildende Prozesse entwässerter Moore. In: Succow, M. & Joosten, H. (eds.): Landschaftsökologische Moorkunde. 2nd edition. Schweizerbart, Stuttgart: 47–57 (Kapitel 2.4.3).

Steinke, C., Wohlgemuth, J., Schoß, G., Kulikowsky, T., Behrens, H., Schmidt, K.-R., Kilian, S., Kuhl, R. & Gräf, R. (1964): Egon und das achte Weltwunder. 85 Min.

Stelte, W. (2011): Fuel pellets from biomass processing, bonding, raw materials. PhD thesis. Technical University of Denmark, Kopenhagen. Danish National Laboratory for Sustainable Energy. 52 p.

Stern, N.H. (2007): The economics of climate change. The Stern review. Cambridge University Press, Cambridge, UK, New York. 692 p.

Sterner, R.W. (2008): On the phosphorus limitation paradigm for lakes. International Review of Hydrobiology 93: 433–445.

Stevens, M. & Hoag, C. (2006): Broad-Leaved Cattail (Typha latifolia L.). USDA NRCS National Plant Data Center & Idaho Plant Materials Center. USDA NRCS National Plant Data Center & Idaho Plant Materials Center. 4 p.

Straka, H. (1960): Über Moore und Torf auf Madagaskar und den Maskarenen. Erdkunde 14: 81–98.

Strasburger, E. & Sitte, P. (1998): Strasburger – Lehrbuch der Botanik. 34th edition. Gustav Fischer Verlag, Heidelberg. 2098 p.

Succow, M. (1988): Landschaftsökologische Moorkunde. Gustav Fischer Verlag, Jena. 340 p.

Succow, M. (2001a): Auswirkungen auf Bewirtschaftbarkeit und Grünlandvegetation. In: Succow, M. & Joosten, H. (eds.): Landschaftsökologische Moorkunde. 2nd edition. Schweizerbart, Stuttgart: 465–469.

Succow, M. (2001b): Kurzer Abriß der Nutzungsgeschichte mitteleuropäischer Moore. In: Succow, M. & Joosten, H. (eds.): Landschaftsökologische Moorkunde. 2nd edition. Schweizerbart, Stuttgart: 404–406.

Succow, M. & Jeschke, L. (1986): Moore in der Landschaft. Entstehung, Haushalt, Lebewelt, Verbreitung, Nutzung und Erhaltung der Moore. Urania, Leipzig, Jena, Berlin. 268 p.

Succow, M. & Jeschke, L. (1990): Moore in der Landschaft. Entstehung, Haushalt, Lebewelt, Verbreitung, Nutzung und Erhaltung der Moore. 2nd edition. Urania, Leipzig, Jena, Berlin. 268 p.

Succow, M. & Joosten, H. (eds.) (2001): Landschaftsökologische Moorkunde. 2nd edition. Schweizerbart, Stuttgart. 622 p.

Succow, M. & Stegmann, H. (2001): Abiotische Kennzeichnung von Moorstandorten. (topische Betrachtung). In: Succow, M. & Joosten, H. (eds.): Landschaftsökologische Moorkunde. 2nd edition. Schweizerbart, Stuttgart: 58–62.

Südbeck, P., Bauer, H.G., Boschert, M., Boye, P. & Knief, W. (2007): Rote Liste der Brutvögel Deutschlands. 4. Fassung, 30 November 2007. Berichte zum Vogelschutz 44: 23–81.

Sundblad, K. & Wittgren, H.-B. (1989): *Glyceria maxima* for wastewater nutrient removal and forage production. Biological Wastes 27: 29–42.

Suyanto, N.K., Sardi, I., Buana, Y. & van Noordwijk, M. (2009): Analysis of local livelihoods from past to present in the central Kalimantan ex-mega rice project area. World Agroforestry Centre, Bogor. 69 p.

Takakai, F., Morishita, T., Hashidoko, Y., Darung, U., Kuramochi, K., Dohong, S., Limin, S.H. & Hatano, R. (2006): Effects of agricultural land-use change and forest fire on N_2O emission from tropical peatlands, Central Kalimantan, Indonesia. Soil Science and Plant Nutrition 52: 662–674.

Tangemann, S. & Cramon-Taubadel, S. (2013): Agricultural policy in the European Union. An overview. Diskussionsbeitrag 1302. Georg-August-Universität, Göttingen.

Department für Agrarökonomie und Rurale Entwicklung.

Tanneberger, F. & Wichtmann, W. (eds.) (2011): Carbon credits from peatland rewetting. Climate – biodiversity – land use. Schweizerbart, Stuttgart. 223 p.

Tanneberger, F., Tegetmeyer, C., Dylawerski, M., Flade, M. & Joosten, H. (2009): Commercially cut reed as a new and sustainable habitat for the globally threatened Aquatic Warbler *Acrocephalus paludicola*. Biodiversity and Conservation 18: 1475–1489.

Tanneberger, F., Flade, M., Preiksa, Z. & Schröder, B. (2010): Habitat selection of the globally threatened Aquatic Warbler *Acrocephalus paludicola* at the western margin of its breeding range and implications for management. Ibis 152: 347–358.

Technologie- und Förderzentrum im Kompetenzzentrum für Nachwachsende Rohstoffe (TFZ) (Ed.) (2009): Miscanthus: Anbau und Nutzung – Informationen für die Praxis. Berichte aus dem TFZ 19. Technologie- und Förderzentrum im Kompetenzzentrum für Nachwachsende Rohstoffe (TFZ), Straubing. 43 p.

Tegetmeyer, C., Tanneberger, F., Dylawerski, M., Flade, M. & Joosten, H. (2007): The Aquatic Warbler – saving Europe's most threatened song bird. Reed cutters and conservationists team up in Polish peatlands. Peatlands International 1: 19–23.

TEEB (2010) The Economics of Ecosystems and Biodiversity: Mainstreaming the Economics of Nature: A synthesis of the approach, conclusions and recommendations of TEEB.

Teh, Y.A., Silver, W.L., Sonnentag, O., Detto, M., Kelly, M. & Baldocchi, D.D. (2011): Large greenhouse gas emissions from a temperate peatland pasture. Ecosystems 14: 311–325.

The Economics of Ecosystems and Biodiversity (TEEB) (2010): Ecological and economic foundation. Edited by Puspham Kumar. Earthscan, London, Washington, D.C. 411 p.

Théroux Rancourt, G., Rochefort, L. & Lapointe, L. (2009): Cloudberry cultivation in cutover peatlands: hydrological and soil physical impacts on the growth of different clones and cultivars. Mires & Peat 5, Art. 6: 1–16.

Thomas, F., Denzel, K., Hartmann, E., Luick, R. & Schmoock, K. (2009): Kurzfassung der Agrarumwelt- und Naturschutzprogramme, Darstellung und Analyse der Entwicklung von Maßnahmen der Agrarumwelt- und Naturschutzprogramme in der Bundesrepublik Deutschland. BfN-Skripten 253. Bundesamt für Naturschutz (BFN), Bonn. 271 p.

Timmermann, T. (1999): Anbau von Schilf (*Phragmites australis*) als ein Weg zur Sanierung von Niedermooren – eine Fallstudie zu Etablierungsmethoden, Vegetationsentwicklung und Konsequenzen für die Praxis. Archiv für Naturschutz und Landschaftsforschung 38: 111–143.

Timmermann, T. (2003): Nutzungsmöglichkeiten der Röhrichte und Riede wiedervernässter Niedermoore Mecklenburg-Vorpommerns. Greifswalder Geographische Arbeiten 31: 31–42.

Timmermann, T. (2009): Biomasse- und Standortskatalog. (Standortpotenzial). In: Wichmann, S. & Wichtmann, W. (eds.): Bericht zum Forschungs- und Entwicklungsprojekt Energiebiomasse aus Niedermooren (ENIM). Universität Greifswald, Greifswald: 37–52.

Timmermann, T., Margóczi, K., Takács, G. & Vegelin, K. (2006): Restoration of peat-forming vegetation by rewetting species-poor fen grasslands. Applied Vegetation Science 9: 241–250.

Tol, R. (2005): The marginal damage costs of climate change: an assessment of the uncertainties. Energy Policy 33: 2064–2074.

Trepel, M. (2004): Zur Wirkung von Niederungen im Landschaftswasser- und -stoffhaushalt. Archiv für Naturschutz und Landschaftsforschung 43: 53–64.

Trepel, M. (2009): Nährstoffrückhalt und Gewässerrenaturierung. Korrespondenz Wasserwirtschaft 2: 211–215.

Trepel, M. (2012): Peatlands and the EU Water Framework Directive: Is the Cynderella syndrome persisting?. IMCG Newsletter 2012/1: 17–18.

Trepel, M. (2013): Moorböden – ein nasser Schatz mit großer Bedeutung. Geographische Rundschau 65: 36–42.

Trepel, M., Davidsson, T.E. & Jørgensen, S.E. (2000): Quantitative simulation of biochemical processes in peatlands a a tool to define sustainable use? Suo 51: 83–93.

Trumper, K., Bertzky, M., Dickson, B., van der Heijden, G., Jenkins, M. & Manning, P. (2009): The natural fix? The role of ecosystems in climate mitigation: A UNEP rapid response assessment. UNEP-WCMC, Cambridge. 65 p.

Tschoeltsch, S. (2008): Reet: Vom Anbau bis zum Dach. Das Reetprojekt aus der Eider-Trene-Sorge Niederung. 61 p.

Turban, E. (1995): Decision support and expert systems. Management support systems. Macmillan, New York, Toronto. 833 p.

Turner, R.K., van den Bergh, J.C.J.M., Söderqvist, T., Barendregt, A., van der Straaten, J., Maltby, E. & van Ierland, E.C. (2000): Ecological-economic analysis of wetlands: scientific integration for management and policy. Ecological Economics 35: 7–23.

Turner, R.K., Georgiou, S. & Fisher, B. (2008): Valuing ecosystem services. The case of multifunctional wetlands. Routledge, London. 240 p.

Tyler, G.A., Smith, K.W. & Burgess, D.J. (1998): Reedbed management and breeding Bitterns *Botaurus stellaris* in the UK. Biological Conservation 86: 257–266.

Ullrich, K. & Riecken, U. (2012): Moorschutzstrategien, -initiativen und -programme in Deutschland. Bog and fen conservation strategies, initiatives and programmes in Germany. Natur und Landschaft 87: 81–86.

Ulrich, A., Malley, D. & Vooran, V. (2009): Peak phosphorus: Opportunity in the making. Why the phosphorus challenge presents a new paradigm for food security and water quality in the Lake Winnipeg basin. MB: Institute for Sustainable Development. International Winnipeg. www.iisd.org/pdf/2010/peak_phosphorus.pdf.

Umweltbundesamt (UBA) (ed.) (2007): Ökonomische Bewertung von Umweltschäden. Methodenkonvention zur Schätzung externer Umweltkosten. Umweltbundesamt (UBA), Dessau-Rosslau.

Umweltbundesamt (UBA) (ed.) (2012): Ökonomische Bewertung von Umweltschäden. Methodenkonvention 2.0 Schätzung von Umweltkosten. Umweltbundesamt (UBA), Dessau-Rosslau. 74 p.

Umweltbundesamt (UBA) (ed.) (2013): National Inventory Report for German Greenhouse Gas Inventory 1990–2011. Submission under the United nations Framework Convention on Climate Change and the Kyoto Protocol 2013. Umweltbundesamt (UBA), Dessau-Rosslau.

Umweltbundesamt (UBA) (2014): Berichterstattung unter der Klimarahmenkonvention der Vereinten Nationen und dem Kyoto-Protokoll 2014 Nationaler Inventarbericht zum Deutschen Treibhausgasinventar 1990 – 2012, Umweltbundesamt Dessau-Roßlau, 963 S. https://www.umweltbundesamt.de/publikationen/berichterstattung-unter-der-klimarahmenkonvention. Last access: 4.10.2015

United Nations Environment Programme (UNEP) (ed.) (2005): Integrated assessment of the impact of trade liberalization: A country study on the Indonesian rice sector. United Nations Environment Programme (UNEP) & Earthprint.

Vadász, C., Német, Á., Biró, C. & Csörgő, T. (2008): The effect of reed cutting on the abundance and diversity of breeding passerines. Acta Zoologica Academiae Scientiarum Hungaricae 54: 177–188.

Valkama, J., Robertson, P. & Currie, D. (1998): Habitat selection by breeding curlews (*Numenius arquata*) on farmland: The importance of grassland. Annales Zoologici Fennici 35: 141–148.

van den Pol-van Dasselaar, A., van Beusichem, M.L. & Oenema, O. (1998): Effects of soil moisture content and temperature on methane uptake by grasslands on sandy soils. Plant and Soil 204: 213–222.

van der Nat, F.-J. & Middelburg, J.J. (2000): Methane emission from tidal freshwater marshes. Biogeochemistry 49: 103–121.

van der Werf, G., Randerson, J.T., Collatz, G.J., Giglio, L., Kasibhatla, P.S., Arellano, A.F., Jr., Olsen, S.C. & Kasischke, E.S. (2004): Continental-scale partitioning of fire emissions during the 1997 to 2001 El Niño/La Niña period. Science 303: 73–76.

van Donkelaar, A., Martin, R.V., Levy, R.C., da Silva, A.M., Krzyzanowski, M., Chubarova, N.E., Semutnikova, E. & Cohen, A.J. (2011): Satellite-based estimates of ground-level fine particulate matter during extreme events: A case study of the Moscow fires in 2010. Atmospheric Environment 45: 6225–6232.

van Eijk, P. & Kumar, R. (2009): Bio-rights in theory and practice. A financing mechanism for linking poverty alleviation and environmental conservation. Wetlands International, Wageningen. 129 p.

van Eijk, P. & Leenman, P.H. (2004): Regeneration of fire degraded peatswamp forest in Berbak National Park and implementation in replanting programmes. Alterra, Wageningen. 83 p.

van Huissteden, J., Petrescu, A.M.R., Hendriks, D.M.D. & Rebel, K.T. (2009): Sensitivity analysis of a wetland methane emission model based on temperate and arctic wetland sites. Biogeosciences 6: 3035–3051.

van Swaay, C., Warren, M. & Loïs, G. (2006): Biotope use and trends of European butterflies. Journal of Insect Conservation 10: 189–209.

van Swaay, C., Cuttelod, A., Collins, S., Maes, D., López Munguira, M., Šašić, M., Settele, J., Verovnik, R., Verstrael, T., Warren, M., Wiemers, M. & Wynhof, I. (2010): European red list of butterflies. IUCN and Butterfly Conservation Europe in collaboration with the European Union, Luxembourg. 48 p.

vanstipdonk (2015): http://vanstipdonk.nl, last access 03 May 2015.

Verheijen, F., Jeffery, S., Bastos, A., van der Velde, M. & Diafas, I. (2010): Biochar application to soils. A critical scientific review of effects on soil properties, processes and functions. EUR. Scientific and technical research series 24099. Publications Office, Luxembourg. 166 p.

Verified Carbon Standard (VCS) (ed.) (2011): Verified Carbon Version 3. http://www.v-c-s.org.

Voigt, R. (1987): Lehrbuch der pharmazeutischen Technologie. 6th edition. VCH, Weinheim. 807 p.

Wagner, A. & Wagner, I. (2003): Leitfaden der Niedermoorrenaturierung in Bayern. Bayerisches Landesamt für Umwelt, Augsburg. 140 p.

Wagner, B. & Czybulka, D. (2012): Das System der Eingriffsregelung. In: Czybulka, D., Hampicke, U. & Litterski, B. (eds.): Produktionsintegrierte Kompensation. Rechtliche Möglichkeiten, Akzeptanz, Effizienz und naturschutzgerechte Nutzung. Erich Schmidt Verlag, Berlin: 13–38.

Wahyunto, S., Ritung, S. & Subagjo, H. (2003): Peta luas sebaran lahan gambut dan kandungan karbon di pulau Sumatra / Maps of area peatland distribution and carbon content in Sumatra, 1990–2002. Wetlands International Indonesia Programme and Wildlife Habitat Canada (WHC), Bogor.

Wahyunto, S., Ritung, S. & Subagjo, H. (2004): Peta sebraran lahan gambut, luas dan kandungan karbon di Kalimantan / Map of peatland distribution area and carbon content in Kalimantan 2000–2002. Wetlands International Indonesia Programme and Wildlife Habitat Canada (WHC), Bogor.

Wahyunto, S., Heryanto, B., Bekti, H. & Widiastuti, F. (2006): Peta-peta sebaran lahan gambut luas dan kandungan larbon di Papua / Maps of peatland distribution, area and carbon content in Papua, 2000–2001. Wetlands International Indonesia Programme and Wildlife Habitat Canada (WHC), Bogor.

Wallot, G., Gusovius, H., Pecenka, R. & Hoffmann, T. (2011): Herstellung von Faserwerkstoff aus pflanzlichen Rohstoffen in einer Scheibenmühle. Landtechnik 66: 100–102.

Walter, H. & Lieth, H. (1960–1967): Klimadiagramm-Weltatlas. Fischer, Jena.

Wang, Z.-P. & Han, X.-G. (2005): Diurnal variation in methane emissions in relation to plants and environmental variables in the Inner Mongolia marshes. Atmospheric Environment 39: 6295–6305.

Wayman, M. (1973): Guide for planning pulpand paper enterprises. Food and Agriculture Organization of the United Nations (FAO). Forest Products Studies, 18.

Weber, C.A. (1901): Über die Erhaltung von Mooren und Heiden Norddeutschlands im Naturzustande, sowie über die Wiederherstellung von Naturwäldern. Abhandlungen des Naturwissenschaftlichen Vereins zu Bremen 15: 263–279.

Weber, S. (2015): Mutterkühe halten? Aber bitte rentabel! Bauernzeitung M-V: 44–45.

Weithäuser, M., Scholwin, F., Fischer, E., Grope, J., Weidele, T. & Gattermann, H. (2010): Leitfaden Biogas. Von der Gewinnung zur Nutzung Gasaufbereitung. 5th edition. Biogasportal.info. Fachagentur Nachwachsende Rohstoffe e. V. (FNR), Gülzow-Prüzen. 272 p.

Westlake, D.F. (1966): The biomass and productivity of *Glyceria maxima*: I. Seasonal changes in biomass. The Journal of Ecology 54: 745–753.

Wetlands International (2005): Ruoergai High Altitude Peatlands. Wetlands International Project Fact Sheet

Series. http://www.wetlands.org/Portals/0/publications/Factsheet/FactSheet_Ruoergai.pdf

Whalen, S. (2005): Biogeochemistry of methane exchange between natural wetlands and the atmosphere. Environmental Engineering Science 22: 73–94.

Wheeler, B.D. & Shaw, S.C. (1995): Restoration of damaged peatlands. With particular reference to lowland raised bogs affected by peat extraction. HMSO, London. 211 p.

White, G. (2009): The future of reedbed management. Information and advice note. Royal Society for the Protection of Birds (RSPB), Version 7. Royal Society for the Protection of Birds (RSPB), Bedfordshire. 11 p. www.rspb.org.uk/Images/Reedbed_management_tcm9-255077.pdf, last access 24 June 2014.

Whiting, G.J. & Chanton, J.P. (1996): Control of the diurnal pattern of methane emission from emergent aquatic macrophytes by gas transport mechanisms. Aquatic Botany 54: 237–253.

Whiting, G.J. & Chanton, J.P. (2001): Greenhouse carbon balance of wetlands: methane emission versus carbon sequestration. Tellus B 53: 521–528.

Wibisono, I., Labueni, S. & Suryadiputra, I.N.N. (2005): Panduan rehabilitasi dan teknik silvikultur di lahan gambut. Proyek Climate Change, Forest and Peatlands in Indonesia. Wetlands International Indonesia Programme and Wildlife Habitat Canada (WHC), Bogor.

Wibisono, I., Silber, T., Lubis, I.R., Rais, D.S., Suryadiputra, I.N.N., Silvius, M., Tol, R. & Joosten, H. (2011): Peatlands in Indonesia`s National REDD+ Strategy. Wetlands International Indonesia / Wetlands International Headquarters, Bogor / Ede. 31 p.

Wichmann, S. (2009): Ernte, Transport und Lagerung. In: Endbericht zum Forschungs- und Entwicklungsprojekt Energiebiomasse aus Niedermooren (ENIM). Universität Greifswald, DUENE e.V., Greifswald: 104–111.

Wichmann, S. (submitted): Commercial viability of paludiculture: a comparison of harvesting reeds for biogas production, direct combustion, and thatching, resubmitted with minor revisions to Ecological Engineering.

Wichmann, S. & Köbbing, J.F. (2015): Common reed for thatching – a first review of the European market. Industrial Crops and Products. Resubmitted (after minor revisions).

Wichmann, S. & Wichmann, W. (2009): Endbericht zum Forschungs- und Entwicklungsprojekt Energiebiomasse aus Niedermooren (ENIM). Universität Greifswald & DUENE e.V., Greifswald. 192 p. www.paludikultur.de/fileadmin/user_upload/Dokumente/pub/enim_endbericht_2009.pdf, last access 24 June 2014.

Wichmann, S., Gaudig, G., Krebs, M., Joosten, H., Albrecht, K.A. & Kumar, S. (2014): Sphagnum farming for replacing peat in horticultural substrates. In: Biancalani, R. & Avagyan, A. (eds.): Toward climate-responsible peatlands management. Food and Agriculture Organization of the United Nations (FAO). Mitigation of climate change in agriculture series 9, Rome: 80–83.

Wichtmann, W. (1999): Nutzung von Schilf (*Phragmites australis*). Archiv für Naturschutz und Landschaftsforschung 38: 211–231.

Wichtmann, W. & Couwenberg, J. (2013): Reed as a renewable resource and other aspects of paludiculture. Foreword to the special issue to the rrr conference at Greifswald University. Mires & Peat 13, Art. 0: 1–2.

Wichtmann, W. & Schäfer, A. (2007): Alternative management options for degraded fens – utilisation of biomass from rewetted peatlands. In: Okruszko, T., Maltby, E., Szatylowicz, J., Swiatek, D. & Kotowski, W. (eds.): Wetlands: Monitoring, Modelling and Management. Taylor & Francis, Leiden: 273–279.

Wichtmann, W. & Succow, M. (2001): Nachwachsende Rohstoffe. In: Kratz, R. & Pfadenhauer, J. (eds.): Ökosystemmanagement für Niedermoore. Strategien und Verfahren zur Renaturierung. Ulmer, Stuttgart (Hohenheim): 177–184.

Wichtmann, W. & Tanneberger, F. (2009): Feasibility of the use of biomass from re-wetted peatland for climate and biodiversity protection in Belarus. Michael-Succow-Stiftung zum Schutz der Natur, Greifswald. 112 p.

Wichtmann, W. & Wichmann, S. (2011): Paludikultur: standortgerechte Bewirtschaftung wiedervernässter Moore. Telma Beiheft 4: 215–234.

Wichtmann, W., Tanneberger, F., Wichmann, S. & Joosten, H. (2010): Paludiculture is paludifuture: Climate, biodiversity and economic benefits from agriculture and forestry on rewetted peatland. Peatlands International: 48–51.

Wichtmann, W., Haberl, A. & Tanneberger, F. (2012): Production of biomass in wet peatlands (paludiculture). The EU-AID project `Wetland energy` in Belarus – solutions for the substitution if fossil fuels (peat briquettes) by biomass from wet peatlands. Bioenergieforum Rostock: 85–96.

Wichtmann, W., Oehmke, C., Bärisch, S., Deschan, F., Malashevich, U. & Tanneberger, F. (2014): Combustibility of biomass from wet fens in Belarus and its potential as a substitute for peat in fuel briquettes. Mires & Peat 13, Art. 6: 1–10.

Wiechert, E. (1947): Die Jeromin-Kinder. Die Furchen der Armen. Vol. 2. Rascher, Zürich. 414 p.

Wiegleb, G. & Krawczynski, R. (2010): Biodiversity management by Water Buffalos in restored wetlands. Waldökologie, Landschaftsforschung und Naturschutz 10: 17–22.

Wild, U., Kamp, T., Lenz, A., Heinz, S. & Pfadenhauer, J. (2001): Cultivation of *Typha* spp. in constructed wetlands for peatland restoration. Ecological Engineering 17: 49–54.

Williams, D.J. (1993): Methane emissions from manure of free-range dairy cows. Chemosphere 26: 179–187.

Wippl, J. & Paar, J. (2015): Motormäher schlägt Traktor. Landwirt: 68–72.

Wirz, A. & Klingmann, P. (2012): Entwicklung von Kriterien für ein bundesweites Regionalsiegel. Gutachten im Auftrag des Bundesministeriums für Ernährung, Landwirtschaft und Verbraucherschutz. Final report. FiBL Deutschland e.V., Frankfurt am Main. 194 p.

Wissenschaftlicher Beirat für Agrarpolitik beim Bundesministerium für Ernährung Landwirtschaft und Verbraucherschutz (WBA) (2007): Nutzung von Biomasse zur Energiegewinnung – Empfehlung an die Politik. Wissenschaftlicher Beirat für Agrarpolitik beim Bundesministerium für Ernährung Landwirtschaft und Verbraucherschutz (WBA), Berlin. 255 p. www.bmel.de/SharedDocs/Downloads/Ministerium/Beiraete/Agrarpolitik/GutachtenWBA.pdf?_blob=publicationFile.

Wissenschaftlicher Beirat für Agrarpolitik beim Bundesministerium für Ernährung Landwirtschaft und Verbraucherschutz (WBA) (2010): EU-Agrarpolitik nach 2013 – Plädoyer für eine neue Politik für Ernährung, Landwirtschaft und ländliche Räume. Wissenschaftlicher Beirat für Agrarpolitik beim Bundesministerium für Ernährung Landwirtschaft und Verbraucherschutz (WBA), Berlin. 38 p.

Wohlgemuth, J. (1962): Egon und das achte Weltwunder. Verlag Neues Leben, Berlin. 413 p.

Wolf, I. & Russow, R. (2000): Different pathways of formation of N_2O, N_2 and NO in black earth soil. Soil Biology and Biochemistry 32: 229–239.

Woo, I. & Zedler, J.B. (2002): Can nutrients alone shift a sedge meadow towards dominance by the invasive Typha x glauca? Wetlands 22: 509–521.

World Bank (2006): Sustainable land management. Challenges, opportunities, and trade-offs. Agriculture and rural development. World Bank, Washington, D.C. 88 p.

World Population Prospects (ed.) (2010): Annual Population 2011–2100. XLS table. United Nations (UN). http://esa.un.org/unpd/wpp/Excel-Data/DB04_Population_ByAgeSex_Annual/WPP2010_DB4_F1B_POPULATION_BY_AGE_BOTH_SEXES_ANNUAL_2011-2100.XLS.

Wulf, A. (2009): Brennstoff-Charakterisierung, Verbrennungstests und Ascheanalysen. In: Wichmann, S. & Wichtmann, W. (eds.): Bericht zum Forschungs- und Entwicklungsprojekt Energiebiomasse aus Niedermooren (ENIM). Universität Greifswald, Greifswald: 53–64.

Wüstemann, H., Hartje, V., Meyerhoff, J., Rühs, M. & Schäfer, A. (2013): Financial costs and benefits of implementing a national strategy on biological diversity in Germany. Land Use Policy 36: 307–318.

Xiao, D.N. & Li, X.Z. (2004): Ecological and environmental function of wetland landscape in the Liaohe Delta. In: Wong, M.H. (ed.): Wetlands ecosystems in Asia: Function and management. Developements in Ecosystems. Elsevier, Amsterdam, Boston, Heidelberg, London, New York, Oxford, Paris, San Diego, San Francisco, Singapore, Sydney, Tokyo: 35–46.

Yamulki, S. & Jarvis, S.C. (2002): Short-term effects of tillage and compaction on nitrous oxide, nitric oxide, nitrogen dioxide, methane and carbon dioxide fluxes from grassland. Biology and Fertility of Soils 36: 224–231.

Yang, Y. (2000): The distribution, degeneration and formation mechanism of peatland in P.R. China. In: Rochefort, L. & Daigle, J.-Y. (eds.): Sustaining Our Peatlands 1. International Peat Congress, Quebec, Canada, 24–26 March 2000. International Peat Society (IPS): 162–169.

Ye, S., Brix, H. & Sun, D. (2013): Large-scale management of Common Reed, *Phragmites australis*, for paper production. A case study from the Liaohe River delta, China. In: International conference "Reed as a renewable resource", Greifswald: 34.

Yli-Petäys, M., Laine, J., Vasander, H. & Tuittila, E.-S. (2007): Carbon gas exchange of a re-vegetated cut-away peatland five decades after abandonment. Boreal Environment Research 12: 177–190.

Yu, Z., Beilman, D.W. & Jones, M.C. (2009): Sensitivity of northern peatland carbon dynamics to Holocene climate change. In: Baird, A.J., Belyea, L.R., Comas, X., Reeve, A.S. & Slater, L.D. (eds.): Carbon cycling in northern peatlands. Geophysical Monograph Series. American Geophysical Union, Washington, D.C.: 55–69.

Zak, D., Wagner, C., Payer, B., Augustin, J. & Gelbrecht, J. (2010): Phosphorus mobilization in rewetted fens: the effect of altered peat properties and implications for their restoration. Ecological Applications 20: 1336–1349.

Zak, D., Augustin, J., Trepel, M. & Gelbrecht, J. (2011): Strategien und Konfliktvermeidung bei der Restaurierung von Niedermooren unter Gewässer-, Klima- und Naturschutzaspekten, dargestellt am Beispiel des nordostdeutschen Tieflandes. Telma Beiheft 4: 133–149.

Zedler, P.H., Anchor, T., Knuteson, D., Gratton, C. & Barzen, J. (2009): Using an ecolabel to promote on-farm conservation: the Wisconsin Healthy Grown experience. International Journal of Agricultural Sustainability 7: 61–74.

Zeitz, J. (1991): Untersuchungen über Filtrationseigenschaften von Niedermoorböden mit Hilfe verschiedener Methoden unter Berücksichtigung der Bodenentwicklung. Zeitschrift für Kulturtechnik und Landentwicklung 32: 227–234.

Zeitz, J. (1992): Bodenphysikalische Eigenschaften von Substrat-Horizont-Gruppen in landwirtschaftlich genutzten Niedermooren. Zeitschrift für Kulturtechnik und Landentwicklung 33: 301–307.

Zeitz, J. (2001): Zur anthropogenen Veränderung der Moore – Randow-Welse-Flußtalmoor. In: Succow, M. & Joosten, H. (eds.): Landschaftsökologische Moorkunde. 2nd edition. Schweizerbart, Stuttgart: 434–436.

Zeitz, J. (2003): Moorkulturen. In: Blume, H.-P. (ed.): Handbuch der Bodenkunde. Wiley-VCH, Weinheim: Kapitel 5.3.2.

Zeitz, J. (2014): Landnutzung der Moore in der Region bis Anfang der 1990er Jahre: Ausgewählte Meliorationsverfahren. In: Luthardt, V. & Zeitz, J. (eds.): Moore in Brandenburg und Berlin. Natur+Text, Rangsdorf: 106–112.

Zeitz, J. & Velty, S. (2002): Soil properties of drained and rewetted fen soils. Journal of Plant Nutrition and Soil Science 165: 618–626.

Zeitz, J., Titze, E. & Kossov, V. (1987): Auswirkungen von tiefen Grundwasserständen auf Standorteigenschaften und Ertrag bei tiefgründigen Niedermooren. Feldwirtschaft 28: 214–216.

Zeller, V., Weiser, C., Hennenberg, K., Reinicke, F., Schaubach, K., Thrän, D., Vetter, A. & Wagner, B. (2011): Basisinformationen für eine nachhaltige Nutzung landwirtschaftlicher Reststoffe zur Bioenergiebereitstellung. Deutsches BiomasseForschungsZentrum (DBFZ), Leipzig. 53 p.

Zeng, T., Borowski, A., Braumann, F., Pollex, A. & Lenz, V. (2013): Verwertungskonzepte zur energetischen Nutzung von geeignetem Grünlandaufwuchs im Naturpark Drömling. In: Nelles, M. (ed.): 7. Rostocker Bioenergieforum. Rostocker Bioenergieforum, Rostock, 20–21 June 2013. Universität Rostock. Schriftenreihe Umweltingenieurwesen 36. Agrar- und Umweltwissenschaftliche Fakultät, Rostock: 73–84.

Zubrycki, K., Venema, H.D., Grosshans, R., Manaigre, J. & Sosa Lerin, E. (2012): Our lake, our solutions. Two years of progress and partnerships. Video. International Institute for Sustainable Development (IISD).

List of contributors

Susanne Abel, Institute of Botany and Landscape Ecology, Greifswald University, Greifswald, Germany

Matthias Ahlhaus, Institute of Renewable Energy Systems, University of Applied Sciences, Stralsund, Germany

Balázs Baranyai, Institute of Botany and Landscape Ecology, Greifswald University, Greifswald, Germany

Alexandra Barthelmes, Institute of Botany and Landscape Ecology, Greifswald University, Greifswald, Germany

Dieter Behrendt, ECOLOG-Institut für sozial-ökologische Forschung und Bildung, Hannover, Germany

Kristina Brust, Dr. Dittrich & Partner Hydro-Consult GmbH, Dresden, Germany

Jörg Burgstaler, Landscape Ecology, Faculty of Agricultural and Environmental Sciences, University of Rostock, Rostock, Germany

John Couwenberg, Institute of Botany and Landscape Ecology, Greifswald University, Greifswald, Germany

Detlef Czybulka, Institute of Constitutional and Administrative Law, Environmental Law and Public Economic Law, University of Rostock, Rostock, Germany

Tobias Dahms, Institute of Botany and Landscape Ecology, Greifswald University, Greifswald, Germany

Steffi Deickert, Institute of Botany and Landscape Ecology, Greifswald University, Greifswald, Germany

Sebastian Dettmann, Landscape Ecology, Faculty of Agricultural and Environmental Sciences, University of Rostock, Rostock, Germany

Ingo Dittrich, Dr. Dittrich & Partner Hydro-Consult GmbH, Dresden

René Dommain, Institute of Botany and Landscape Ecology, Greifswald University, Greifswald, Germany

Frank Edom, Hydrotelm, Dresden, Germany

Klaus Fischer, Institute of Zoology, Greifswald University, Greifswald, Germany

Theo Fock, Departement of Agriculture and Food Sciences, University of Applied Sciences, Neubrandenburg, Germany

Thomas Frase, Institute of Zoology, University of Rostock, Rostock, Germany

Dariusz Gątkowski, The Polish Society for the Protection of Birds, OTOP/BirdLife Poland, Marki, Poland

Stephan Glatzel, Departement of Geography and Regional Research, University of Vienna, Vienna, Austria

Sebastian Görn, Institute of Zoology, Greifswald University, Greifswald, Germany

Richard Grosshans, International Institute for Sustainable Developement (IISD), Manitoba, Canada

Anke Günther, Landscape Ecology, Faculty of Agricultural and Environmental Sciences, University of Rostock, Rostock, Germany

Andreas Haberl, Michael Succow Foundation, Greifswald, Germany

Bernhard Hasch, p2m berlin GmbH, Berlin, Germany

Thomas Hering, Thüringer Landesanstalt für Landwirtschaft, Jena, Germany

Benjamin Herold, life Schreiadler Project Schorfheide Chorin, Angermünde, Germany

Henning Holst, LedA-Landentwicklung durch Agrarkultur, Guest, Germany

Bettina Holsten, Institute for Ecosystem Research, University of Kiel, Kiel, Germany

Till Holsten, Institute for Geography and Geology, Greifswald University, Greifswald, Germany

Stefan Horn, Faculty of Agricultural and Environmental Sciences, University of Rostock, Rostock, Germany

Vytas Huth, Landscape Ecology, Faculty of Agricultural and Environmental Sciences, University of Rostock, Rostock, Germany

Christian Jantzen, Institute of Renewable Energy Systems, University of Applied Sciences, Stralsund, Germany

Hans Joosten, Institute of Botany and Landscape Ecology, Greifswald University, Greifswald, Germany

Gerald Jurasinski, Landscape Ecology, Faculty of Agricultural and Environmental Sciences, University of Rostock, Rostock, Germany

Martin Kaltschmitt, Institute of Environmental Technology and Energy Economics, Technische Universität Hamburg-Harburg, Hamburg, Germany

Vladimir Kapitsa, International Sakharov Environmental University Minsk, Belarus

Silke Kleinhückelkotten, ECOLOG-Institut für sozialökologische Forschung und Bildung, Hannover, Germany

Jan Felix Köbbing, Institute of Botany and Landscape Ecology, Greifswald University, Greifswald, Germany

Laura Kölsch, Institute of Constitutional and Adminstrative Law, Environmental Law and Public Economic Law, University of Rostock, Rostock, Germany

Ulrich König, MRG Blautonwerk Friedland, Friedland, Germany

List of contributors

Hellmuth-Hans Kranemann, Kranemann GmbH, Blücherhof, Germany

Jarosław Krogulec, The Polish Society for the Protection of Birds, OTOP/BirdLife Poland, Marki, Poland

Vera Luthardt, Faculty of Landscape Management and Nature Conservation, Eberswalde University for Sustainable Development, Eberswalde, Germany

Uladzimir Malashevich, APB BirdLife Belarus, Belarus, Minsk

Michael Manthey, Institute of Botany and Landscape Ecology, Greifswald University, Greifswald, Germany

Jürgen Müller, Faculty of Agricultural and Environmental Science, University of Rostock, Rostock, Germany

Horst-Peter Neitzke, ECOLOG-Institut für sozial-ökologische Forschung und Bildung, Hannover, Germany

Anke Nordt, Institute of Botany and Landscape Ecology, Greifswald University, Greifswald, Germany

Rainer Nowotny, Hanffaser Uckermark eG Prenzlau, Germany

Claudia Oehmke, Institute of Botany and Landscape Ecology, Greifswald University, Greifswald, Germany

Jenny Piegsa, Ministry of Rural Affairs and Consumer Protection Baden-Württemberg, Stuttgart, Germany

Steffi Raabe, Institute of Botany and Landscape Ecology, Greifswald University, Greifswald, Germany

Michael Rühs, Institute of Botany and Landscape Ecology, Greifswald University, Greifswald, Germany

Achim Schäfer, Institute of Botany and Landscape Ecology, Greifswald University, Greifswald, Germany

Claudia Schröder, Faculty of Landscape Management and Nature Conservation, Eberswalde University for Sustainable Development, Eberswalde, Germany

Christian Schröder, Institute of Botany and Landscape Ecology, Greifswald University, Greifswald, Germany

Philipp Schroeder, Institute of Botany and Landscape Ecology, Greifswald University, Greifswald, Germany

Paul Schulze, Albrecht Daniel Thaer-Institute of Agricultural and Horticultural Sciences, Humboldt-Universität zu Berlin, Berlin, Germany

Michael Succow, Michael Succow Foundation, Greifswald, Germany

Weert Sweers, Faculty of Agricultural and Environmental Science, University of Rostock, Rostock, Germany

Franziska Tanneberger, Institute of Botany and Landscape Ecology, Greifswald University, Greifswald, Germany

Nina Tanovitskaya, Institute for Management of Natural Resources, National Academy of Sciences of Belarus, Minsk, Belarus

Michael Trepel, Ministry of Energy, Agriculture, the Environment and Rural Areas, Schleswig-Holstein, Kiel, Germany

Hannes Wagner, Institute of Environmental Technology and Energy Economics, Technische Universität Hamburg-Harburg, Hamburg, Germany

Andreas Wahren, Dr. Dittrich & Partner Hydro-Consult GmbH, Dresden, Germany

Sabine Wichmann, Institute of Botany and Landscape Ecology, Greifswald University, Greifswald, Germany

Wendelin Wichtmann, Michael Succow Foundation, Greifswald, Germany

Denny Wiedow, Landscape Ecology, Faculty of Agricultural and Environmental Sciences, University of Rostock, Rostock, Germany

Simone Witzel, Hochschule Neubrandenburg, University of Applied Sciences, Neubrandenburg, Germany

Anne Wollert, Center for Innovation and Education Hohen Luckow e.V. (IBZ), Hohen Luckow, Germany

Jutta Zeitz, Albrecht Daniel Thaer-Institute of Agricultural and Horticultural Sciences, Humboldt-Universität zu Berlin, Berlin, Germany

Luisa Zielke, Institute of Botany and Landscape Ecology, Greifswald University, Greifswald, Germany

Index

abandonment 94, 178, 204, 229
abatement costs 135, 140f.
Abura = *Mitragyna ledermannii*
abusus 136
Acacia 11, 217ff., 222
Açaí Palm 221
acceptance 109, 129, 164, 168
accessibility 73, 76, 180
acidification 200
acid sulphate soils 8, 11
Acmella oleracea 221
Acrocephalus arundinaceus 94
Acrocephalus paludicola 94, 97, 205ff., 213
Acrocephalus schoenobaenus 94, 97
activating interview 166
adaptation 109, 198
added value 43, 174
added value creation 132f., 152
additional costs 155
administration 172
administrative approval 194
adoption 168
aerenchyma 80
aesthetic and ethical values 161
afforestation 154, 218
Africa 200
Agathis borneensis 221
agricultural activity 145
agricultural land 150f., 232
agricultural land management 143
agricultural policy 109, 149, 204
agricultural products 150f.
agri-environmental programme 96, 116f., 119, 137, 145, 149, 155, 170, 207f.
agri-environment measures 150ff.
agri-environment subsidies 150
agroforestry 219, 221
Agrotherm GmbH 174
Aichi Targets 199
air-fuel ratio 48
air quality requirements 143
Alan (*Shorea* spp.) 220
Alcon Blue = *Maculinea alcon*
Alder = *Alnus glutinosa*
alleviation of poverty 217
Alnus glutinosa 31, 37f., 51, 130, 140, 187
Alseodaphne coriacea 222
Alstonia spp. 214
Alstonia pneumatophora 222
ammonium 106
amortisation 140
Amur Silver-Grass = *Miscanthus sacchariflorus*

Anas querquedula 97
Andiroba = *Carapa guianensis*
Angelica archangelica 38
Angus 42
Anhui 224
animal farming 116
animal husbandry 39, 42, 109
animal welfare 40
Annex II species 148
annuals 96
annuity 115, 140
anoxic conditions 79
antifascist 161
Ant Rattan = *Korthalsia flagellaris*
Aphidina 95
appreciation 13, 141
aquatic biodiversity 105
Aquatic Warbler = *Acrocephalus paludicola*
aquifer 103
Aquila clanga 205
Araneae 95
archaeological artefacts 161
Archanara geminipuncta 96
archive 20
area availability 194
area selection 179
Argentina 39
Aronia melanocarpa 38
art 157
arthropods 95
ash 45, 49
ash content 36
ash melting temperature 49f.
Asia 200
associations and representatives 172
atmospheric deposition 106
atmospheric residence 80
Auchenorrhyncha 95
Australia 39
Austria 23
avifauna 97
avoidance costs 140
awareness raising 231
Awnless Barnyard Grass = *Echinochloa colona*
Bacopa monnieri 221
Badhamia lilacina 38
Baillon's Crake = *Porzana pusilla*
bales 36, 68, 71f., 113f.
Bali 217
balloon tyres 62ff., 73
Baltic Sea Action Plan 199
ban on ploughing grassland 149, 195
Baruth Valley 182
Bathyphantes setiger 38
Bavaria 14

Bearded Reedling = *Panurus biarmicus*
bearing capacity 7, 12, 59ff., 69, 110
beaver 148
beetles 95
Belangeran = *Shorea balangeran*
Belarus 201, 205ff., 212
Belgian Blue 42
Bell Heather = *Erica tetralix*
beneficiaries 163f.
benefit-to-cost relationship 141
best practice 142, 231, 233
Biebrza 213
Biebrza National Park 207f.
biochar 52, 228
biodiversity 13f., 19f., 41, 94, 110, 132, 134f., 149, 154, 161, 205, 207, 218, 229, 231
biodiversity conservation 137, 143
biodiversity value 79
bioenergy 44, 226f.
bioenergy crops 143, 164
bioenergy production 140
bioethanol 55
biofuels 49, 54
biogas 35f., 54f., 71, 111, 114, 140, 169, 207
Bioland 129
biomass 19, 197
biomass boiler 174
biomass combustion 113
biomass fuels 132
biomass harvesting 90
biomass removal 71, 75, 77, 91, 94, 114
biomass supply costs 113
biomass yield 75, 114
biomethane 57
Biopark 129
biosphere reserves 148
Birds 95f.
Birds Directive 145, 199
Black Alder = *Alnus glutinosa*
Black Chokeberry = *Aronia melanocarpa*
black fen cultivation 4
Black Grouse = *Tetrao tetrix*
Black-tailed Godwit = *Limosa limosa*
Black Tern = *Chlidonias niger*
Blauer Engel 120, 129f.
Blindower Wiesen 139
Bluethroat = *Luscinia svecica*
Bogbean = *Menyanthes trifoliata*
bog breathing 104
Bog Myrtle = *Myrica gale*
Bogie tracks 64
bonus-malus system 153

Index

Borneo 217
Botaurus stellaris 94, 98
bottom-up process 169
Brandenburg 14
Brazil 39
briquettes 35f., 47, 49, 51f., 69, 83, 151, 207, 212
briquetting 53, 206
Broadleaf Cattail = *Typha latifolia*
bugs 96
Building panel 83
Bulbalus arnee = Water Buffalo
bulk density 8, 53, 71ff.
Bundesbodenschutzgesetz 136, 142
bundles 68, 72f., 111
buoyancy 8
burnout phase 48
business organisation 109
butterflies 95f.
Cacari = *Myrciaria dubia*
cajeput oil 219
Calamagrostis canescens 206
Calamus spp. (Rattan palm trees) 222
Calthion 99
Camargue 40
Camocamo = *Myrciaria dubia*
Camu Camu = *Myrciaria dubia*
Canada 201, 226
CAP 133, 149ff., 155
capillary fringe 10
capillary groundwater 11
capillary rise 12
CAP reform 154
Carabidae 96
Carapa guianensis 221
carbohydrates 55
carbon dioxide 79, 89
carbon markets 152
carbon sequestration 79, 90, 149, 154, 217
carbon sink = carbon sequestration
carbon stocks 161
carbon storage 154, 218
carcinogenic 203
Carex spp. 30, 35, 37, 66, 99, 106f., 148, 150, 200
Carex acuta 24, 28, 37
Carex acutiformis 24, 37, 41, 91, 100
Carex disticha 99
Carex elata 206
Carex riparia 24, 37, 206
Cariru = *Talinum fruticosum*
Casuarina equisetifolia 220
catchment 184
Cattail 30, 35f., 43, 61, 80, 83, 90f., 98, 115, 119, 130, 148, 150, 170, 182, 203, 207, 216, 226
CBD 199
cellulose 55
certificate 120f., 129
certification 109, 120, 129
Ceylon Spinach = *Talinum fruticosum*
CH_4 emissions 89f., 92f.

chaff = chopped biomass
Charolais 42
China 197f., 203, 215f., 223, 225
Chinese Celery = *Oenanthe javanica*
Chinese Swamp Cypress = *Glyptostrobus pensilis*
Chlidonias leucopterus 97
Chlidonias niger 98
chopped biomass 68, 71, 73, 75, 114
chopping 122
CHP 57
Cicadas 95
Circus aeruginosus 94, 146
Circus cyaneus 97
Circus pygargus 97
citizens' jury 167
citizens' panel 166f.
Cladium mariscus 29, 93, 208
climate 135, 160
climate change 13, 109, 135, 154, 161, 199
climate change adaptation 102
climate change mitigation 79, 135f., 140, 152, 154, 174, 218, 229
Climate, Community and Biodiversity Standard 223
climate effect 134
climate policy 154
climate protection 119, 137, 169, 222
climate protection subsidies 152
climate regulation 15
Climate Smart Agriculture 231
Cloudberry = *Rubus chamaemorus*
CO_2-equivalent 80
CO_2 sequestration 80
coalification 52
coastal flood mires 41
coastal marshes 41
Cockspur = *Echinochloa crus-galli*
Coix lacryma-jobi 221
Coleoptera 95
colmation 7
colonisation 157, 159f.
combined heat and power (CHP) generation 57
Combretocarpus rotundatus 222
combustion 45, 48ff., 113f., 207
combustion properties 48
combustion technology 114
commodification 15
Common Agricultural Policy 133, 149, 157
Common Barnyard Grass = *Echinochloa crus-galli*
Common Ironwood = *Casuarina equisetifolia*
Common Reed = *Phragmites australis*
Common Snipe = *Gallinago gallinago*
common strategic framework 133
communication 162, 164
community forestry 219
compaction 7, 9, 45, 69, 76, 92, 200
compensation 14, 146ff., 195

compensatory measures 148
compensatory payments 137, 142, 146
competition for land 162, 179
competitive 140
competitiveness 133
complex amelioration 3f., 7
compost 207
compression 8
concentration camps 161
conditioning 43
conductivity 8
conflicts of interest 157, 162, 165f.
consensus 167
conservation 157, 160ff., 229, 231
conservation area 144f.
conservationists 132
conservation objectives 145
conservation status 180
consolidation 7ff.
constructed wetlands 36, 226
construction 109
construction panel 83
consultants 172
consumer demand 168
contribution margin 110, 118
control mechanisms 152
convective gas flow 90
conventional cattle grazing 116
Convention on Biological Diversity 198
conversion 44, 46
conversion efficiencies 57
conversion of machinery 70
conversion of permanent grassland 148f.
cooling effect 102f.
Corn Crake = *Crex crex*
corporate management 109
cost accounting 110
cost-benefit analysis 109, 134ff., 141f.
costs 109, 116, 230
Cotton Grass = *Eriophorum*
Cowberry = *Vaccinium vitis-idaea*
Crab Wood = *Carapa guianensis*
Cradle2Cradle 130f.
Cranberry = *Vaccinium*
credibility 129
Crex crex 97
cropland 154
crop rotation 162
cross-compliance 151, 155, 171, 230
crossing 70, 77
CSF 133
cultivars 115
cultivation 157, 159
Cultivation Act 3f.
cultivation methods 162
cultural diversity 175
cultural heritage 175
cultural services 18, 20
Cumbungi = *Typha domingensis*
Cup Plant = *Sylphium perforatum*
Curlew = *Numenius arquata*
cutter bars 68

Index

Cyperus papyrus 55, 220
Dactylorhiza majalis 99
damage 166
damage costs 135, 141 f.
Danube Delta 23, 64
Darß 118
Dawn Redwood = *Metasequoia glyptostroboides*
decentralised heating plants 113
decision makers 163 f.
decision making process 162
Decision Support System 185
decision tree 186
decomposition 3, 9, 106
deep ploughing 25
deep plough sand cover cultivation 6
deforestation 3, 217
degradation 161, 200
degrading peatland 210
delta tracks 64, 87
demand for land 197
demonstration projects 152
denitrification 80, 89, 104, 106 f.
Denmark 23
depreciation 111, 232
desertification 200
desiccation 7, 229
desulfurization 55
deterministic calculations 110
dieback of trees 161
diffusion 168 f.
direct costs 111
direct marketing 111
direct payments 133, 137, 149 ff., 204, 230, 231 f.
disc mower 122 f.
dislocation 7
dissemination 168
dissolved organic carbon 9
district heating 174
diversification 133, 151
DOC = dissolved organic carbon
Dokudovskoye 206
double wheels 86
drainage 4 f., 7 ff., 20, 59, 79 f., 102, 106, 131, 157, 199, 207, 217, 229
drainage costs 10
drainage depth 12
drainage ditches 77, 144
drainage system 190
Droserae herba 39
Drosera intermedia 39
Drosera rotundifolia 38 f., 81
Dryad = *Minois dryas*
dry fermentation 54
DSS-TORBOS 185
Durian = *Durio carinatus*
Durian paya 220
Durio carinatus 214, 219 f.
Dusky Large Blue = *Maculinea nausithous*
Dyera lowii 214
Dyera polyphylla 220, 222

EAFRD = European Agricultural Fund for Rural Development
early adopter 168
early compensatory measure 147
earthified 10, 12
Eastern Prussia 160
Echinochloa colona 221
Echinochloa crus-galli 221
eco-account 147
Ecolabel 120
ecological balance 144
economic benefits 136, 162
economic development 132, 175
economic evaluation 134
economic utilisation 161
economic valuation 135
ecosystem functions 134, 147
ecosystem goods 16
ecosystem services 3, 13, 15 f., 19 f., 79, 119, 134 ff., 139, 153, 161, 175, 218, 223, 225, 229 ff.
ecotourism 132
educational institutions 172
efficiency 76, 136, 140 f.
Elaeis guineensis 11, 217 ff., 222
Elbe 40
electricity generation 132
El Niño 217
emission 48, 54
emission factor 93, 198, 204, 217
emission reduction 93, 205, 222
employment opportunities 152
Emslandplan 161
encroachment 207 f.
energy autarchic bioenergy villages 132
energy balance 102
energy crops 169
energy efficiency 45
energy generation 36, 109
energy production 114
energy yield 227
enrichment planting 218
ensilage 40, 42, 55 f., 71, 73
Entandrophragma palustre 220
entrained-flow gasifiers 57
environmental benefits 120
environmental consciousness 160
environmental education 19, 132
environmental impact assessment 194 f.
environmental integrity 131
environmental performance 120
environmental policy 161
environmental protection 161
Erdniedermoor 9, 12
Erica tetralix 38
Eriophorum 93, 159
Erythrina excelsa 220
essential oils 125
establishment 115
ethanol 55, 228
EU Biodiversity Strategy 199
EU Ecolabel 129 f.

EU Habitats Directive 105, 145, 148, 194, 199
Euro-Leaf 120
Europe 199
European Agricultural Fund for Rural Development 133, 150 f.
European Innovation Partnership 152
European Union 199, 231
European Water Framework Directive 14, 105, 151, 194 f., 198
Euryale ferox 220
Euterpe oleracea 221
eutrophication 13, 207, 226
evaluation 177
evaporation 15, 102 f., 191
evapotranspiration 102
exemption 147, 149
exhaust gasses 174
existence values 134
ex-Mega Rice Project 214
expenditures 111
experts 163
extensification 168
external costs 136 f., 140
external effects 133 ff., 153
fallow 98
FAO = Food and Agricultural Organisation
farmers 172
fascines 4, 76
favourable conservation status 145
Federal Soil Conservation Act 153, 155
Federal Soil Protection Act 142
fermentation 54 ff., 114
fermentation residues 169
fertilisation 3, 9, 11, 13, 89 f., 107, 170, 217
fertility management 116 f.
fibers 226
Fibraurea tinctoria 220
Filipendula ulmaria 38
filter and buffer function 19
fire 218, 222, 229
first mover 168
First Pillar 149 f., 152, 154
Fischer-Tropsch synthesis 57
fixed-bed gasifiers 57
flail mower 73
Flame Wainscot 146
flooding 26
Flooding Pampa 39
flood protection 103, 105, 194
floods 229
Floods Directive 105
floodwater mires 4
flotation wheels 64
fluidized bed gasifiers 57
fodder 19, 39 ff., 176
fodder production 19, 176
Food and Agricultural Organisation 199
food production 158

food provision 38
food security 217
forage quality 41
forest management 154, 223
forestry 13, 154, 205
Forest Stewardship Council (FSC) 119, 129 f.
form bodies 83
Fox Nut = *Euryale ferox*
Frederick the Great 3 f., 157, 159 f.
Frederick William I 4
freezing 9
Friedländer Große Wiese 160
FSC = Forest Stewardship Council
fuel characteristics 114
fuel costs 113
fuel peat 3
fuel price 112
full-cost accounting 111
funding opportunities 154
GAK 150 f.
Gallinago gallinago 94, 97
Gallinago media 97
Galloway cattle 42
Garden Angelica = *Angelica archangelica*
Garganey = *Anas querquedula*
GDR 160 f.
Gemor = *Alseodaphne coriacea*
general development plan 193
generation of heat 132
Genjer = *Limnocharis flava*
German Democratic Republic 160
GEST = Greenhouse gas Emission Site Type 93
GHG balance 89 f.
GHG emissions 89, 91, 93, 135, 139, 140, 204 f.
glass-liquid transition 47
glass-transition temperature 53
global warming potential 80
Glyceria maxima 35, 40 f., 99
Glyptostrobus pensilis 220
Gonystylus bancanus 220 f.
good practice 136, 145, 147 f., 150 f., 153, 155
goods and services 134 f.
gramineous biomass 114
Grashopper Warbler = *Locustella naevia*
grasshoppers 95
grassland 107
gravity drainage 12
grazing 41 f., 90, 92, 99 f., 116, 154
Great Bittern = *Botaurus stellaris*
Greater Pond Sedge = *Carex riparia*
Greater Spotted Eagle = *Aquila clanga*
Great Reedmace = *Typha latifolia*
Great Reed Warbler = *Acrocephalus arundinaceus*
Great Snipe = *Gallinago media*
green gas label 131

greenhouse gas emissions 12, 14, 20, 79 f., 110, 129, 131, 164, 199, 229, 231
Greenhouse gas Emission Site Type (GEST) 92 f.
greening 151, 153, 155
green politics 160
Grey Sedge = *Lepironia articulata*
Grias peruviana 221
gross calorific value 34, 227
ground beetles 95 f.
ground pressure 60, 62 ff.
groundwater 13, 102 f., 144
growing media 37
Grünes Gas Label 129 f.
Gut Darß 118
Gute fachliche Praxis 136
Gwydir Wetlands 39
Gypsywort = *Lycopus europaeus*
habitat management 206
habitat protection 145 f., 194 f., 207, 230, 232
Habitats Directive = EU Habitats Directive
hardwoods 222
harmful impact 144
harvest 59, 71 ff., 90 f., 111
harvester 62 ff., 66, 68, 86, 121
harvest frequency 71
harvest time 22, 33
Havelländisches Luch 4, 182
hay 71 f.
haze 12
health hazards 203
Heck cattle 41
Heilongjiang 224
HELCOM 199
Hemiptera 95
Hen Harrier = *Circus cyaneus*
Herb of Grace = *Bacopa monnieri*
Heritiera utilis 220
heterogeneity 8
Hierochloe odorata 130
high intensity grassland 99, 107
Highland 42
homeland 158
homogeneity 45
Honey Myrtles = *Melaleuca cajuputi*
hot spot approach 198
Hubei 224
humid meadow 40
humification 7, 9 f.
humus balance 154
Hunan 223 f.
Hungary 23
Hutan Desa 221
hydraulic conductivity 7, 102 f.
hydrogenetic mire type 179 f., 181
hydrological effects 195
hydrology 15, 102 ff.
hydromorphic soils 154
hydrophobic 8, 12
hydrothermal carbonisation 52
identification 132, 159

Ilex cymosa 220
implementation 157, 168, 179, 229
incentive 150, 152 f., 168 f.
income generation 119
income policy 152
Indian Pennywort = *Bacopa monnieri*
Indonesia 199, 203, 214, 217, 219, 222 f.
information 162, 164
infrastructure 73, 76 f., 138, 183, 190, 195, 230
initiators 132, 162
Inner Mongolia 215 f., 223 f.
innovation 109, 168, 172 f.
insulation 8, 36, 43, 82 f., 131
insulation boards 36
intensification 95
intensive livestock farming 169
interdisciplinary approach 172
inter-generational justice 175
inter-geographical justice 176
Intergovernmental Panel on Climate Change 198
internalisation 120, 137
International Sustainability and Carbon Certification (ISCC) 120, 129 f.
intervention and compensation regulation 144 f., 194
inundation 40, 192, 221
investment 115, 138
investment aid 149
IPCC = Intergovernmental Panel on Climate Change
Ipomoea aquatica 221
Ireland 23
Iris pseudacorus 100
iron 106
irrigation 7
ISCC = International Sustainability and Carbon Certification
Isopoda 95
ISO standard 14020 120
Ixobrychus minutus 97
Jambu = *Acmella oleracea*
Japanese Parsley = *Oenanthe javanica*
Java 217
Jelutong = *Dyera polyphylla*
Jiangsu 224
Jilin 224
Job's Tears = *Coix lacryma-jobi*
Juncus gerardii 101
Jungle Rice = *Echinochloa colona*
Kalimantan 217 ff., 222
Kaliningrad Oblast 160
Kangkong = *Ipomoea aquatica*
Kans Grass = *Saccharum spontaneum*
Kanya = *Pentadesma butyracea*
Kauri = *Agathis borneensis*
Kempas = *Koompassia malaccensis*
key players 132
Khas Kha = *Vetiveria zizanioides*

knowledge transfer 171 f., 231
Komplexmelioration 7, 160
Koompassia malaccensis 222
Korthalsia flagellaris 220
kuda-kuda system 223
Kyoto Protocol 154, 198
labelling 129
labour camps 161
labour costs 110
Lac Alaotra 200
lactic acid 40, 56
lactobacteria 56
Lake Winnipeg 226
land acquisition 138 ff.
land cultivation 144
land reclamation 217
land reserve 160
landscape 155, 162, 164
landscape ecology 161
landscape elements 155
landscape protection area 144
land use 80
land use change 143, 154
land use intensity 110
Land Use, Land Use Change and Forestry (LULUCF) 154
land use sector 154
Lapwing = *Vanellus vanellus*
latex 222
leaching 7, 9, 106
legal framework 110, 143, 230
legal protection 145 f.
legislation 132, 146, 150, 161 f., 170
Lepidoptera 95
Lepironia articulata 220
Lesser Pond Sedge = *Carex acutiformis*
Lesser Reedmace = *Typha angustifolia*
liability 153
Liaohe Delta 223, 225
Liaoning 224 f.
licences 111
Lida 207
ligno-cellulose 55
liming 11, 217
Limnocharis flava 221
Limosa limosa 94, 97
Limousin 42
Lioaning 223
liquid fuels 57
liquid manure 169
literature 157
litter 37, 40, 95
litter meadows 40
Little Bittern = *Ixobrychus minutus*
Little Crake = *Porzana parva*
livestock density 41 f., 100 f., 116
load-bearing capacity = bearing capacity
loading capacity 69
loading point 77
loam 82
local climate 102 f.

Locustella luscinioides 94, 97
Locustella naevia 97
logistic 59, 70, 72, 76 f., 230
Lotus = *Nelumbo nucifera*
lower heating value 34 ff.
Lower Saxony 14
low-intensity grazing 116
LULUCF = Land Use, Land Use Change and Forestry
Luscinia svecica 97
Lycopus europaeus 38
Macaranga pruinosa 223
Macquarie Marshes 39
Maculinea alcon 96
Maculinea nausithous 96
Madagascar 200
maize 54 f., 176
Malaysia 199
Malchin 174
management 95, 98, 229
management costs 116
management plan 145
Mangifera = *Mangifera havilandii*
Mangifera havilandii
Mango (*Mangifera*) 219
Manitoba 226
marginal costs 134
Marine Stewardship Council 120
Marine Strategy Framework Directive 105
marketability 230
market price 116, 136
markets 170
market value 110
Marsh Harrier = *Circus aeruginosus*
Material processing 131
Maulwurfsfräsdränung 5 f.
meadow birds 100
Meadowsweet = *Filipendula ulmaria*
meat production 116
Mecklenburg-West Pomerania 14, 44, 73, 114, 137, 140, 146, 164, 180, 204 f., 212, 231
medical plants 38
Mega Rice Project 218, 222
Melaleuca cajuputi 219 f.
melioration 160 f.
Menyanthes trifoliata 38, 81, 99
Meranti = *Shorea*
Metasequoia glyptostroboides 220
methane 55, 80, 89, 92, 169
methanol 57
Metroxylon sagu 218, 220 f.
MICCA = Mitigation of Climate Change in Agriculture
microrelief 8, 12
microtopography 217
milk production 117
Millennium Ecosystem Assessment 15
mineralisation 3, 7, 9, 80, 102, 104, 106 f., 229
minerotrophic 102
Minois dryas 96

mire conservation 161
Miscanthus 50, 54, 72, 113, 151
Miscanthus sacchariflorus 220
mitigation 109, 135, 198
Mitigation of Climate Change in Agriculture (MICCA) 198
Mitragyna ledermannii 220
moist meadows 107, 189
moisture class 10
moisture content 53
mole pipe drainage = Maulwurfsfräsdränung
Molinia caerulea 66
Molinietum caeruleae 40
monetisation 15, 92, 134
Mongolia 200
monitoring 120
Montagu's Harrier = *Circus pygargus*
Monte Carlo simulation 110 f., 113, 116
monuments 160
Mooratmung 9
Moordammkultur 4, 6 f.
MoorFutures® 139, 154
mosaic 189
mosquitoes 166
moths 95
mowing 95, 100, 122
MSC = Marine Stewardship Council
mulcher 68
mulm 11, 60
Mulmniedermoor 9, 12
multi-fuel furnaces 48
multifunctionality 231
multi-level contribution margin accounting 110 f.
multipliers 163 f., 172
multisectoral demands 231
MV tut gut 129
Myrciaria dubia 221
Myrica gale 81, 125
N_2O 90, 92
Narrowleaf Cattail = *Typha angustifolia*
Nasturtium officinale 221
National Biodiversity Strategy 140 f.
national parks 148
National Park Vorpommersche Boddenlandschaft 118
national willingness to pay 141
Natura 2000 145, 149, 186, 195
natural resources 175
nature conservation 33, 41, 98, 119, 143 f., 146 ff., 160, 194
nature conservation schemes 117
natureplus 129 f.
nature reserve 144, 148, 161, 187
Naturland 130
Naturland Wald-Zertifikat 129
Nelumbo nucifera 221
Neptunia oleracea 221
net calorific value 34, 45, 49
net energy of lactation (NEL) 41
Netherlands 23

network 173
New Guinea 217
Niangón = *Heritiera utilis*
Niedermoorschwarzkultur 4, 6f.
Nipa Palm = *Nypa fruticans*
nitrate 80, 106
nitrogen 11, 80, 89, 105ff.
N-Leaching 107
Noctuidae 96
non-timber forest products 217f., 221f.
non-use values 135
North America 226
North Carolina 203
NO_X emissions 49
Numenius arquata 97
Nuthe floodplain 183
nutrient availability 36
nutrient balance 104, 106
nutrient recovery 226
nutrient removal 96
nutrient retention 105ff., 231
nutritional energy content 41
Nypa fruticans 220
Oberes Rhinluch 7
Obstacles 170
Oderbruch 160
Oenanthe aquatica 38
Oenanthe javanica 220
Oil Palm = *Elaeis guineensis*
ombrotrophic 102
operating costs 111
operating income 135
opinion leaders 168
opponents 163f.
opportunity costs 134, 140
opposition 162, 183
organic farming (EU) 129f., 168f.
origin of agriculture 229
Orthoptera 95f.
oscillation 9, 12
otter 148
Otto gas engines 57
Ottomeyer 161
overuse 200
owlet moths = Noctuidae
ownership 136
oxidation 7, 200
paludiculture 1, 144
Pandanus = *Pandanus tectorius*
Pandanus tectorius 220
Pantanal 39
Panurus biarmicus 94
paper 43, 216, 223, 225
Papua 217
Papyrus = *Cyperus papyrus*
paradigm 171f.
paradigm change 1, 232f.
Pardosa sphagnicola 38
participation 162, 164
pastoral management 42
payment entitlements 151
payment for ecosystem services 230, 232
peak discharge 15, 104

peat briquettes 207
peat crumb horizon 9
peat degradation 4, 7, 19
peat extraction 13f., 19, 205
peat fires 12, 203f., 206, 217f.
peat formation 3, 15, 36
peatland conservation 14, 140f.
Peatland Rewetting and Conservation 154
peat loss 5
peat mineralisation 6, 11
Peatmoss (*Sphagnum* spp.) 32, 37, 119, 130
peat oxidation 4, 7, 11
peat swamp forest 217f., 221
pedogenesis 10
Peene Valley 39, 61, 95f., 99
pelleting 47, 53, 84, 113, 207
pellets 35f., 47f., 51ff., 69, 71, 84, 206ff.
penetrometer 59ff.
Pentadesma butyracea 220
perception 157, 159, 162
percolation mire 4, 8, 97ff., 103
performance 71, 73, 110, 112
permanent crops 150f.
permanent grassland 149f.
permanent grazing 41
permissibility 144
permission 194
perverse subsidies 153
Phak Khan Chong = *Limnocharis flava*
Phalaris arundinacea 24, 29, 35ff., 40f., 50f., 66, 93, 130, 148, 150f., 200, 206
Philomachus pugnax 95, 97
Phoenix reclinata 220
phosphorus 104ff., 226f.
phosphorus release 106, 199
Phragmites australis 22, 24, 28, 35, 41, 43, 50, 55, 80, 82, 84, 86, 91, 93f., 96, 111ff., 115, 121, 124, 143, 146, 148, 150f., 170, 206, 216, 223, 226
pilot projects 152, 157, 230
pioneers 163f., 173
pioneer species 98
Plagefenn 160
plan approval procedure 144, 193ff.
planning and construction costs 138f.
planning security 143, 230
planning support 185f.
plantation 143, 217f.
plant biodiversity 98
planting 115
plaster 43f.
Poggeln 5
Poland 23, 203, 207
polder 4, 8, 11
Polder Kieve 139
Poleski National Park 208
policy agreements 198

polls 162
pollutants 45
polluter pays principle 138, 142, 152f. 155, 231f.
pontoon bridge 127, 184
population growth 158, 197, 217
porosity 8, 102
Porzana parva 98
Porzana porzana 97, 146
Porzana pusilla 97
post 2020 climate regime 154
potential areas 180
poverty 133, 222, 229
power generation 140
practicability 109
precision farming 70
preferences 135
preprocessing = processing
preselection of sites 179
prices 135
primary air 49
priming effect 9
private land ownership 136
processing 46, 70, 72
product 120
product diversification 133
production chain 120
production facility expansion 133
production factors 134
production function 13f., 229
production method 111
productivity 9, 12f., 71, 161
profitability 109ff., 119, 132, 136
profit margin 116
project layout 193
project proponents 162
property rights 135, 141, 152
protected areas 148, 180, 186
protected habitat 146
provisioning services 8, 13, 15, 19f.
proxy 93
public acceptance 162
public participation 194
pulp production 218
purchase of the land 138
Purple Small-Reed = *Calamagrostis canescens*
Purun = *Korthalsia flagellaris*
pyrolysis 52
radiative energy 103
Raffia Palm = *Raphia*
raised bog 102f., 217f.
Rallus aquaticus 97
Ramin = *Gonystylus bancanus*
Ramsar Convention 161, 198
Randow Valley 182
Raphia hookeri 220
ratrak 208
Rattan palm trees = *Calamus* spp.
reclamation 158ff.
Reclamation Edict 159
recovery 72
recreation 19
REDD+ 222

Index

Red List 97
Redshank = *Tringa totanus*
reed = *Phragmites australis*
Reed Canary Grass = *Phalaris arundinacea*
reed control 42
reed cultivation 113, 225
reed for thatch 110f., 119
Reedmace = *Typha*
Reed Manna Grass = *Glyceria maxima*
reed mats 215
Reed Warbler = *Acrocephalus scirpaceus*
refugees 158, 161
regional economic development 133
regional hydrology 161
regional labels 129
regional products 111, 132
regional value creation 132
regulating services 19f.
regulation 229
regulation functions 13
regulatory frameworks 143
regulatory policy 153
Reich Conservation Act 161
relaxation 47, 53
reload distance 75
reloading 69
remote sensing 179
renewable energy 132, 170
Renewable Energy Act 157, 169, 230
Renewable Energy Directive 120
research institutes 172
reservations 164, 170
residential areas 161
resistance 162, 186
restoration 20, 89, 97, 147, 161f. 223, 228
retailer 120
revenues 109, 111, 116
rewettability 8
rewetting 14, 20, 59, 79, 89, 91, 97, 98, 104, 105, 109, 115, 135f., 138ff., 150, 152, 154, 161f., 165f., 186ff., 197f., 204f., 230
rewetting costs 139
Rhinluch 25, 183
Rhynchospora alba 38
Rimpau 4, 6f.
riparian zones 103
risk assessment 194
risks 170
root exudates 90, 92
root penetration 4, 8
rosette plants 96
rotary mowers 68
rotational grazing 41
Round-leaved Sundew = *Drosera rotundifolia*
rubber tracks 87
Rubus chamaemorus 38, 81
Ruff = *Philomachus pugnax*
Ruoergai peatlands 200

rural development 132, 139, 149, 151
Russia 203
saccharification 55
Saccharum spontaneum 220
Sachamangua = *Grias peruviana*
Sago Palm = *Metroxylon sagu*
Salix spp. 206
salt marsh 100
salt meadow 101
sand cover cultivation 4f., 7
Sanddeckkultur 4f.
saturation deficit 102
Savi's Warbler = *Locustella luscinioides*
Saw Sedge = *Cladium mariscus*
Schleswig-Holstein 14
Scirpus spp. 93, 99f.
Screwpine = *Pandanus tectorius*
sea level rise 12
sealing 7
sea water levels 11
Sebangau National Park 214f.
secondary air 49
secondary pedogenesis 3, 7, 9f., 12f., 59
Second Pillar 149ff. 154
section-wise harvest 147f., 151, 155
sedges = *Carex* spp.
Sedge Warbler = *Acrocephalus schoenobaenus*
sedimentation 104, 106
segregation 9
Seiga 23, 61, 62ff., 73, 86
self-interest 136
self-regulation 15, 218
self-sufficiency 160f., 173
semi-natural biotopes 94
Senegal date palm = *Phoenix reclinata*
setting 8
settlement 158
sewage treatment 105
Shandong 224
shear strength 59ff.
shear vane tester 60f.
Shorea = *Shorea belangeran*
Shorea belangeran 214, 220f.
shorelines 144
short rotation coppice 140, 143, 178
shredder 52, 85
shrinkage 7, 9, 11, 200
shrinking 102f.
shunt 93
silage 55f.
sintering 49f., 54
sleigh 128
Slender Tufted-Sedge = *Carex acuta*
slime mold 38
smallholder plantations 11, 219
smouldering 12
snow groomer 62f., 66f., 70, 86ff., 208
social aspects 157
social damage 231

social inclusion 133
socialism 160
socialistic land improvement 161
social welfare 109
societal appreciation 140
socio-demographic groups 162
soil compaction 9, 26, 59
soil compression 9
Soil Conservation Act 136, 154, 230
soil moisture classes 92
soil penetration resistance 59ff.
soil preservation 231
soil productivity 7
Soil Protection Act = Soil Conservation Act
solid biofuels 45, 49, 52
solid fermentation 55
Southeast Asia 11, 200, 203
Southern Cattail = *Typha domingensis*
Specialists 164
species diversity 98
species protection 147f., 161
Sphagnum spp. 37, 39, 119
Sphagnum farming 32, 37
Sphagnum palustre 32, 37
Sphagnum papillosum 32, 37
spiders 38, 95
Spondias mombin 221
Sporava 206
Spotted Crake = *Porzana porzana*
Spreewald 182
stakeholder analysis 163
stakeholders 133, 162ff.
starch 55
steam-powered ploughs 161
storage ability 71
storage capacity 192
storage coefficient 103
straw 50, 54, 72, 112ff.
subsidence 7ff., 26, 102, 107, 162, 200, 217
subsidies 116, 135, 150, 155, 162, 168, 170
substitutes 110
subsurface drainage 4f.
succession 95, 100
Succisa pratensis 99
suckler cows 101, 116f., 168
suitable area 182
Sulawesi 217
sulphur 55
Sumatra 217ff., 222
summer harvest 95f., 98, 114
Sundew = *Drosera* spp.
supply costs 112
sustainability 136, 175f.
sustainable land management 171f., 175f.
Sweet Grass = *Hierochloe odorata*
swelling 9, 60, 102f.
Sylphium perforatum 40
Symposia 164
synergies 105, 199

Index

Synthetic Natural Gas 52, 57
Talinum fruticosum 221
target species 96, 115
taxation 153
Tea Tree = *Melaleuca cajuputi*
Tengkawang nut trees = *Shorea* spp.
termination of lease 139
Terra-wheels 86
terrestrialisation mire 106, 160
territory protection 145
Tetrao tetrix 97
Thalictrum aquilegifolium 99
thatch 43, 45, 82, 111, 114
thatching 114
thatching reed 45, 111
thawing 9
thermal conductivity 102
thermal utilisation 54
thermo-chemical gasification 57
Thurbruch 167
Thyme-leafed Gratiol = *Bacopa monnieri*
Tibet 203
tillage 9
timber 219, 221
total economic value 134
tourism 132
tracked vehicle 64, 73, 122
tracking 120
tracks 88
trafficability 3, 8, 12, 59f., 64, 68ff., 75f., 226, 230
trailer 122
trampling 91f., 100
transferability 148
transfer of knowledge 171f.
transfer payments 105, 136f.
transfer point 184
transhipment 75
translocation 9
transmigration 217
transport 72ff.
transportability 45, 133
transport capacity 71, 75f.
transport distance 71, 110
transport of the biomass 70
transport tracks 76
transport trips 75
Trapa natans 221
Tringa totanus 97
tropical peatlands 11
True Sago Palm = *Metroxylon sagu*
Tufted Sedge = *Carex elata*
Turkey 23
Twin-spotted Wainscot = *Archanara geminipuncta*
twin tyres 64, 123
Typha spp. 24, 35, 43, 80, 119, 226f.

Typha angustifolia 24, 35, 206, 226
Typha domingensis 221
Typha latifolia 24, 35, 91, 170, 221, 226
Typha x glauca 24, 35, 226
Ucker Valley 139, 183
Ukraine 23
Umweltbundesamt 137
UNFCCC = United Nations Framework Convention on Climate Change
United Kingdom 23
United Nations Framework Convention on Climate Change (UNFCCC) 140, 154, 198
upper heat value 34
uptake 55, 68f., 77, 91, 114, 122, 188
Urbarmachungsedikt 3f., 159
usus 136
usus fructus 136
utilisation 157
Vaccinium angustifolium 38
Vaccinium macrocarpon 38
Vaccinium oxycoccos 38, 81
Vaccinium vitis-idaea 38
valley mires 106
value added 136
value creation 20, 109, 132f., 135, 137, 141f.
Vanellus vanellus 94
variable costs 110
VCS = Verified Carbon Standard
vegetation dieback 8
vegetation form 93
venturi effect 90
verification 120
Verified Carbon Standard 154
Vetiveria zizanioides 220
vibration 60
vicious cycle of peatland utilisation 10f., 153
volatile matter 48
voluntary carbon market 119, 154, 223
wasteland 157
wastewater 36, 226
water authorities 144
water balance 102, 192
Water Buffalo (*Bulbalus arnee*) 31, 41f., 100f., 116ff., 130f.
Water Chestnut = *Trapa natans*
water content 71
Watercress = *Nasturtium officinale*
water demand 192
Water Dropwort = *Oenanthe aquatica*
Water Framework Directive 14, 105, 151, 194f., 199
water holding capacity 103

Waterhyssop = *Bacopa monnieri*
water law 144
Waterleaf = *Talinum fruticosum*
waterlogging 4, 9, 10, 26, 229
water management = water regulation
Water Mimosa = *Neptunia oleracea*
water pollution 106
water protection area 144
water quality 20, 119
Water Rail = *Rallus aquaticus*
water reed 146
water regulation 4, 8, 12, 144, 190
water repellency 9
water retention 7, 105, 149, 154, 192, 203
water rise mires 4
Water Spinach = *Ipomoea aquatica*
weak sustainability 175
weight distribution 60
welfare effects 135f., 140ff., 152
Western Marsh Harrier = *Circus aeruginosus*
wet fermentation 54f., 114
wetland conservation 161
Wetland Drainage and Rewetting 154, 198
wet meadows 107, 195
White-beak Sedge = *Rhynchospora alba*
white peat 37
White Samet = *Melaleuca cajuputi*
White-winged Tern = *Chlidonias leucopterus*
wild boars 98
wilderness 13, 19, 157
Wild Rice = *Zizania* spp.
willingness to pay 110, 119, 133f., 141f.
Willows (*Salix* spp.) 51, 130, 207
wind erosion 4, 8, 11
Winnipeg 216
winter harvest 94ff., 98, 113, 143, 146ff.
winter harvest of round bales 113
winter mowing 145ff.
working width 71, 75
World Trade Organisation 129
World Wars I and II 158, 161
Xinjiang 223f.
Yellow Mombin = *Spondias mombin*
Yellow Sawah Lettuce = *Limnocharis flava*
yield 12, 24, 35, 41, 73, 75, 110, 150, 223f., 227
yield reduction 150
Zehlaubruch 160
Zizania spp. 38, 220